TRENDS IN AGRICULTURE AND SOIL POLLUTION RESEARCH

TRENDS IN AGRICULTURE AND SOIL POLLUTION RESEARCH

JAMES V. LIVINGSTON
EDITOR

Nova Science
New York

For permission to use material from this book please contact us:
Telephone 631-231-7269; Fax 631-231-8175
Web Site: http://www.novapublishers.com

NOTICE TO THE READER

LIBRARY OF CONGRESS CATALOGING-IN-PUBLICATION DATA:
Available Upon Request

ISBN 1-59454-325-9

Published by Nova Science Publishers, Inc. ✢ New York

CONTENTS

Preface		**vii**
Chapter 1	Detection of Tributyl and Dibutyl Tin in the Environment with a Bioluminescent Bacteria: From the Bioassay to the Biosensor *M. J. Durand, M. S. Dubow, H. Horry, P. Picart, P. H. Daniel, S. Lebeau and G. Thouand*	**1**
Chapter 2	Fate of Petroleum Hydrocarbons in the Subsurface Environment: Abiotic Processes *Ishai Dror*	**43**
Chapter 3	Exposure Assessment to Dioxins of People Living in the Seveso (Italy) Area *Elena Fattore and Roberto Fanelli*	**79**
Chapter 4	Chemical Degradation of Phenylurea Herbicides in the Soil *Elio Coppola, Stefano Salvestrini, Giuseppe Santoro and Sante Capasso*	**95**
Chapter 5	Adsorption of Trace Metals onto Natural Surface Coatings *Deming Dong, Fan Yang, Yu Li, Xiuyi Hua, Rong Zhang, Xiaojun Lü, Jingjing Zhang, Botao Zhang, Na Zheng and Liang Liu*	**121**
Chapter 6	Desorption of Lead and Cesium from Kaolinite and Illite Using Cationic Surfactants *Zhaohui Li, Christine D. Reardon and Christine V. Evans*	**145**
Chapter 7	Deriving Critical Loads for India, Where Do We Stand Today? *Ajay Taneja and Gur Sumiran Satsangi*	**165**
Chapter 8	Urban Expansion: A Threat to Crop Land. A GIS Based Study from a Predominantly Agrarian Economy (India) *Shahab Fazal*	**191**

Chapter 9 Eco-Friendly Molluscicides, Piscicides and Insecticides from Common Plants **205**
Ajay Singh, S. K. Singh, R. P. Yadav, V.K. Srivastava, D. Singh and S. Tiwari

Index **231**

PREFACE

Pollution is undesirable state of the natural environment being contaminated with harmful substances as a consequence of human activities so that the environment becomes harmful or unfit for living things; especially applicable to the contamination of soil, water, or the atmosphere by the discharge of harmful substances. In addition to the harm, either present or future and known or unknown, to living beings, pollution cleanup and surveillance are enormous financial drains of the economies of the world. This volume details outstanding research covering agriculture and soil pollution.

In: Trends in Agriculture and Soil Pollution Research
Editor: James P. Livinston, pp. 1-41
ISBN: 1-59454-325-9

Chapter 1

DETECTION OF TRIBUTYL AND DIBUTYL TIN IN THE ENVIRONMENT WITH A BIOLUMINESCENT BACTERIA: FROM THE BIOASSAY TO THE BIOSENSOR

M. J. Durand[1], M. S. Dubow[2], H. Horry[1,3], P. Picart[4], P. H. Daniel[5], S. Lebeau[1] and G. Thouand[1,*]

[1] UMR CNRS 6144 GEPEA / ERT CBAC, Campus de la Courtaisière, IUT de la Roche sur Yon, Dpt Génie Biologique, 18 Bd G Defferre, 85035 La Roche sur Yon- France

[2] Institut de Génétique et de Microbiologie, Bâtiment 409, Université Paris Sud, 91405 Orsay Cedex, France.

[3] Biolumine SA, Plate-Forme Vendée Recherche, parc d'activité de la croix verte, 85600 Boufféré, France.

[4] ENSIM, rue Aristote, 72085 Le Mans Cedex 9, France, LAUM, UMR CNRS 6613, Avenue Olivier Messiaen, 72085 Le Mans Cedex 9, France

[5] LPEC, UMR CNRS 6087, Avenue Olivier Messiaen, 72085 Le Mans Cedex 9, France

ABSTRACT

The measurement of compounds constitutes a major tool for regulatory agencies as well as for the control and the monitoring of processes in industrial areas. Organotin compounds such as tributyltin chloride (TBT) have been extensively used as biocides for over 50 years and are currently present in marine and terrestrial environments. Tributyltin is toxic at nanogram.L-1 levels and led, for example, to the settling of oyster spat and causing oyster shell thickening and growth inhibition. The detection of organotin compounds in water and sediments is based on chromatographic separation after extraction from the matrix. These techniques are very sensitive, but time-consuming, expensive and require extensive technical competence. They also have the disadvantage that they do not take into account the bioavailability of the compounds.

* Tel : 00 33 2 51 47 84 41; Fax : 00 33 2 51 47 84 51; E-mail : gerald.thouand@iut-laroche.univ-nantes.fr

The development of new detection methods like bioassay and biosensors becomes desirable.

We have developed since 2000 a bioassay and a biosensor for the detection of TBT and DBT using an *Escherichia coli*::*lux*AB strain. This strain called TBT3 has been isolated after a random insertion of the *lux*AB (promoter-less) genes of *Vibrio harveyi* into the *Escherichia coli* DH1 chromosome. The genetic construction and the physiology of the strain are presented. The performance of the bioassay include the determination of the specificity, the sensitivity, the repeatability and the reproducibility. For an application of this bioassay in the environment, some results on contaminated sand are demonstrated.

A comprehensive procedure is also used in our Lumisens 1 range biosensor for on-line and automatic detection of pollutants like TBT in a technical and an application point of view. This review presents the major results of our laboratory and the future developments of our bioassays and biosensors.

1. INTRODUCTION

In 2001, the European Community published a list of prioritary substances in the field of water policy in which tributyltin (TBT) compounds are classified as "dangerous priority substances (Table1). Tributyltin (TBT) was used as a biocide in antifouling paints on ships and in the wood protection. Due to its toxicity around one ng.L-1 to oyster and other marine species, legislation in many countries banned the application of TBT-based paints to vessel less than 25 m (EC directive 89/677/EEC). Since the 1st January 2003, the International Marine Organization (IMO) prohibited the use of organotin compound as active biocide in antifouling paint for all vessels (last date for the application). By the 1st of January 2008, organotin compound should be eliminated for all the vessels (Champ, 2000; Champ, 2003). Therefore sources of TBT will continue to be generated until at least 2008.

The reduction of the levels of TBT in water occurred since the 1982 regulation, the TBT concentration was 4.6 ng.L-1 on average in the French coastal water in 1997. Nevertheless, in 1997, the campaign conducted by IFREMER established that among 36 stations tested, 10 of them were TBT contaminated at concentration above 100 ng.L-1 (Michel and Averty, 1999).

Due to its hydrophobic character, TBT binds to dissolved or suspended material. As a consequence the sediment is recognized to be a tank for TBT. In 1996, Ruiz et al reported that 1.2 ton TBT as Sn was still stored in sediment of the Arcachon Bay in France. Resuspension of contaminated sediments will lead to enhance concentration in overlying water phase.

Organotin pollution of sediment is of global concern. Concentrations of organotin in municipal wastewater, sewage sludge and sediment were reported by Hoch (2001). Marine sediment from Hawaii (Pearl Harbor: 4 to 2830 ng.g-1 dry wt), from Hambourg in Germany (5200 ng.g-1) and from Vancouver, Canada (10780ng.g-1) were the most polluted sites. Concentration of TBT in North Sea marina ranged from 80 to 720 ng.g-1 and between 570 and 17000 ng.g-1 (dry weight) and in Baltic Sea marina (Biselli et al 2000). Antartic sediment are also highly contaminated, a concentration of 2290 ng.g-1 was found in a station of Ross sea (Negri et al, 2004). TBT contamination is also reported in all the Asian coastal areas (Shim et al, 1999, Cheung et al, 2003, Midorikawa et al 2004).

The persistence and behaviour of TBT in aquatic environment is well documented (Fent 1996, Ruiz et al 1996, Ceulemans et al 1998). Degradation of TBT proceeds via successive dealkylation from TBT to DBT and finally MBT. In aquatic ecosystems, photodegradation of

TBT by UV radiation is not significant, biodegradation is the major breakdown pathway of TBT. Biodegradation depends on the environmental conditions and only a limited number of microorganisms such as bacteria, fungi and algae are able to degrade TBT.

In natural water, TBT has a short residence time with a half –live in a range of several days to week.

Nevertheless, due to anoxic condition, the estimated half lives are between 1 to 5 years in freshwater and marine sediments.

Table 1. List of priority substances in the field of water policy (EECn° 2455/2001/EC)

CAS[(1)] Number	Name of priority substance	CAS Number	Name of priority substance
15972-60-8	Alachlor	7439-97-6	Mercury and its compounds
120-12-7	Anthracene	91-20-3	Naphthalene
1912-24-9	Atrazine	7440-02-0	Nickel and its compounds
71-43-2	Benzene	25154-52-3	Nonylphenols
n.a.[(2)]	Brominated diphenylethers	104-40-5	(4-(para)-nonylphenol)
7440-43-9	Cadmium and its compounds	1806-26-4	Octylphenols
85535-84-8	-13-Chloroalkanes	140-66-9	(para-tert-octylphenol)
470-90-6	Chlorfenvinphos	608-93-5	Pentachlorobenzene
2921-88-2	Chlorpyriphos	87-86-5	Pentachlorophenol
107-06-2	1,2-Dichloroethane	n.a.	Polyaromatic hydrocarbons
75-09-2	Dichloromethane	50-32-8	(Benzo(a)pyrene),
117-81-7	Di(2-ethylhexyl)phtalate (DEHP)	205-99-2	(Benzo(b)fluoranthene),
330-54-1	Diuron	191-24-2	(Benzo(g,h,i)perylene),
115-29-7	Endosulfan	207-08-9	(Benzo(k)fluoranthene),
959-98-8	(alpha-endosulfan)	193-39-5	(Indeno(1,2,3-cd)pyrene)
206-44-0	Fluoroanthene	122-34-9	Simazine
118-74-1	Hexachlorobenzene	**688-73-3**	**Tributyltin compounds**
87-68-3	Hexachlorobutadiene	**36643-28-4**	**(Tributyltin-cation)**
608-73-1	Hexachlorocyclohexane	12002-48-1	Trichlorobenzenes
58-89-9	(gamma-isomer, Lindane)	120-82-1	(1,2,4-Trichlorobenzene)
34123-59-6	Isoproturon	67-66-3	Trichloromethane (Chloroform)
7439-92-1	Lead and its compounds	1582-09-8	Trifluralin

(1) CAS : Chemical Abstract Services ;

(2) na : Not applicable

The measurement of such compounds constitutes a major tool for regulatory agencies as well as for the control and the monitoring of processes in industrial areas.

The present paper focuses on in-depth description of our research from the "bacteria" to the biotools used to detect in a new way, organotin compounds such as TBT and DBT. An introduction is devoted to the presentation of chemical measurement used to monitor TBT and then some chapters useful to understand the three main topics *ie*: the description of the bacteria *Escherichia coli*::*lux*AB TBT3 (*E. coli* TBT3) physiology, its application to a bioassay and a biosensor.

2. Conventional Measurements Methods of TBT/DBT

Analytical procedures developed for organotin species analysis are based on chromatographic separation, gas chromatography (GC) or liquid chromatography (LC) after extraction from diverse matrices (Abalos et al, 1997 for review)

GC methods are widely used, because organotin are mostly polar and not volatile, derivatization is required before their separation.

The preparation of environmental sample is rather complexe and requires 5 steps :

(1) pre-treatment : addition of internal standard(s) into sample to correct losses in extraction steps and derivatization.

(2) extraction by solvents such as hexane, benzene, toluene for TBT. DBT and MBT need complexation or acidification of the sample before extraction. Extraction could be done also by microwave or sonication (Gómez-Ariza et al 2001 for review).

(3) derivatization to produce volatile compounds, two alkylation reactions can be performed using Grignard reagents or sodium tetraethylborate (NaBEt4).

(4) column clean-up using adsorbents like Florisil, alumina-silica, florisil silica.

(5) For the detection a large kind of detectors are used including : Flame ionisation detection (FID), electron-capture detection (ECD), atomic absorption spectrometry (AAS), Flame photometric detection (FPD, tin selective), atomic emission detection (AED) and Mass Spectrometry (MS).

Pulsed flame photometric detection could be used instead of conventional FPD for a better sensitivity (Bancon-Montigny et al 2000, 2004). Procedure using liquid chromatography inductively coupled plasma mass spectrometry (LC-ICP-MS) was described by White et al (1998). Detection limit of TBT in sediment was 7 nmol.kg-1 to 170 nmol.kg-1 TBT cation according to the method used (Sturgeon et al 2003).

In order to assess the quality of the extraction of TBT from sediment, and the quantification for low concentration an exercise of intercalibration was organized by the Institute for National Measurement Standards of National Research Council of Canada. The authors concluded that determinations were acceptable for the majority of participants (Sturgeon et al 2003). Nevertheless they emphasised that attention should be given to the impact of extraction efficiency. The use of reference material could allow the evaluation of the extraction quality and measurement (Quevauviller and Morabito 2000).

A reference procedure by gas chromatography for the evaluation of organotin in sediment is proposed by AFNOR (Association Française pour la Normalisation, NF T90-250), for a range of concentration from 20 µg Sn/kg to 2000µg Sn/kg (dry sediment), extraction procedure was not specified.

3. A New Strategy for the Detection of TBT/DBT

In order to increase the number of measurements and to be in the meantime cost effective, we propose a two-staged strategy underneath (Figure 1). Each sample could be firstly tested with a biotool (bioassay or biosensor) allowing a fast and inexpensive measurement. If TBT is detected, the strategy is ended. Conversely, each negative response meaning no TBT or TBT concentration under the detection limit of the biotool would lead to the use of the sensitive chemical measurement.

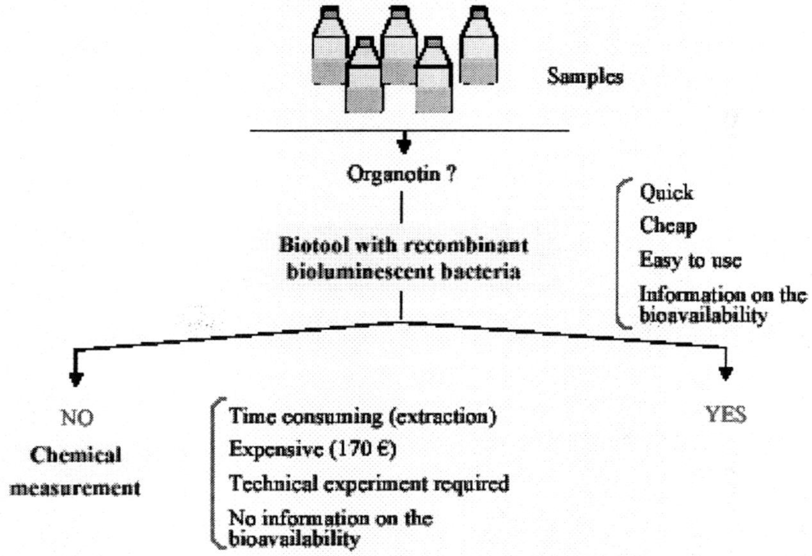

Figure 1. Proposal of a new strategy for the detection of TBT in an environmental sample

As this strategy mentions the term bioassay it is important to separate from biosensor (Figure 2). A bioassay is a simple, easy to use set of measures performed either on tube or on microplate. Usually, after the measurement, the bioassay is disposable. Conversely, a biosensor is a device, in which the bioelement (the specific biological part) is in close contact with the transducer able to transform the biological signal (produced after contact of the bioelement with the analyte) in an easily detectable electric signal (Kress-Rogers, 1997). Thanks to this configuration, a biosensor is reusable and consequently, is able to perform on line measurements. Since a living biological element is the sensitive part of both biosensor and bioassay, they are able to sense only the bioavalable part of TBT in a sample *i.e* : the concentration of substance that causes a toxic effect. This is a major difference with a chemical measurement.

In a technical point of view, the development of a bioassay is simpler and faster than a biosensor since the latter needs the use of electronic and time for the immobilisation of the bioelements to the transducer.

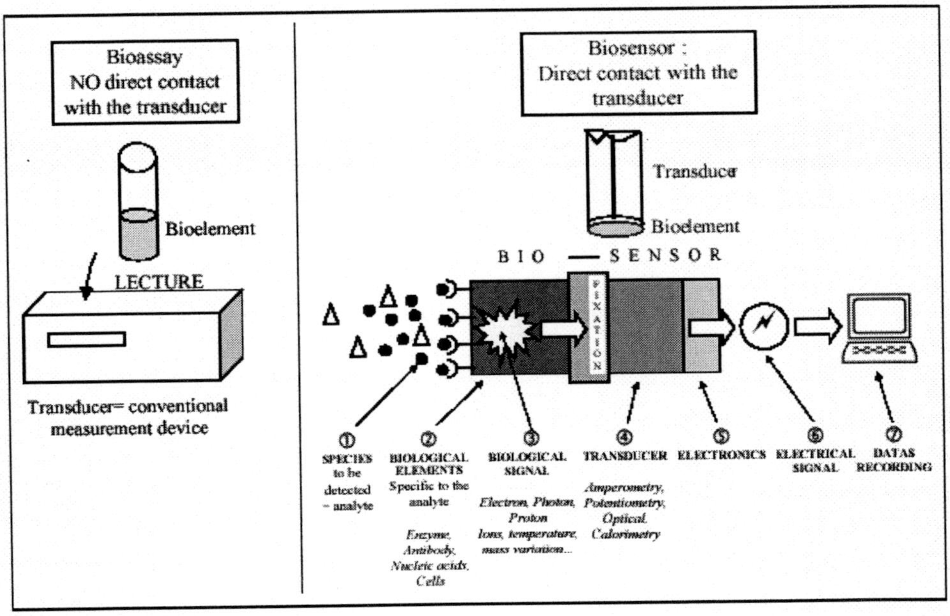

Figure 2. Difference between a bioassay and a biosensor.

4. The Use of Microorganisms to Sense TBT/DBT

Few studies have reported the use of bacteria for the detection of organotin compounds. A luminescent *Pseudomonas fluorescens* was constructed for a test based on the decrease of luminescence in the presence of the organotin compounds (Bundy et al, 1997). The EC50 value after 15 min contact time was 2.62 µM but this assay was found non specific and was less sensitive than the Microtox® assay (which uses bioluminescent *Vibrio fischeri* bacteria). Cha and Kim (2002), fused a promotorless *lac*Z operon to a DNA-damage inducible promoter (*rec*A, *din*D, *uvr*A) cloned in a plasmid and transformed in *Escherichia coli* . The bacteria was induced with a relatively high level of TBT (from 0.01 to 1 mg.L-1 depending on the genetic construction) and lacked specificity for organotins. In 1996, Briscoe et al. reported the isolation of an *Escherichia coli* namely "TBT3" strain containing a chromosomale *lux*AB transcriptional fusion using *lux*AB from *Vibrio harveyi* as the reporter gene (see chapter for the genetic construction).

In 1995, Miller et al isolated TBT resistant bacteria harboring a plasmid opening a possibility to build new bioluminescent strains after insertion of a bioluminescent operon. Recently, Roy et al (2004) isolated from high contaminated sites several strains as *Pseudomonas* sp able to degrade TBT. A strain identified as *Pseudomonas diminuta* was isolated from polluted river water and was able to degrade TBT at initial water concentration of 4-40 µg Sn.L-1 in 24 hours (Kawai et al 1998). TBT-resistant bacteria was also isolated from polluted sediments, the mechanism involved in the resistance of *Pseudomonas stutzeri* is associated to the operon tbt ABM, homologous to the resistance- modulation-cell division (RDV) efflux pumps (Jude et al 2004). Despite those new strains and insights towards the TBT biodegradation understanding, no bioluminescent strains were actually constructed.

5. BIOLUMINESCENCE

Bioluminescence is the light produced by some organisms mediated by the enzyme luciferase or a photoprotein mediated an oxidation reaction (Blum, 1997). Bacterial luminescence has been found in 11 bacterial species, from 4 genera (*Vibrio*, *Photobacterium*, *Shewanella* (*Altermonas*) and *Photorhabdus* (*Xenorhabdus*). The first three genera are found in marine habitats and the latter is terrestrial (Ulitzur, 1997). The molecular biology of these genes and enzymes has been well studied in *Vibrio harveyi*, *Vibrio fischeri*, *Photobacterium phosphoreum* and *Photobacterium leiognathi* and reviewed by Meighen (1994). In these bacteria, the emitted light (490 nm) is generated after a complex biochemical reaction summurized as follow.

$$FMNH_2 + RCHO + O_2 \xrightarrow{\text{Bacterial luciferase}} FMN + H_2O + RCOOH + \text{light (490 nm)}$$

In the above reaction, an aldehyde (decanal, dodecanal or tetradecanal) is oxydized into an organic acid in the presence of oxygen and reduced flavine mononucleotide. The light is emitted in the form of a glow and with a small quantum yield of light (0.1). Bacterial luciferase is a heterodimeric enzyme of 78 kDa encoded by the *lux* A and *lux* B genes. Three other genes (*lux* C, *lux* D and *lux* E) are associated and are involved in the production and turnover of the aldehyde (Meighen 1994).

A reporter reaction like bioluminescence is often coupled to gene expression to allow a fast, sensitive and non-destructive signal. The most common reporter genes come from the promoter-less *lux* operon of *Vibrio harveyi* and *Vibrio fischeri*. Strains are either transformed with the complete *lux*CDABE genes or the *lux*AB genes. In this case, the aldehyde must be added exogeneously.

Many bioassays have been developed for the detection of chemical compounds (Köhler et al. 2000, Daunert et al, 2000, Nivens et al, 2004) or bacteria in food (Thouand et al. 2001).

The detection of bioluminescence can be performed using different procedures:

(i) The most simple is the use of the naked eye or the exposure of a colony or cells to photographic or X-ray film (Figure 3)

(ii) Camera CCD (charge coupling device) is very sensitive and can be used for monitoring bioluminescence in real time (Sternberg et al, 1997)

(iii) Commercial luminometers are available either with single assay tubes or for use with microplates (see Ellis and Wright, 1999; Stanley, 1997; Yotter et al, 2003 for review)

For construction, a biosensor can be built using a photomultiplier tube coupled to a fiber optic probe (see chapter 8 below).

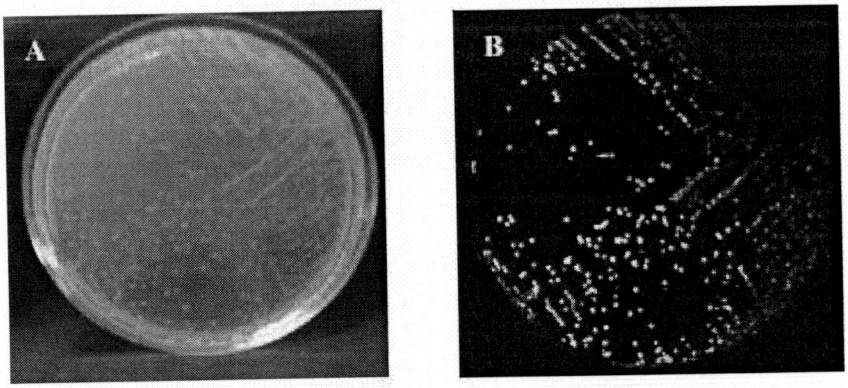

Figure 3. Light production from single colonies of a constitutive bioluminescent *Escherichia coli* XL1 pBVF*lux*CDABE on petri dish (LB medium, tetracycline, ampicilline) incubated 12H at 30°C. **A** : the plate observed in our laboratory. **B** : The same plate observed in a dark room (Momochromatic CCD camera, integration time 10 seconds) Thomas Charrier, not published data.

5.1. Measurement of the Spectral Emission of Bioluminescent Cells

In order to develop a biosensor or a bioassay using recombinant bioluminescent strains the choice of optical equipment, like lenses, optical fibers and photomultiplier tubes depends on the spectral characteristics of the bioluminescent strains. We developed a method using a highly sensitive CCD (Charge Coupled Device) of a Raman spectrometer (Thouand et al, 2003a). We taken advantage of the high sensitivity and the rapidity of the device. Because of the multichannel configuration, the total spectrum takes only 5-6 minutes, which allow to avoid all dynamical problems and signal variation. Additionally in high frequency range the quantum efficiency of the CCD is approximately of an order of magnitude 3, compared to the detection with a classical spectrometer using a photomultiplier that would need 210 hours to obtain a comparable spectrum!.

5.2. Description of the Set Up for the Measurement of a Spectral Emission

The luminescence spectra are recorded in a dedicated stainless-steel, autoclavable, mini-bioreactor (stainless-steel 317L enriched with molybdenum). This mini-bioreactor (100 mL working volume) for the measurement (namely MR) is made up of a hollow wall with a water input and output that allowed thermostatic control with water circulation (30°C or 25°C for *Vibrio* strains). Agitation is performed with a magnetic stirrer (300 rpm). The bioluminescence signal emitted from bacteria in the bioreactor is focussed via a convex lens (f=15mm) at the entrance of a glass bundle optic fibers (LOT ORIEL, N°77526, length = 1.219m, Diameter max = 3.2mm, numerical aperture = 0,56; diameter= 3mm). The lens is fixed at the intake of a pipe screwed to the lid of the bioreactor. The other part of the fiber optic is connected with an entrance slit (10mm x 1mm) to the Raman spectrometer.

Before each spectral measurement, the mini-bioreactor is sterilized by autoclaving (120°C, 20 min) and the fiber-optic is fixed on the lid and connected to the Raman spectrometer (Figure 4).

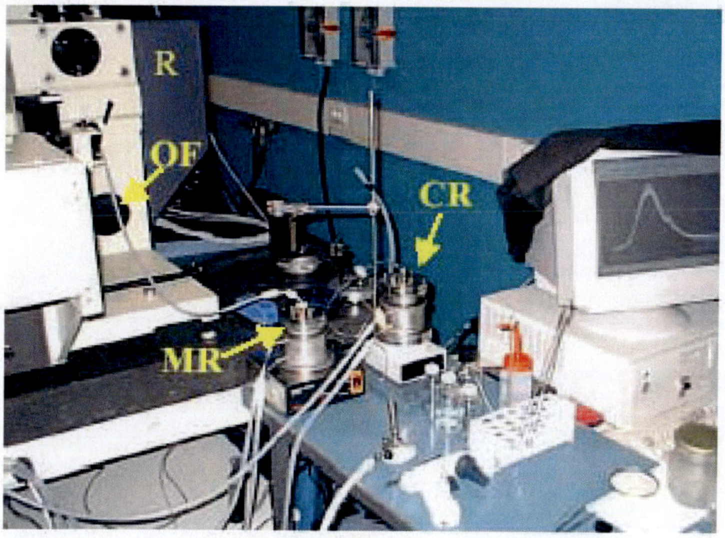

Figure 4. Representation of the experimental set up for the measurement of spectrum from Bioluminescent strains. Jobin-Yvon T64000 Raman Spectrometer.

For each experiment, strains are grown in a separate culture reactor (CR) similar to the measurement reactor described above. At each studied conditions, 10 mL of the bacterial culture is harversted and transfered in the MR bioreactor. The spectrum acquisitions are started immediately after the manual addition of a decanal stock solution to reach a 30μM final concentration for the *lux*AB recombinant strains. For *E.coli* TBT3, cells are pre-induced for one hour with 0.2 μM TBT in the CR bioreactor prior to recording its spectrum (see Thouand et al 2003 for a description of the Raman CCD calibration). The spectra are recorded between 400 nm up to 750 nm divided in windows of 61.25nm. Light intensity was expressed in arbitrary units (a.u.). Spectra are corrected from the background and related to the optical density of the solution.

Figure 5 reported spectral emission measurement of the wild type naturally luminescent *Vibrio harveyi* and *E.coli* TBT3 harboring the *lux*AB genes from *Vibrio harveyi*. Thanks to the Raman CCD it was able to confirm a main peak at about 490 nm (blue light) and moreover, a second peak at 590 nm that was never described because of the poor sensitivity of traditional spectrometer using photomultiplier as detector device. *E.coli* TBT3 strain produced the same spectrum though the light intensity was lower conducting to a difficulty to clearly identify the weaker peak at 590 nm. From other experiments, we demonstrated that the spectra was similar to other strains Gram positive or negative and whatever the growth conditions (growth phases and growth medium). From these studies, the choice of optical devices as well as photomultiplier was optimised and common to every bioluminescent strain.

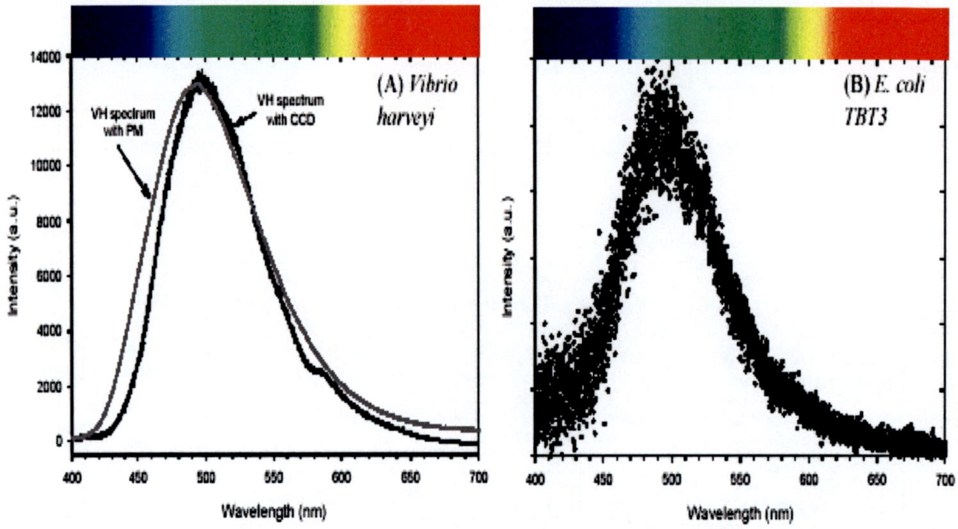

Figure 5. A : Comparison of the spectral emission of *Vibrio harveyi* with a CCD chip detector from a Raman spectrometer and with a photomultiplier (PM) detector (slits 500 microns, step 1 nm, 10 sec per point, 2 accumulations, acquisition time of 1 hour). Spectral emission from culture at A_{620nm} = 3.86 were corrected from the quantum efficiency (QE) of each apparatus (two repetitions). The strain *Vibrio harveyi* was grown in DSMZ N°6904 at 25°C in batch conditions aerated and agitated at 300 rpm. (B) : Spectral emission of a culture of *E.coli* TBT3 at A_{620nm} = 0.3 (induced with 0.2 µM of TBT) and grown at 30°C in batch condition in Glucose medium. (adapted from Thouand et al, 2003a)

6. The Strain *E. Coli* TBT3 is the Specific Bioelement in the Bioassays and Biosensors

6.1. Molecular Genetic of the *E.coli* TBT3 Strain

In order to identify *Escherichia coli* genes that are transcriptionally regulated by cellular exposure to organotins, a *luxAB* genes fusion library was constructed (Figure 6) and screened in the absence and presence of these compounds incorporated in Luria Bertani Agar plates. Using this approach, we found a clone whose luminescence was augmented, in a dose-dependant manner, with increasing concentrations of 0,1 and 10 µg.mL-1 TBT, or 0, 100, 250 and 500 ng.mL-1 DBT. This clone was called TBT3.

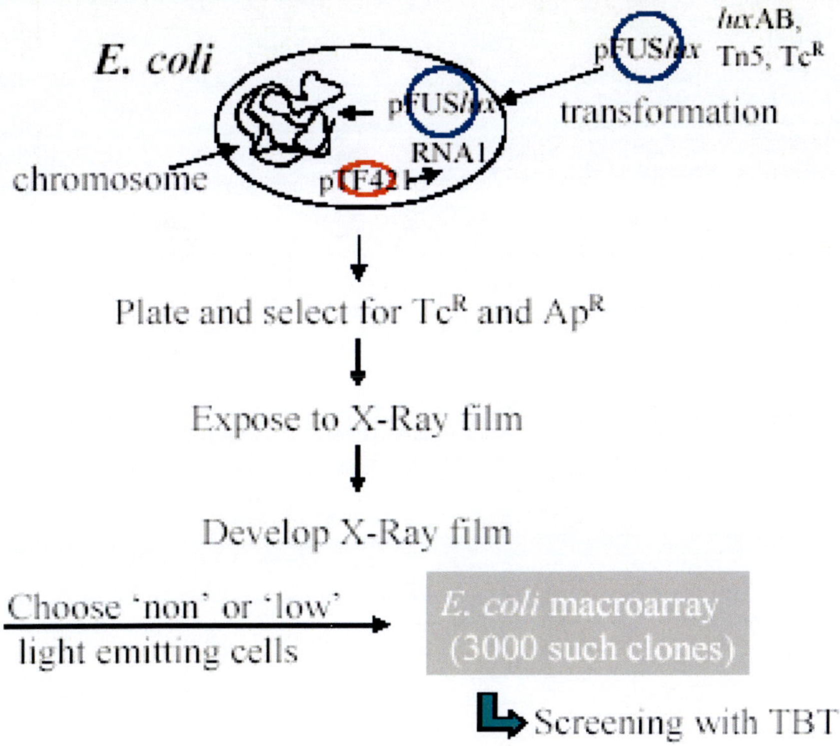

Figure 6. Protocol used for the construction of the *luxAB* genes fusion library and selection with TBT (Guzzo and DuBow, 1991)

Following clone identification, the *luxAB* insertion site was cloned (Figure 7A), the junction site sequenced, and identified as being 75bp from the translational end of the *stpA* gene (Figure 7B).

Its protein product, StpA, is a histone-like protein homologue of HN-S. It was determined that the promoterless reporter genes *luxAB,* is inserted in the opposite direction of *stpA* transcription. The gene(s) downstream of *stpA* was found to be *ygaP*, an orphan gene, which appears to form an operon with a second orphan gene, *b2667* (Figure 8). These two unstudied genes appear to have the same transcriptional orientation as the reporter gene, suggesting that luminescence may be produced upon induction of transcription of these unknown open reading frames into the 3' end of the *stpA* gene.

The gene *b2667*, is 300 bp in length and produces a putative protein of 99 amino acids. Bioinformatic approaches suggest that this gene may be a DNA binding protein containing a helix-á - turn-á -helix motif. The second gene, *ygaP*, is 525 bp in length and produces a putative protein of 174 amino acids, with unknown function.

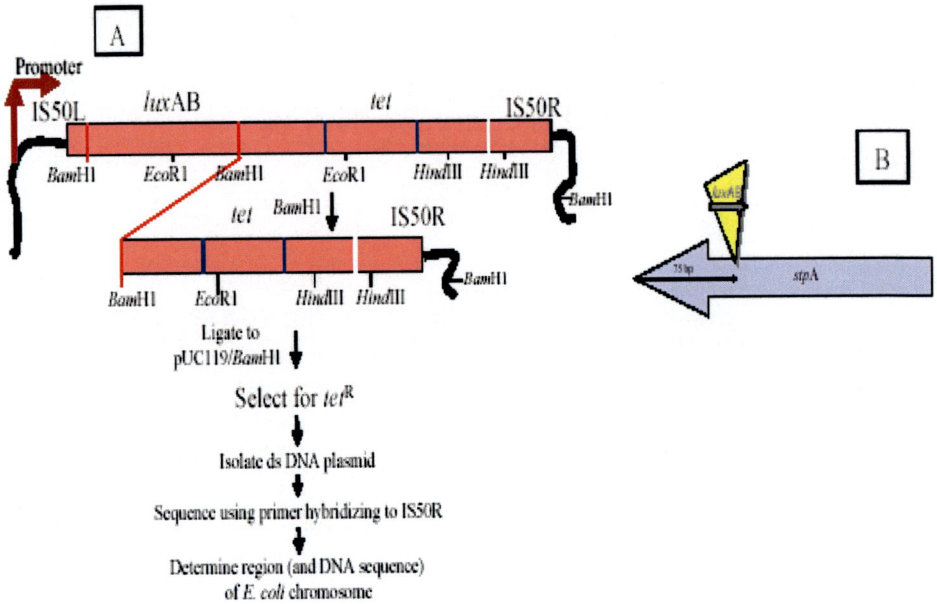

Figure 7. Cloning of the *lux*AB insertion site (A) and localization (B). Not published results

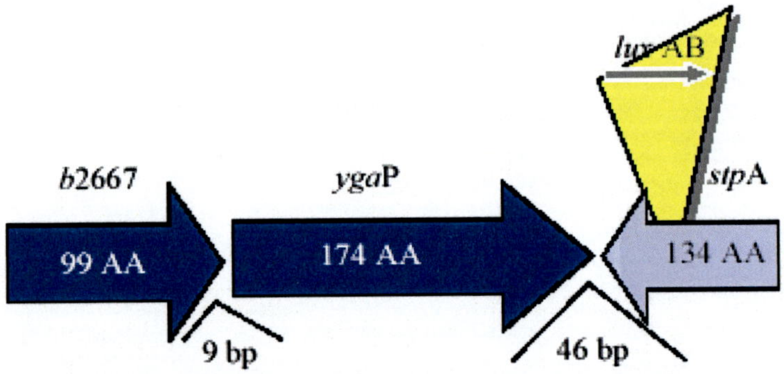

Figure 8. Final localization of the *lux*AB genes into the *stp*A gene of *E coli* TBT3. Not published results

6.2. Typical Induction Response of the *E.coli* TBT3 Strain with TBT

The figure 9 report the typical response of the strain *E.coli* TBT3 after induction with TBT-Cl from 0.001 to 10 μM (see below for a full description of the protocol). Three domains have to be considered. When concentration is less than 0.01 μM, no induction was recorded (domain A). From the first detection limit (DL1) corresponding to 0.02 μM TBT, the strain produced light but not linearly with the pollutant concentration. The linear domain started from 0.08 μM and finished one log later (domain B). When TBT concentration increased over 1.5 μM, a toxic effect was recorded leading to a complete light inibition at 10 μM.

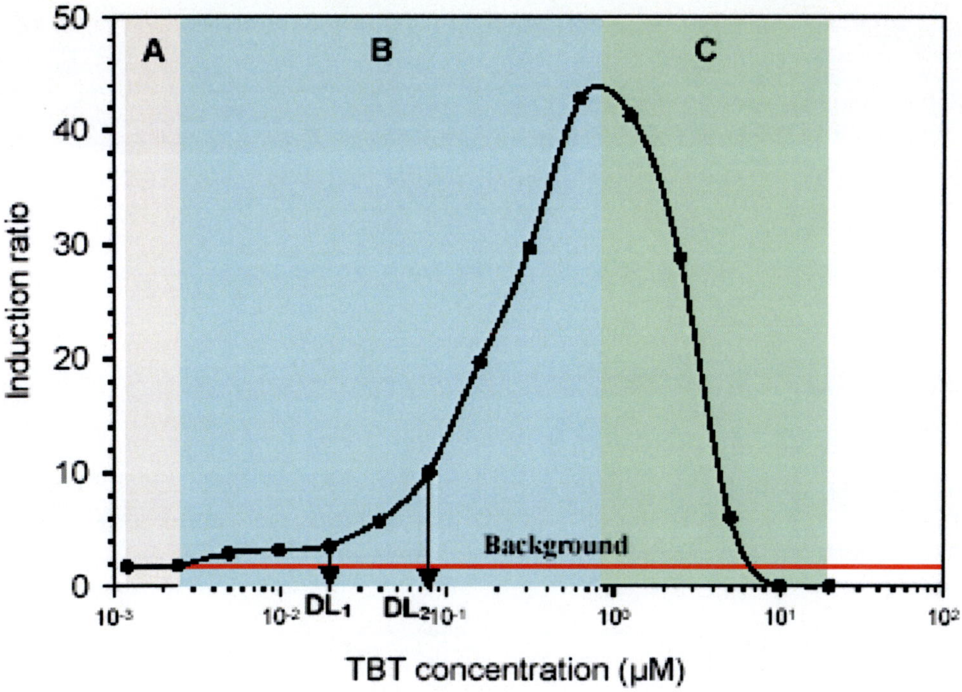

Figure 9. Induction of strain *E.coli* TBT3 with TBT (tributyl tin monochloride) diluted in $MgSO_4$. Each point shown are the averages of data from four replicates. Standard errors are below 1% and hence are not shown. For comparative assays, two other experiments were repeated and the trends were found to be identical. Adapted from Durand et al, 2003. The strain were first grown over night in glucose medium, diluted to OD 0.05 in $MgSO_4$ 0.01 M and then induced with several concentration of TBT. Bioluminescence was recorded after addition of decanal at 30 µM (final concentration). The induction ratio is the ratio between the light produced with TBT on the light produced without the compound.

6.3. Growth of *E. coli* TBT3 in Batch and Continuous Culture

Before to use a strain both for a bioassay or a biosensor, several preliminary steps should be considered. Among them, the most important is to evaluate the growth of the strain and the effect of the culture medium on the bioluminescence and the detection limit of the compounds.

6.3.1. Growth of E. coli TBT3 in Batch Culture and Selection of a Growth Medium

A selection of 33 representative publications from 1990 to 2001 about bioluminescent bioassays development showed that the rich Luria Bertani medium (LB) was often preferred for the growth of the cells than synthetic medium before induction with the target compounds (Table 2). Moreover, the growth phases from which the strain are harversted for induction was seldom mentioned.

We investigated the effect of the growth medium on the main growth parameters, on the bioluminescence and on the detection limit for TBT measurement.

Table 2. Media currently used for growth of recombinant *Escherichia coli* in microbial luminescent bioassays. Selection of 33 publications since 1990 to 2001

		Growth phases used before induction			
	TOTAL %	Not stated	Exponential phase		Stationary phase
			Early	Middle	
LB (rich)	82	74	19	7	0
Synthetic*	18	83	17	0	0

(*) : M9 : Na_2HPO_4, K_2HPO_4, NaCl, NH_4Cl, $MgSO_4$, $CaCl_2$, Glucose, Proline, Thiamine (Atlas, 1997) ; TY : bacto tryptone, bacto yeast extract, NaCl (Atlas, 1997) ; MM : R medium with 10 trace elements and pyruvic acid as carbon source instead of glucose. (Atlas, 1997).

Media and Methods

Luria-Bertani medium (LB) was prepared according to Atlas (1997) and adjusted to pH 7 before sterilization by autoclaving (120 °C, 20 min). Glucose medium with a C/N/P ratio (w/w/w) of 100/10/1 was prepared as described in Thouand et al, 2003b.

Batch growth experiments were performed in a 2 L reactor (New Brunswick, Toulouse, France) with a 1 L working volume of LB or glucose medium. Each batch was inoculated with 5 % (v/v) of a preculture of the bacteria grown for 15 h in the same medium. Dissolved oxygen and pH were monitored on-line with sterilizable probes (Mettler Toledo). Dissolved oxygen was feed-back regulated with the agitation according to the manufacturer's recommendation.

The respiration activity of the bacteria was evaluated by the measurement of INT reduction into formazan according to a modified protocol previously described by Bitton et al. (1986). The acetic acid concentrations (kit Boehringer Mannheim) were measured on the filtrate of a bacterial suspension according to the manufacturer's recommendation. The amount of intracellular ATP was measured on the pellet of a two milliters of a bacterial suspension that was first treated with 800 µL of a sterile cracking buffer. Lysed cells were centrifuged and the intracellular level of ATP was measured from the supernatant fluid with the ATP bioluminescent kit (Sigma, FLAA) according to the manufacturer's recommendation. Bioluminescence was monitored with a luminometer (Microlumat LB 96V, EG&G Berthold) in white opaque 96-well microtiter plates (Nunc). Each well contained 100 µL of the bacterial suspension sample initially exposed to several concentrations of TBT for 1 h at 30 °C (150 µL final volume). Bioluminescence was recorded after the automatic injection of 25 µL of a 210 µM decanal stock solution.

Growth of *E. coli* TBT3 strain was faster on LB medium than on glucose medium (respectively 1 ± 0,23 h-1 and 0,31 ±0,035 h-1). The rich medium allowed a higher cell production regardless of the CFU and OD production (Figure 10). Besides, one of the most significative difference is on the respiration activity and the ATP level. Initial level of oxygen dropped rapidly in less than 400 min for LB medium instead of 600 min for the synthetic glucose medium. The respiration activity (INT) increased and decreased dramatically for LB as well as the energetic level (ATP) in the early growth stage. Conversely, for glucose medium, these two parameters were more stable than for LB.

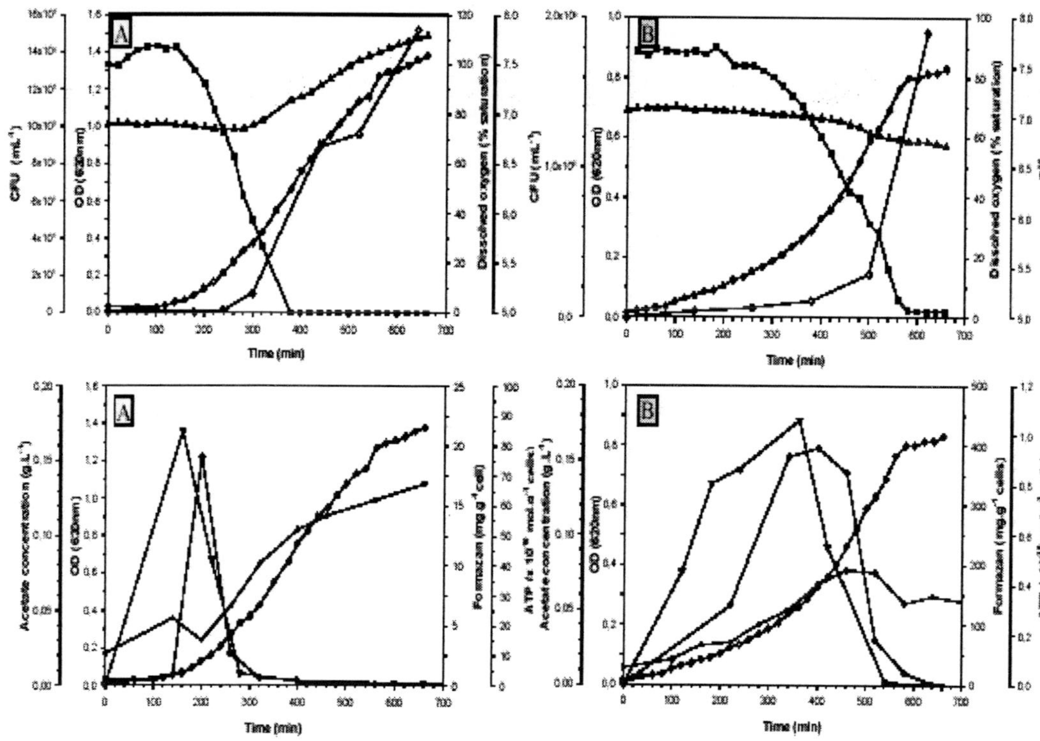

Figure 10. Growth of *E. coli* TBT3 in batch condition on a 2-L reactor using either LB medium (A) or Glucose medium (B). Conditions : 30°C, 250 rpm, $pH_{initial}$ 7, 3 vvm - volume of air/culture volume/min). Not published figures. Optical density, dissolved oxygen, pH, CFU, INT, ATP, acetate

The two media was then used to assess their effect on the detection limit of the strain for TBT. The strain *E. coli* TBT3 was grown in either LB or Glucose medium, harvested in early exponential phase, and was subsequently induced with TBT. Induction of *E. coli* TBT3 bioluminescence after growth in Glucose medium began at 0.02 μM and the light increased strongly up to 1.5 μM of TBT. After growth in LB medium, the shape of the figure was similar but the detection limit shifted to 1.5 μM. Considering the above results, all further experiments were performed in Glucose medium and besides the growth stage from which bacteria are harvested for subsequent bioassays was important to take into account. Indeed, the bioluminescence is growth dependent as reported by Figure 11. The light output is maximal in the acceleration and exponential growth phases and decreased with time when cells entering the stationary phase.

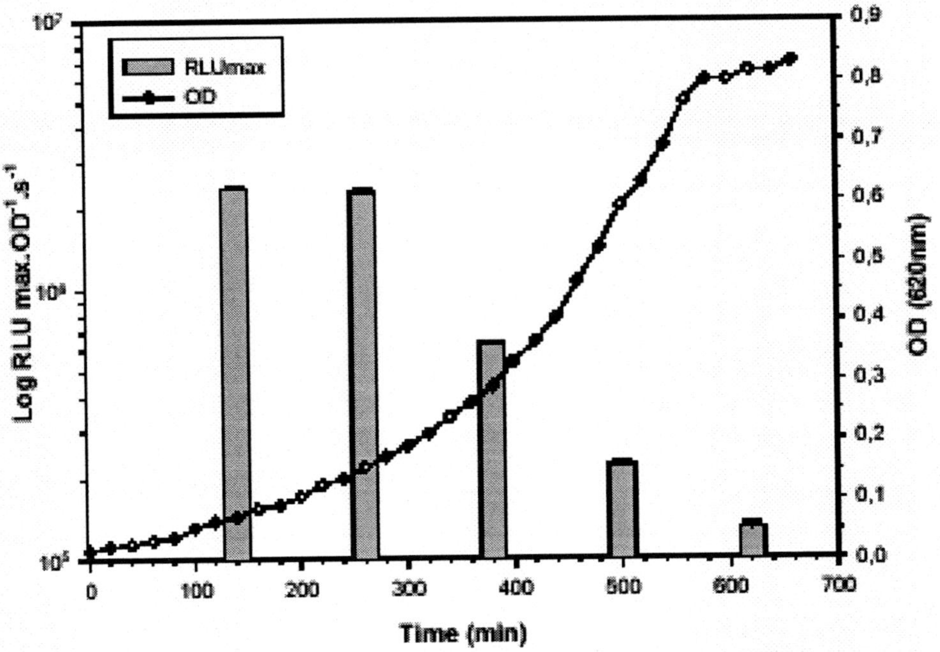

Figure 11. Growth stage dependence of bioluminescence in batch reactor of the strain E. coli TBT3 in glucose medium (30 °C, 250 rpm, pHinitial 7, 3 v.v.m air). The growth curve is similar to Figure 10B and was added for clarity. Bioluminescence corresponds to values from the most inducible TBT concentration. Experiments were performed three times. (Data are the average of three replicates with a standard error of less than 5 %). Adapted from Thouand et al, 2003b.

6.3.2. Effect of the pH of the Media on the Bioluminescence Of E. coli TBT3 Strain

The pH both influence the bacterial activity and the speciation of the TBT. In alcalin conditions, organotin compounds are not charged as in acid conditions the tin atom is a cationic group (Fent, 1996). Hence, these behaviors justified a survey aimed to verify the effect of the pH on the bioluminescence of the *E. coli* TBT3 strain as well as the usefulness of a buffer to maintain the pH of the glucose medium during the growth of strains. The table 3 shows results from several batch cultures on glucose medium with and without buffer. The effect on the cell production was also investigated.

Both two buffers are suitable at 50 mM for the maintenance of pH around 7 unit. Nevertheless, when the buffer concentrations increased, a toxic effect led to a decrease of cell production, especially for the PIPES buffer. We also verified that the presence of the buffer did not influence both the bioluminescence production as well as the detection limit due to a possible adsorption of the TBT to the Tris molecule (Table 3). Interestingly, glucose medium without buffer allowed data close to medium supplemented with Tris-HCl, nevertheless, due to potential long time use of the biosensor, we decided to buffer glucose medium.

Table 3. Effect of the buffer in the glucose medium on the pH and the cell production (Xmax) as measured with the optical density ($OD_{620\ nm}$). DL : detection limit. Not published data

	Control (no buffer)	PIPES (mol.L-1)			Tris-HCL (mol.L-1)		
	0 0.05	0.5	1	0.05	0.5	1	
pHi	7.2 7.1	7.08	7.17	7	7.06	7.1	
pHf	4.77 7.08	7.12	7.17	6.81	7	7.26	
. pH	-2.43	-0.02	0.04	0	-0.19	-0.06	0.16
Xmax ()	0.97	1.11	0.22	0.001	1.01	0.705	0.004
$RLU.OD^{-1}$ max	1.7 106				2.4 106		
DL (µM TBT)	0.08				0.08		

6.3.3. Effect of the Oxygenation on the Bioluminescence of E. coli TBT3 Strain

Oxygen is a key parameter both for the growth and the bioluminescence reaction of the strain. Simple aeration with bubbling was adopted for the growth of the bacteria and was then tested. Three batch cultures were conducted in 2 L bioreactors with exactly the same ratio between working volume and air space as those of the biosensor. Table 4 shows the results for the culture regulated with 50 % (B_{50}), 10 % oxygen regulation (B_{10}) or unregulated oxygen (B_{NR}) which is close to the conditions adopted for the biosensor. As expected, B_{10} offered the poorest conditions of growth with a small growth rate and a pH variation of 2.53 units due to acetate production (0.153 g.L-1), and the results from B_{50} and B_{NR} batches were similar. Interestingly, the bioluminescence produced by bacteria from B_{50} was significantly different from that in the B_{10} and B_{NR} reactions as revealed with a Student test (respectively $t_{B50,B10}$= 47, p = 1.25.10-6; $t_{B50,BNR}$ = 32, p = 5.4.10-6, dll = 4, $t_{student}$ = 2.776).

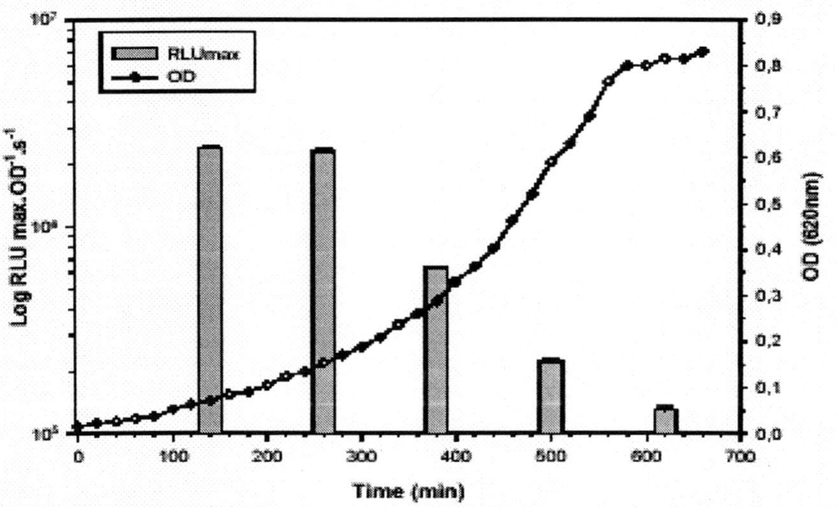

Figure 12. Effect of the oxygenation mode on a *E.coli* TBT3 strain culture continuously fed with glucose medium. Experiments were conducted in the mini-reactor (100 mL working-volume), 30°C, 250 rpm, pH initial 7, D = 0.6 h-1. B50 mode was regulated with pure oxygen bubbling. Bioluminescence was measured after manual addition of 30 µM decanal (final concentration). Not published data.

Table 4. Effect of several modes of oxygenation on the growth parameters of the *E.coli* TBT3 strain grown in batch conditions on glucose medium (30°C, pHinitial7). (adapted from Thouand et al, 2003b)

	Oxygen		
	Regulated		Not regulated
	B_{50} [(a)]	B_{10} [(b)]	B_{NR} [(c)]
	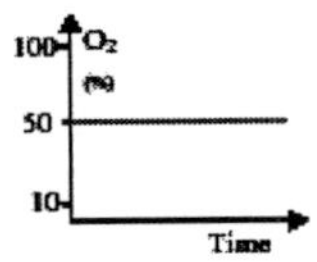	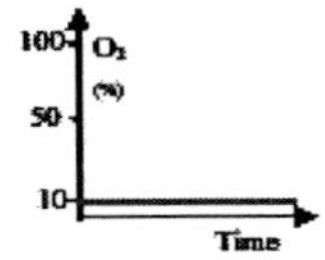	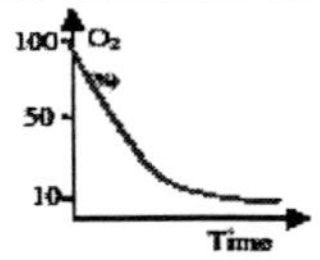
pH variation [(d)]	0.83	2.53	0.24
Glucose consumption (%)	91	74	91
Acetate production ($g.L^{-1}$)	0.055	0.460	0.153
OD_{max} [(e)]	1.04	0.38	0.98
Cells production ($CFU.L^{-1}$)[(f)]	$2.79x10^{11}$	$2.3x10^{10}$	$2.09x10^{11}$
Growth rate (h^{-1})	0.491	0.226	0.493
Maximum light production ($RLU.OD^{-1}.s^{-1}$)	$6.7x10^5$ ($2.2x10^4$)	$1.7x10^6$ ($3.3x10^4$)	$1.89x10^6$ ($6.8x10^4$)

[a] Oxygenation was regulated with agitation from 250 to 350 rpm, 3 v.v.m air

[b] Oxygen was removed at the beginning with N2 and then, growth was performed without air bubbling at 200 rpm.

[c] No regulation of the oxygenation throughout the course of growth. Constant agitation and bubbling (250 rpm, 3 v.v.m air)

[d] pH variation from the beginning to the end of the culture

[e] Optical density at the end of the culture

[f] Colony Forming Units in Agar glucose medium with antibiotics

The unregulated mode (BNR) was subsequently assessed with a continuous culture of the *E.coli* TBT3 strain in the biosensor at D = 0.6 h-1 (Figure 12). Decanal was manually added in the reactor (30 μM, final concentration) and the background level of bioluminescence was then measured. TBT was not added. When the culture was fully aerated at 5 mgO2.L-1 (50 %), bioluminescence increased sharply to 5000 RLU.s-1 followed by a rapid decrease due to decanal consumption. The same pattern was recorded for the unregulated mode that led to increased luminescence (approximately 7000 RLU.s-1). Thus appears that the unregulated mode did not affect the bioluminescence reaction and thus was suitable for the biosensor.

6.4. Growth of *E.coli* TBT3 in Continuous Culture Condition in the Lumisens1 Bioreactor

In the biosensor Lumisens 1, the strain *E.coli* TBT3 is continuously fed with the glucose medium. We attempted to investigate the effect of the rate of glucose feeding (dilution rate) on the bioluminescence.

All the experiments were performed in the mini-bioreactor of Lumisens 1 as a part of the biosensor described below (Chapter 8). Growths were started as for batch culture. Continuous feedings of sterile glucose medium were started when exponential growth phases were reached (OD620nm = 0.5). Dilution rates ranged from 0.3 to 0.9 h-1. Samples were taken at steady-state (10 folds the hydraulic residence time) from the outlet of the reactor to measure the descriptors of each culture (OD, CFU, INT, ATP) as well as the bioluminescence after induction with the most inducible TBT concentration. The pH, oxygen, temperature and agitation were recorded as for the biosensor (see below).

Figure 13A plots light output after TBT induction and the culture descriptors (INT and ATP levels) as Figure 13B plots the biomass descriptors as well as glucose and acetate concentrations inside the reactor. When dilution rate increased, biomass, glucose and acetate decreased confirming the continuous culture theory (Kubitschek, 1970). Interestingly, light emission was clearly influenced by the dilution rate. Bioluminescence increased approximately ten-fold when the dilution rate was raised from 0.3 h-1 to 0.9 h-1. The explanation could be multifaceted, since at high growth rates (high dilution rates) the main components (RNA, proteins and DNA) are augmented (Schaechter et al. 1958), the envelope composition is modified (Shokri et al. 2002) and the availability of both decanal and TBT could be changed. One another explanation could be that at D=0.9 h-1, close to the .washout. level, the cells consumed merely 12 % of the substrate and overall metabolism dropped. Hence, it is possible that oxygen, normally used for metabolism, should then become available for the bioluminescence reaction. The overall cellular activity (INT and ATP levels) increased with the dilution rate up to 0.8 h-1. After that, a decrease was mainly observed for ATP. This overall activity drop did not affect both the bioluminescence and the response of the strain for induction with TBT.

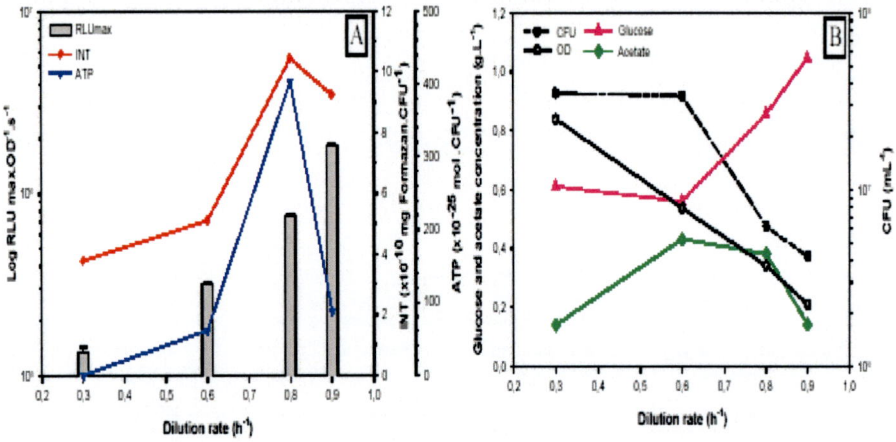

Figure 13. Comparison of the growth and activity of the *E.coli* TBT3 strain fed with glucose medium (30 °C, 250 rpm, pHinitial 7, 3 v.v.m air) for several dilution rates. Experiments were performed twice. (Data are the average of three replicates with a standard error of less than 5 %)
INT : (2-p(iodophenyl)-3(p-nitrophenyl)5-phenyltetrazolium chloride) represents the overall respiration activity; ATP : adenosine triphosphate; CFU : colony forming unit in agar glucose medium with antibiotics; OD (absorbance) : optical density at 620 nm. (Adapted from Thouand et al, 2003b).

7. Development of a Bioassay for the Screening of TBT in Water and in Sediments

Many bacterial bioassays have been developed for toxicity evaluation of water, wastewater, sediment eludriats, soil, some of them used recombinant bacteria (Billard and Dubow, 1998, Kölhler et al 2000). Performances of bioassay include specificity of the strain for compounds, detection limit, repeatability and reproducibility.

7.1. Experimental Conditions of the Bioassay with Liquid Samples

Cells were grown over night in glucose medium (14 h) at 37°C with appropriate concentration of ampicillin and tetracycline (Durand et al, 2003). The cells were diluted to an OD620 nm (absorbance) of 0.075 in fresh medium. Culture aliquots (100µl) were added to each well of an opaque white 96 microwell plate (NuncTM, ref.236108) containing dilution series of compound to be tested in 50µl sea water or spring water. After 1 hour induction (30°C), luminescence was followed using a temperature controlled (30°C) microtiter plate luminometer (Microlumate L96V, EG§G Berthold) after automatic injection of decanal to reach a final concentration of 30 µM. The Induction Ratio (IR) was calculated as follow : IRi = (RLU.s-1.DO-1)i / (RLU.s-1.DO-1)o where (RLU.s-1.DO-1)i is the luminescence after induction with organotin and (RLU.s-1.DO-1)o is the background luminescence. We consider that a compound induces the bacteria if IR value is more than 2,5.

Organotin stock solutions (600µM) were prepared with 60 ml ethanol (70%) adjusted to 100 ml with pure water. From stock solutions serial dilutions were prepared in synthetic sea water pH 8.2 (Instant Ocean® aquarium system, Sarreboug, France) or spring water pH 7,4 (Prystel®, Vendée, France).

7.2. Specificity and Sensitivity of the Bioassay

The shape of bioluminescence curve was found to be almost gaussian (Figure 14) and whatever the medium used for the dilution of TBT, solution of MgSO4 (10-2 M), spring water (pH 7.3) or sea water (pH 8.2). Bioluminescence increased when TBT was diluted in sea water. This result was not due to a difference of pH between sea water and spring water (see below).

All the commercial formulations of chloride form of organotin compounds were tested in order to assess the specificity of the strain *E.coli* TBT3. As it can be seen in figure 14, DBT and TBT significantly induced the bacteria. The detection limit was 0.1 nM for DBT (0,03 µgL-1) and 0.08 µM (26µgL-1) for TBT. For the other form, MBT, TPT and TBTox, a decrease of light output was observed, the maximum induction ratio (less than 6) was observed with 5µM of TBTox and 10 µM of MBT, no induction was observed for TPT. The induction ratio was not found to be proportional to the number of butyl group bound with tin. The substitution of butyl group with phenyl group led to a loss of sensitivity of the strain.

As TBT could be bound to bromide, chloride, iodide, the effect of the halogen group on the induction of the strain was studied. The three halogen forms of TBT induced the stain. TBTBr was the most potent inducer with the detection limit of 0.02 µM, for the two other forms, the detection limit was the same (0.08 µM) (see Durand et al 2003 for details).

The presence of tin atom was also found to be important for the light production of the strain. Organogermanium and organomercury were tested in same conditions as organotin, no induction was observed when tin was substituted by germanium or mercury.

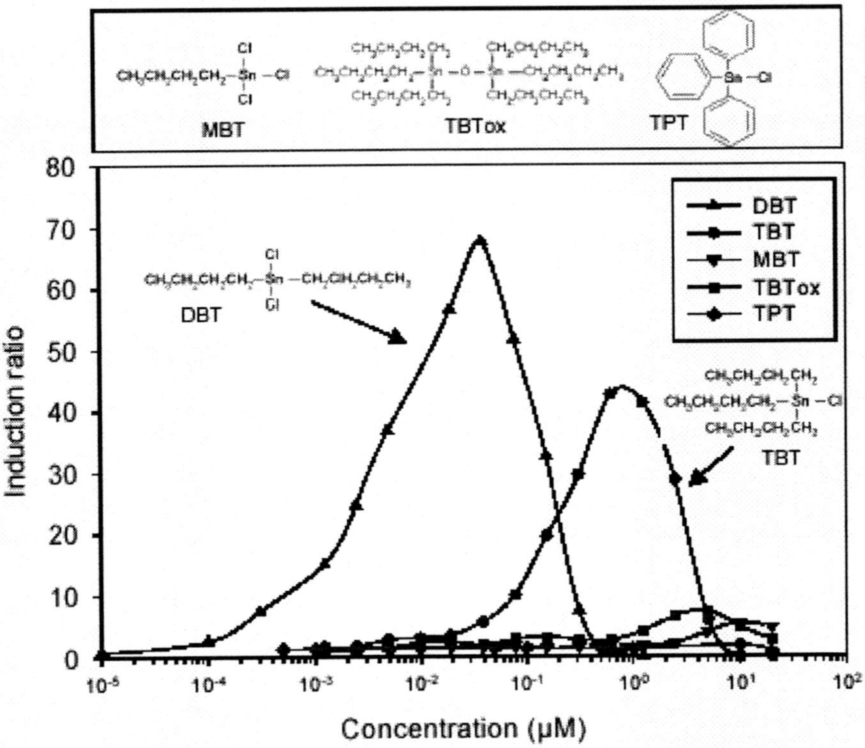

Figure 14. Specificity of strain *E.coli* TBT3 for several organotin compounds. Each point shown are the averages of data from four replicates. Standard errors are below 1% and hence are not shown. Organotin were diluted in synthetic sea water. (TBT : tributyl tin monochloride; DBT : dibutyl tin chloride; MBT : monobutyl tin trichloride; TBTox : Bis(tributyl tin) oxide; TPT : triphenyl tin monochloride.) From Durand et al 2003.

The repeatability of the bioassay was evaluated. Forty assays were performed with TBT or DBT under the same conditions with the same inoculum. The coefficient of variation between these experiments was 8-9%. The reproducibility was also assessed after four independent experiments performed on different days with different experimenters. The lowest TBT concentration detected with the strain occurred at a mean value of 0.076 µM with a coefficient of variation of 14%, while

the highest TBT concentration that eliminated the light (toxic concentration) was 7.25 µM with a coefficient of variation of 13%.

7.3. Influence of the pH of Spring Water

Organotin compounds are dissociated in water. The dominant TBT species at pH<pKa (6.51) is cation (TBT+ + Cl-). At pH 8.0, TBT predominates as the neutral TBTOH species. Fent (1996) reported a pH dependant toxicity of TBT on *Daphnia magna*, due to the fact that TBTOH can penetrate biomembranes much more easily than the cation form.

In order to test the influence of the pH of the water on the induction of the bacteria, three different pH on spring water (Prystel®, Vendée, France) were tested : the natural pH of the water (pH 7.3), acid water (adjusted to 5.5 with HCl 1M) and basic water (adjusted to 8.6 with NaOH 1M). Bioluminescence intensity was influenced by the pH of the water. The shape of bioluminescence curve was found to be the same whatever the pH of the water used for the dilution of TBT. Moreover, the detection limit for TBT was not affected (Figure 15).

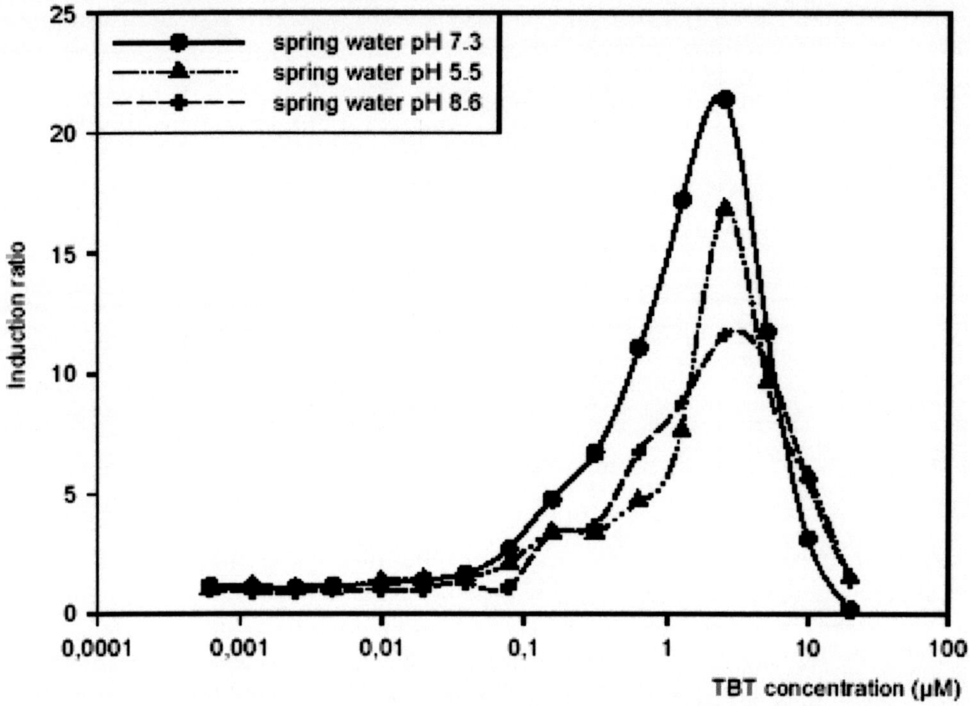

Figure 15. Influence of the pH of spring water on the induction of strain *E.coli* TBT3. Each point shown are the average of data from four replicates. Standard errors are below 5% and hence are not shown. (not published data)

7. 4. Application of the Bioassay to Solid Matrix

Generally, concentrations of TBT in coastal water range from 0.2 to 16 ng Sn.L-[1] (0.55 to 44 ng TBT-Cl L-1). From these data only the highest concentration could be detected by *E.coli* TBT3 in water samples. Nonetheless, concentrations of TBT and DBT in sediments are in the range of the detection limit of the bioassay (Figure 16).

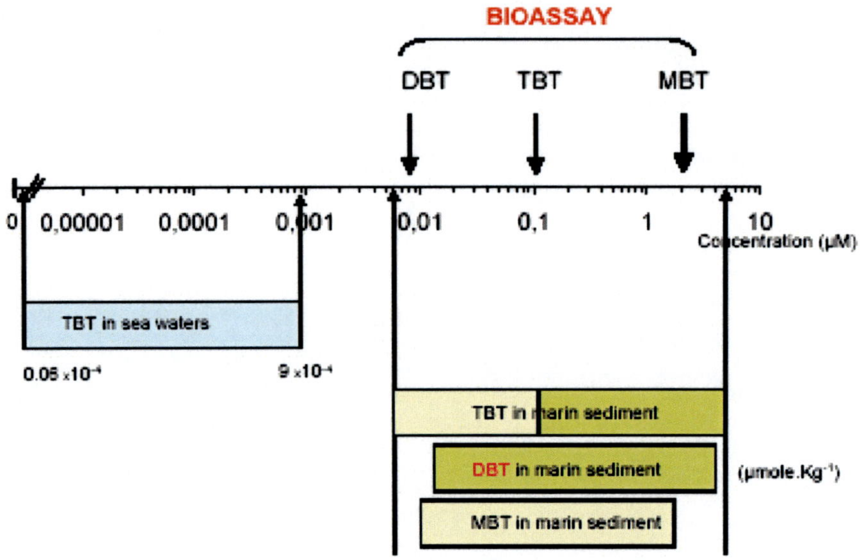

Figure 16. Concentration ranges of organotin in aquatic ecosystem and sensitivity of the TBT bioassay.

Chemical analysis results are precise but the bioavailability of the pollutant is not taken into account due to the extraction procedure. The development of solid-phase contacts assays consist to a direct contact between the bioelement and the matrix (soil or sediment). Several of these assays were used in order to assess toxicity of sediment (Microtox® bioassay) or for the measurement of bioavailability of pollutant such as heavy metals (Ivask et al, 2004) or organic compound (Brandt et al, 2002). In order to apply the detection of TBT in a solid matrix with the strain *E.coli* TBT3, sand was contaminated with TBT sea water solutions. Spikes of TBT were added in samples of sand form Fontainebleau (dry sand < 500 µm, Labogros, France). A mass of 20 g (dry) sand was contaminated by 10 ml of sea water with TBT and mixed at 300 rpm during 2 hours .

Bioassay was conducted as follow: 1 g of wet contaminated sand received 1ml of bacteria ($OD620_{nm}$ 0.2). Tubes were incubated 2 hours at 30°C (300 rpm). After incubation samples were settled 5 minutes at room temperature. Luminescence was analysed with 500 µL of the supernatant with a MiniLumat luminometer (LB9506, Berthold) after addition of 100 µl decanal (300 µM).

Results ploted in Figure 17 shows that luminescence increased for concentrations above 0,01 µg.g-^{1} of TBT. In the case of concentrations over than 10 µg.g-1 a decrease of luminescence was observed due to a toxic effect of the sand. Nevertheless for the concentrations 0,001 and 0.01 µg.g-1 luminescence was not dose response related, perhaps due to the heterogeneity of the sample The preliminary results showed that luminescence should be related to the concentration of bacteria (for example Colony Forming Unit).

As adsorption mechanisms of TBT and DBT are dictaded by mineralogical and chemical composition of the sediment (Hoch et al 2003, Behra et al 2003), we are going to study the detection of TBT and DBT with the strain in different matrix as for example benthonite and kaolonite.

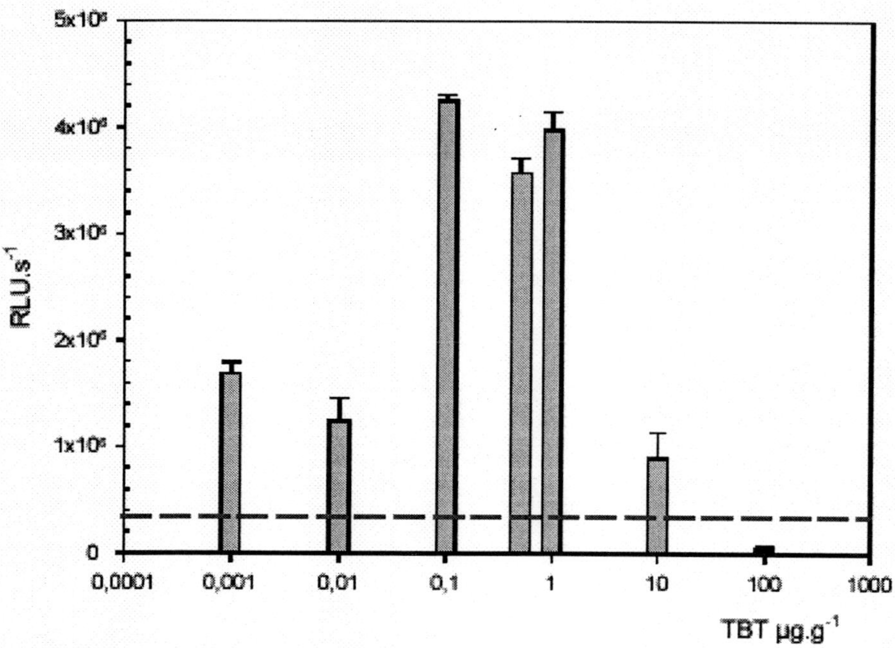

Figure 17. Application of the strain E. coli TBT3 to a solid phase contact bioassay. Bacteria were incubated 2H with a TBT contaminated sand and then the supernatant was used for bioluminescence measurement (----) Background of luminescence (bacteria with uncontaminated sand)

7. 5. Conservation of the Strain

The main concern for a microbiologist is the way to conserve the strains. The conservation methods should allow an inexpensive way and it should be easily handled for a field analysis. In order to match these conditions, we tested several protocol of conservation with inclusion in agarose and with lyophilization.

Immobilization in Agarose

Inclusion of *E.coli* TBT3 bacteria in agarose was conducted as follow. Low melting agarose (SoBiGel, Hendaye, France) solution was prepared by dissolving 2% agarose (w/v) in MgSO4 (10-^{2}M) at 70°C. After homogenisation, one volume of agarose solution was kept at 37°C and mixed with one volume of bacteria (OD620 nm 0.4). Aliquots (100µl) of this solution were casted in microtiter–wells (NuncTM). After solidification, the microplates were sealed and kept at 4°C. In order to analyse the conservation effect, the bacteria were induced by TBT solution prepared in sea water. Induction of the strain was done with solution of TBT as describe above. After 2 hours of incubation (30°C), 25µl of *N*-Decyl aldehyde solution (700µM) were injected in each well and the luminescence was recorded.

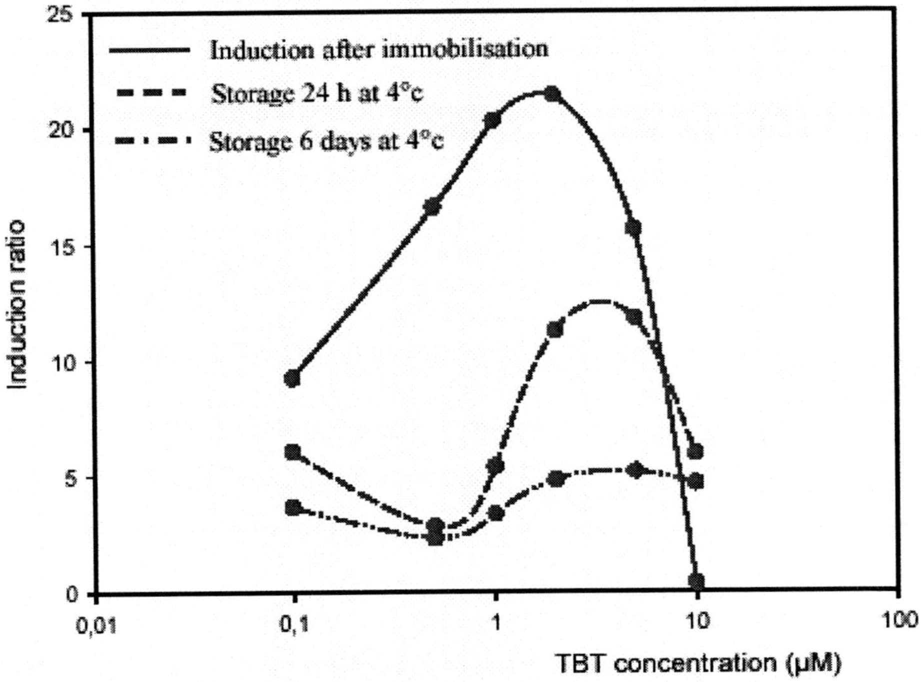

Figure 18. Induction of the strain *E.coli* TBT3 after immobilization in agarose (1% final concentration) with TBT in sea water. Induction was evaluated just after immobilization and after storage at 4°C. Each point shown are the average of data from twelve replicate. Standard errors are below 10%. (not published data)

Figure 18 shows that just after immobilization, the induction of bacteria were preserved, nevertheless after 24 hours of conservation at 4°C, a strong decrease of induction was observed. This protocol of conservation in agarose was not appropriate for a long term conservation.

Lyophilization of the Cells

The selected protocol was adapted from Wagner and Van Dyk (1998). One liter of glucose medium contained in a 2 liters bioreactor (New Brunswick, Discovery 100) was inoculated with 10 % of an over-night preculture of *E.coli* TBT3 strain. The use of a bioreactor with speed, oxygen and pH controls is one of the main conditions for the success and the reproducibility of the conservation. Cells were grown for 14 hours to the stationary phase (OD_{620nm} = 0.8) at 37°C. Oxygen (50 %) was regulated by feedback regulation with aeration and agitation. The pH was regulated to 7 via the addition of 0.2 M HCl or NaOH solutions sterilized by autoclaving. Cells were then centrifuged and resuspended in a mixture containing one volume of fresh glucose medium plus one volume of a sucrose solution at 24 % (W/V). The final optical density was adjusted to 0.1 unit. One hundred microliters were then added in each well of a 96 well-microtiter plate, frozen for 18H at -80°C and then lyophilized for 24 H in a Christ Alpha 1-2 lyophilizer. Before the bioassay, the lyophilized bacteria were reconstituted with 100 µL of pure water for 30 min at 30°C. Figure 19 shows the results of *E.coli* TBT3 strain induction with TBT before and after 4 and 8 months of conservation at –20°C. The response of the lyophilized strain was very close in comparison to the strain before lyophilization. Both the detection limit and the maximum induction ratio were comparable.

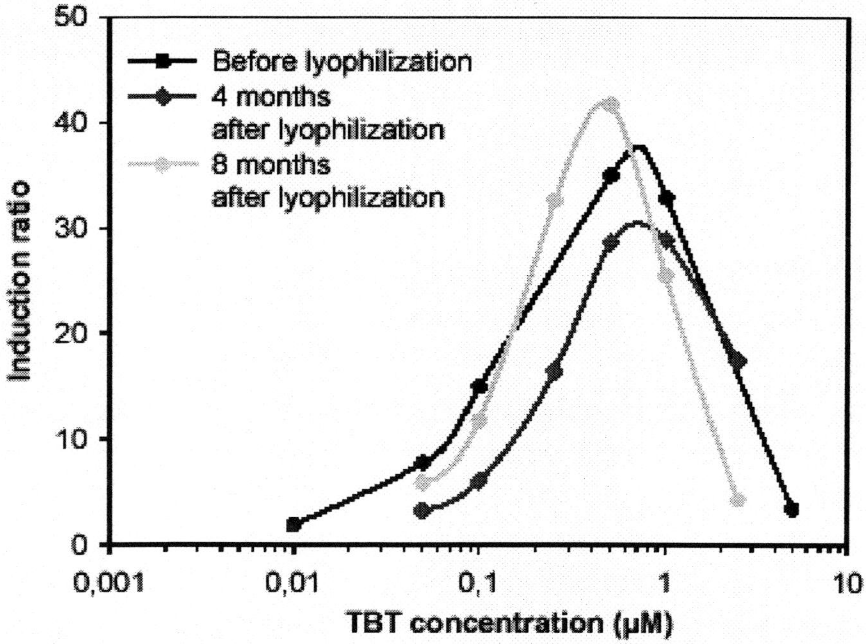

Figure 19. Induction of the strain before and after lyophilization with TBT in sea water after 4 and 8 months of conservation at –20°C. Each point shown are the averages of data from twelve replicates. Standard errors are bellow 5 % and hence are not shown. (Not published data)

8. LUMISENS : A RANGE OF BIOSENSORS FOR ON LINE MONITORING TBT/DBT

Biosensors can be divided in two categories: (1) bioluminescent cells immobilized in a matrix (Choi and Gu 2002; Bolton et al. 2002; Polyak et al. 2001; Ikariyama et al. 1997; Heitzer et al. 1994) or (2) bioluminescent cells in suspension in a growth medium (Gu et al. 1996, Thouand et al, 2003b). In our approach, the latter category was chosen for the development of a biosensor in order to detect on line pollution of TBT in continuous culture conditions. For the development of our biosensor, several features were first chosen before to start its construction (Table 5). LUMISENS was the name of all the biosensor developped. As other biosensor are currently developed in our laboratories, we added a number to each prototype. We present here the range LUMISENS 1.

Here, we will describe in a first part the different kind of biosensors as well as the overall set up needed to the on line measurement of TBT or DBT. In a second part, we will focuse on the description of the bioreactor of Lumisens 1B and the transducer and the third part will illustrate the validation of the biosensor for the measuremnt of TBT.

Table 5. Initial features wished for the development of the biosensor Lumisens 1

Characteristics	Features whished for LUMISENS 1 development
Transportable	Should be transportable in other sites or fields
Size	The working volume should not exceed 100 mL to allow economic use of reactant
Autonomous	Should be at least autonomous for 7 days with one or two control per day
Automation and control	Should allow automatic oxygen, pH, temperature, optical density and bioluminescence measurements.
Oxygenation and agitation Simple	air bubbling and magnetic stirrer
Light measurement	With a highly sensitive photomultiplier chosen thanks to the spectral measurement of bacterial bioluminescence
Material	Should allow sterilization with autoclave Lightproof material

8.1. Overall Set Up of the LUMISENS 1 Biosensors

It is interesting and important to consider that to match the features mentioned in Table 5, three years were necessary in order to produce a first biosensor ("ancestor") to the most developed LUMISENS 1B. The figure 20 shows also that it is possible to construct a simple device with key elements including a stainless-steel reactor with temperature controller, a motor for agitation, a photomultiplier connected to an amplifier and a computer for acquisition of data. Here we used a home-made software on LAB VIEW 5.1.

Thanks to the first prototype, Lumisens 1A was developed. We were able to improve the reactor by addition of probes and a glass jacket avoiding the contact of the bacteria with the steel. The PM was included in a lightproof box to avoid light disturbance. In order to avoid biofilm formation onto the lens, the admission pipe from which the fiber optic conveys light to the PM was on the cover. A set of home-made transducers was added in order to move the biosensor outside of the laboratory (see below). Finally, Lumisens 1B was developed in order to measure both the bioluminescence and the cell density (optical density) inside the reactor, without contact and disturbance of the cell suspension (French patent N°2831182). The feed of glucose as well as the outlet of the medium in the continuous culture mode are allowed by a peristaltic pump with at least 8 rollers (see Thouand et al, 2003b and Picart et al, 2004 for a complete technical description).

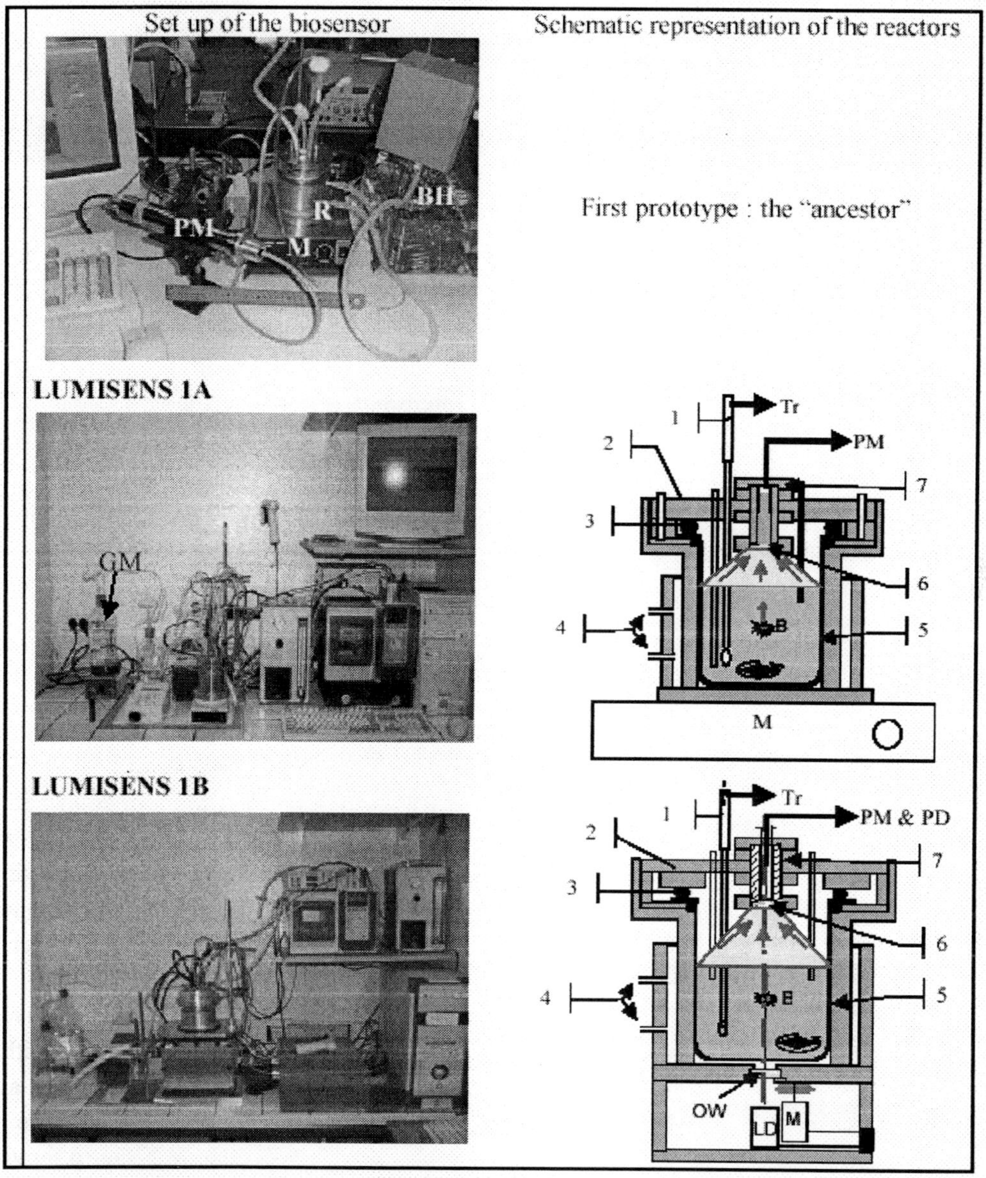

Figure 20. Photographies of the 3 biosensors developped during 3 years by our research team and drawings of the related mini-reactors (Not published photographies)
B: bioluminescent bacteria (*E.coli* TBT3 in this case), **BH**: bath heater; **GM**: Glucose medium, **LD**: laser diode; **M** : motor, **OW**: optical window, **PD**: photodetector, **PM**: photomultiplier tube, **R** : mini-reactor, **Tr** : transducer for oxygen, pH and temperature sensors - **1**: sensors (oxygen, pH and temperature), **2**: cover (the screws were not mentioned in this draw for clarity), **3**: o-ring seal, **4** : temperature control with BH (optional), **5** : glass lining containing bacteria, **6** : lens, **7** : fiber optic admission pipe.

8.2. Overview of the Lumisens 1B Biosensor

Lumisens 1B was built to improve the autonomous of the device. The aim was to avoid the manual sampling in the reactor for the cell density measurement. The latter is important to measure in the maintime with bioluminescence since these two parameters are cross dependent. Thus, the bioluminescence could decrease either when pollution is present or when the cell density decreases, leading to the misinterpretation of the toxicity of an environmental sample.

8.2.1. Principle of the Simultaneous Measurement of Bioluminescence and Cell Density

The cell density was measured according to the absorbance principle based on the Beer-Lambert's law, and the bioluminescence was measured by light acquisition with a photo-detector (Figure 21). A laser beam (ë = 670 nm) crosses the growth medium and is focused on to a fibre optic through the light admission pipe, with the bioluminescent signal emitted from the cells. The two signals both are conveyed equally in two twin fibers optic. One towards the PMT and the other towards the photodiode. Then, the bioluminescence signal is isolated from the laser signal using a short pass filter that cut any signal over than 650 nm and is directed towards a highly sensitive photomultiplier for the measurement of bioluminescence. The laser beam signal is discriminated from the bioluminescence signal thanks to their own physical property. The former signal is a frequence as the latter is a continuous phenomenon. Consequently, the photodiode, sensitive to a change of frequency, takes only into account the laser diode beam.

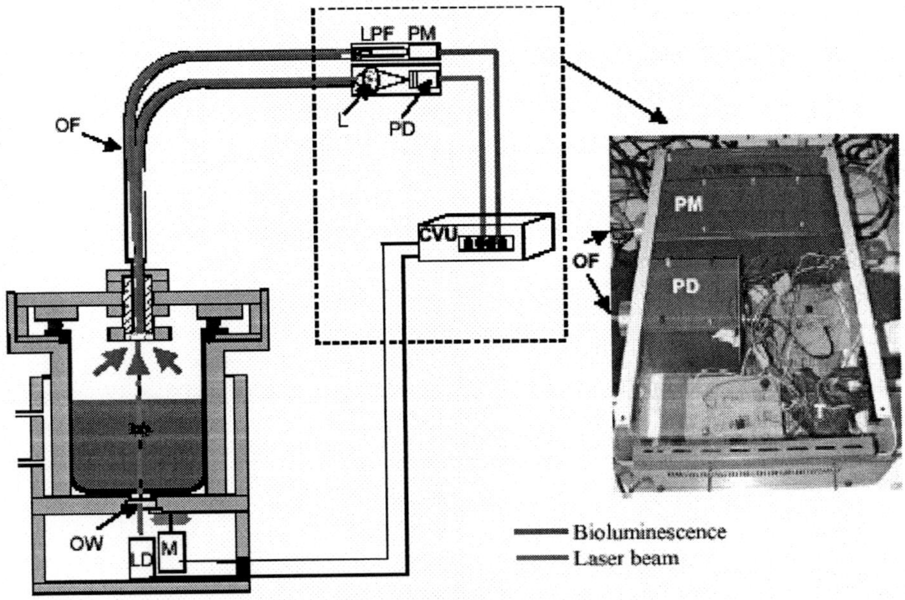

Figure 21. Schematic representation of the setup necessary for the on line and *in situ* measurement of bioluminescence and cell density in the Lumisens 1B biosensor. **CVU**: control voltage unit, **L**: lens, **LD**: laser diode, **LPF**: low pass filter, **M**: motor, **OF**: optical fibers, **OW**: optical window, **PD**: photodiode, **PM**: photomultiplier tube. Not published photographies.

The system has been validated with the naturally luminescent *Vibrio fischeri* strain. Results have shown strong correlations between manual measurements using commercially available equipment, and automatic measurements using the biosensor (Picart et al, 2004). As a consequence, LUMISENS I, in accordance with its original design, can use any *lux*-engineered or naturally luminescent bacteria for monitoring pollutants.

8.2.2. Automatic Bioluminescence and Cell Density Measurements

As example, the figure 22 shows the automatic monitoring of both the optical density (OD) and the bioluminescence (BL) of the strain *E.coli* TBT3 fed continuously with glucose medium without and with decanal. The latter is one substrate of the luciferase, the key enzyme of the bioluminescence reaction.

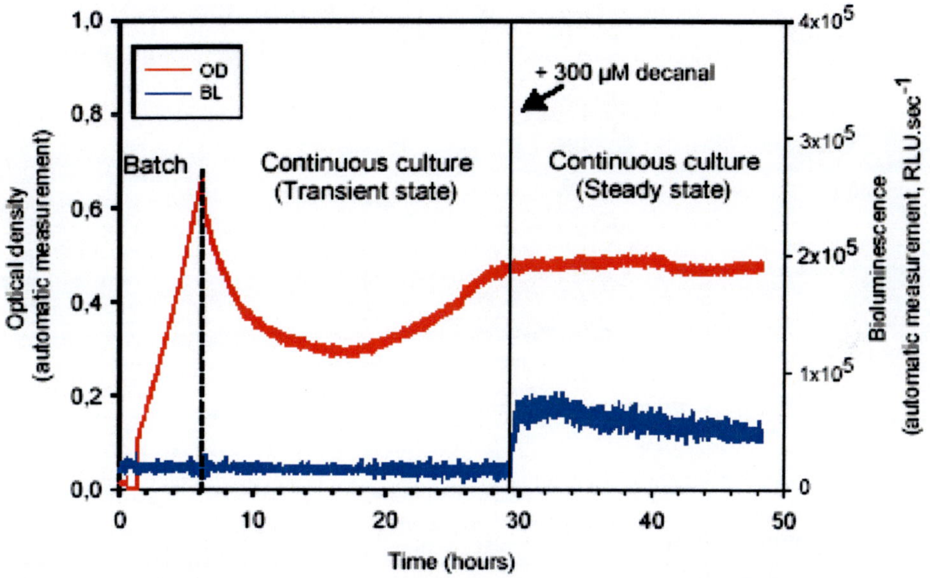

Figure 22. Automatic monitoring of bioluminescence (BL) and cell density (optical density) with Lumisens1B during batch and continuous cultivation of *E.coli* TBT3 strain in Glucose medium with TrisHcl (50 mM). D = 0.9h-1, 30 °C, 250 rpm, $pH_{initial}$ 7, 3 v.v.m air. Not published results.

Thanks to the automatic measurement, it was easy to distinguish the batch time (beginning of every continuous culture) followed by a transient state of about 20 h and then the steady state with glucose+decanal. From this stage, the measurement of TBT can be assessed.

8.3. On Line Detection of TBT

When remembering the genetic construction of the strain *E.coli* TBT3, it appears that an exogenous aldehyde must be added in the bacterial culture to measure bioluminescence since the strain harbors only the *lux*AB genes from *Vibrio harveyi*. The latter produce a large aldehyde (tetradecanal, Ulitzur and Hastings, 1979) but its low solubility is a major inconvenience and decanal is commonly preferred (Meighen 1994). The external addition of decanal was then undertaken and two main possibilities were examined : a manual or a continuous addition of the aldehyde in the bioreactor. In the meantime, the culture descriptors were followed in order to assess the possible toxic effect of decanal on the strain (Figure 23). In the former case, decanal was injected twice at 30 µM in a continuous culture of *E.coli* TBT3 fed with glucose medium at D= 0.9h-1. *i.e.* before TBT injection to record the background bioluminescence (constitutive level of luciferase) and just after induction with the compound (2 μM). This two-staged mode was tested and allowed a significant bioluminescence output that levelled off 10 hours after injection of the aldehyde (Figure 24).

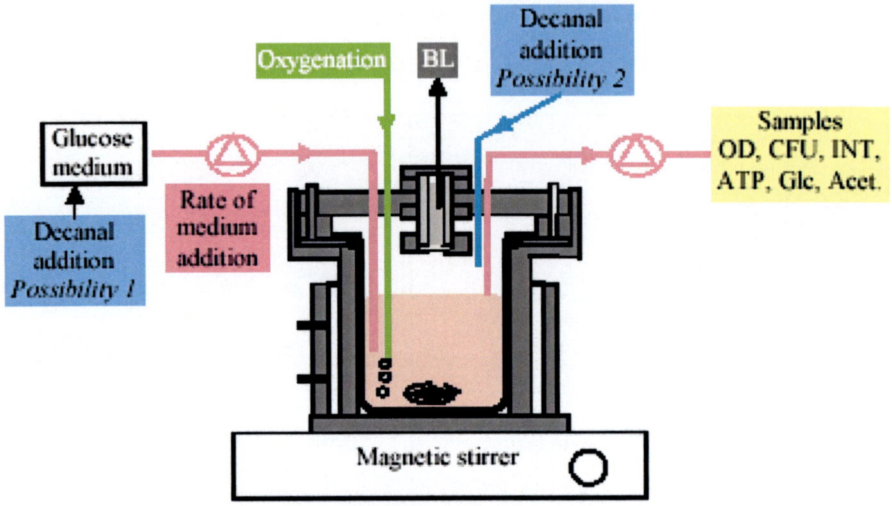

Figure 23. Schematic representation of the mini-reactor of Lumisens 1A used to evaluate the influence of decanal addition mode. (BL: bioluminescence, OD: optical density, CFU: colony forming unit, INT : iodonitrophenyltetrazolium, ATP : adenosine triphosphate, Glc : glucose, Acet.: acetate). (Decanal addition : Possibility 1 = continuous, Possibility 2 : two stages)

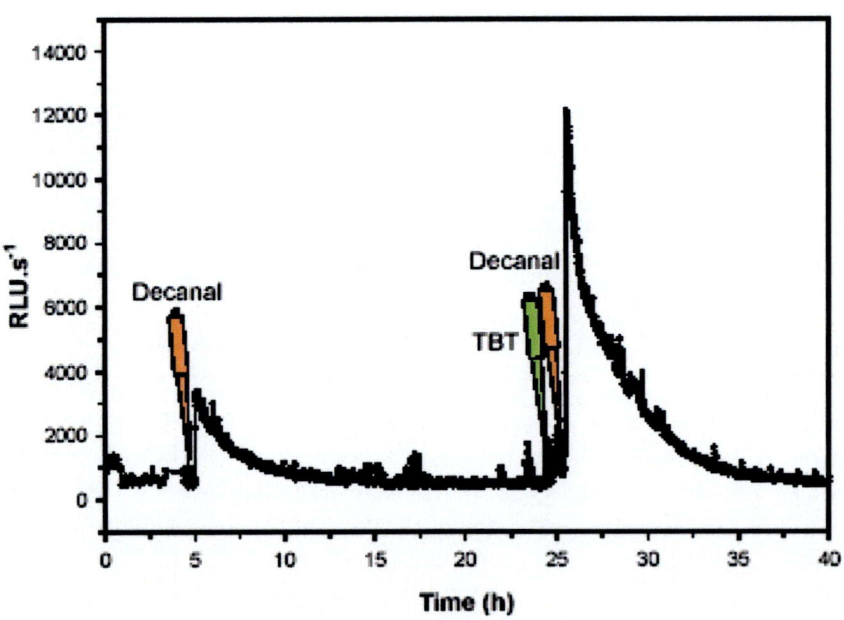

Figure 24. Bioluminescence of *E.coli* TBT3 in Lumisens1A in the two stages mode of decanal addition. Cells were grown at D=0.9 h^{-1} in glucose medium. TBT was added at 2 μM in the reactor (30 °C, 250 rpm, $pH_{initial}$ 7, 3 v.v.m air). Decanal was added twice at 30 μM (final). (Modified from Thouand et al, 2003b)

A second possibility is the continuous supply of the aldehyde. Decanal was introduced in the glucose medium in order to reach a final concentration of 30 and 300 µM and then the mixture was fed to the bacteria at D = 0.9 h^{-1} or 0.6 h^{-1}. One example of results is displayed with Figure 25 and all the results are gathered on Table 6. Clearly, decanal can be added on line in the reactor during time with the feeding medium but both the dilution rate and the decanal concentration played a role on the bioluminescence level.

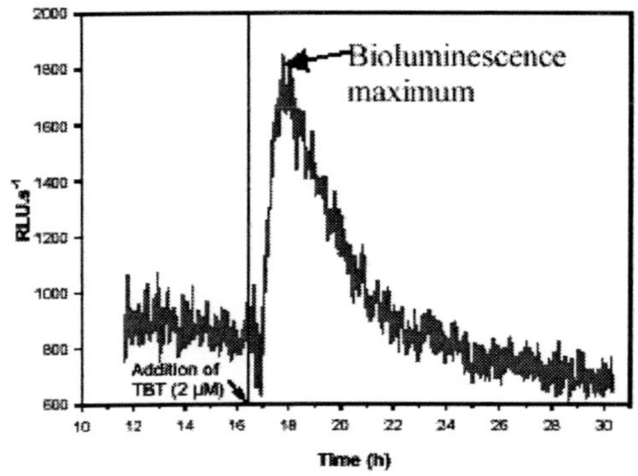

Table 6. Maximum bioluminescence output before and after TBT induction (2µM) in four conditions of decanal and dilution rate.

	Decanal 30 µM		Decanal 300 µM	
D (h^{-1})	-TBT	+TBT	-TBT	+TBT
0.6	400	400	900	1200
0.9	350	600	900	1800

Figure 25. Example of a continuous culture of *E.coli* TBT3 fed with glucose medium complemented with 300 µM of decanal at D = 0.9h^{-1}. The data in Table 6 are coming from the "Bioluminescence maximum" showed in the graph. Same culture conditions as above. (Adapted from Thouand et al, 2003b).

Indeed, the lower bioluminescence was monitored at a low dilution rate and at a low decanal concentration. Nevertheless, the background level (without TBT) is also influenced by the decanal concentration and not the dilution rate (Table 6). The best conditions allowing a high bioluminescence after induction with TBT is a high dilution rate (D=0.9 h-1) and a high decanal concentration (300 µM). One of the main concern was to know the effect of decanal aldehyde addition on the bacterial culture activity after a long time addition. Indeed the in vitro bioluminescence reaction is inhibited at aldehyde levels exceeding 100 µM (Holzman and Baldwin, 1983). Moreover, the aliphatic aldehydes are known to form Schiff bases with protein amino group (Holzman and Baldwin 1983) and the use of 300 µM aldehyde concentration could have been toxic for the culture. This was not observed and the aldehyde was not found to be toxic throughout the experiment (Table 7). The ATP and INT levels increased dramatically meaning probably that the overall bioluminescence reaction requires both energy and a high respiratory activity (INT). Surprisingly, glucose was better consumed (about 3 times) and the by-product acetate increased (about 3.6 times) when decanal was introduced in the culture. Nevertheless, the yield of glucose to acetate conversion did not change significantly. Clearly, decanal boosted the main metabolic pathways of the cells but as glucose was consumed, the carbon flux was not used to form new cells.

The yield of biomass production decreased from 4.5 to $1.5x10^8$ UFC.g glucose^{-1}. The optical density rise should be related to the cell volume increase.

Table 7. Effect of continuous addition of decanal on several descriptor of the *E.coli* TBT3 culture fed with glucose medium (GM) or glucose medium and decanal (300 μM). Adapted from Thouand et al, 2003b.

Before TBT induction	After TBT induction		
	GM only	**GM + decanal 300μM**	
	0.19 (0.04)a	0.3 (0.06)	0.25 (0.02)
CFU (106 mL-[1])	139.5	173	150
Glucose (%)[b]	22	67	75
Acetate (g.L-[1])	0.07 (0.01)	0.25 (0.03)	0.26 (0.03)
$YX/_S$ (x108 UFC/g Glc)	4.5	1.9	1.5
Glc (g/g) %	23	27	25
ATP (10-[25] mol.CFU-[1])	99 (5)	565 (28)	663 (33)
INT (10-[10] mg.CFU-[1])	0.29 (0.03)	1.64 (0.2)	1.75 (0.2)

[a] Standard deviation; bglucose consumed

Lumisens 1 was used to monitor TBT addition with the followed optimised operating conditions: 300 μM of decanal incorporated into a glucose medium, with simple atmospheric air bubbling and at D = 0.9 h-1. Both bioluminescence and optical density were automatically measured and one example is showed in Figure 26. The injection of 1.25 μM of tributyltin (final concentration) led to a strong increase of bioluminescence after one hour. Then, the light emission slowly decreased to reach the baseline while the cell density (Optical Density, OD) remained constant at about 0.50 OD units. Similar experiments were conducted twice and allowed us to evaluate the characteristics of Lumisens 1 (Table 8). The response time is merely 1.45 H as the recovering time is about 12 H meaning that only two detections could be performed per day. This latter parameters should be optimised in the future. Moreover, we evaluated the live time of the biosensor ie : the time during which measurements can be performed. It was possible to conduct continuous culture of *E.coli* TBT3 over 14 days but we observed 8 days after the beginning the formation of biofilm both on the glass jacket and the glass window from which the laser beam is introduced (Figure 27). A biofilm of 106 to 250.106 CFU.cm-2 were usually counted depending on the dilution rate and experimental conditions (temperature, pH,..). Then, the life time of Lumisens 1 was limited to 7 days in a first approach. We attempted to limit this phenomenon with glass coating with silanization or with Parylene®. The latter was successful but expensive since the coating should be reitarate after two or three use of the biosensor.

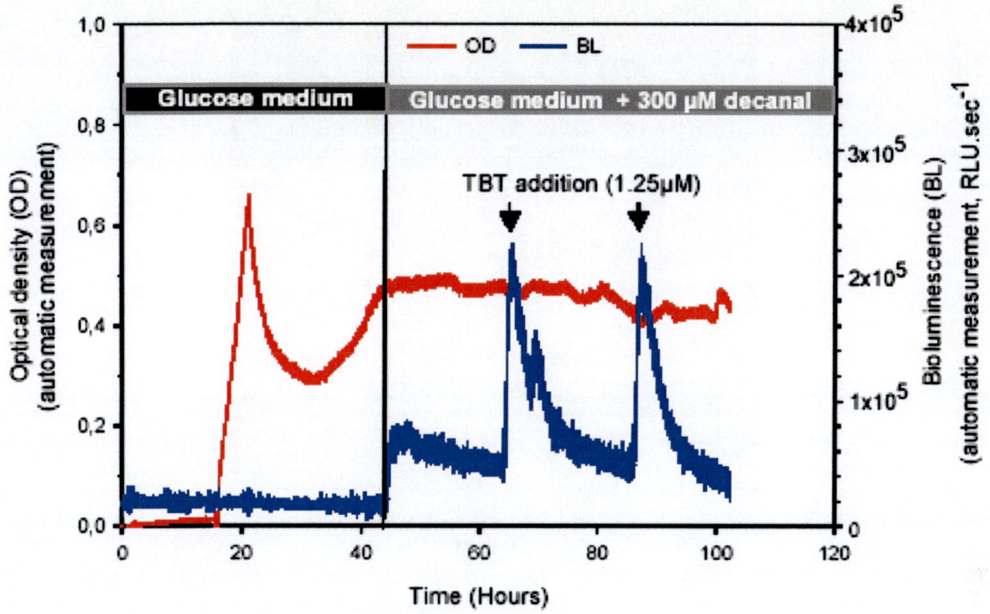

Figure 26. Typical example of Lumisens 1B behavior during the on line detection of TBT. The bioelement *E.coli* TBT3 was cultivated in continuous culture mode in glucose medium with TrisHcl (50 mM). D = 0.9h-1, 30 °C, 250 rpm, $pH_{initial}$ 7, 3 v.v.m air. Not published results.

Table 8. Characteristics of Lumisens 1B biosensor for the detection of TBT with *E.coli* TBT3 continuously fed with glucose medium at D = 0.9h-1, 30°C, 300 µM decanal. Results from two similar experiments. Values in parenthesis are standards deviations

Induction ratio [a]	Response Time (h)	Recovery time (h)	Repeatability [b]	Reproducibility [c]	Life Time
3.1 (1)	1.45 (0.07)	12 (0.6)	98 % (1.4)	70 %	7 days

(a) : Ratio between the bioluminescence with TBT versus bioluminescence without TBT

(b) : calculated from the induction ratio after two identical TBT addition in a same experiment.

(c) : calculated from the induction ratio after two identical TBT addition from two separate experiments.

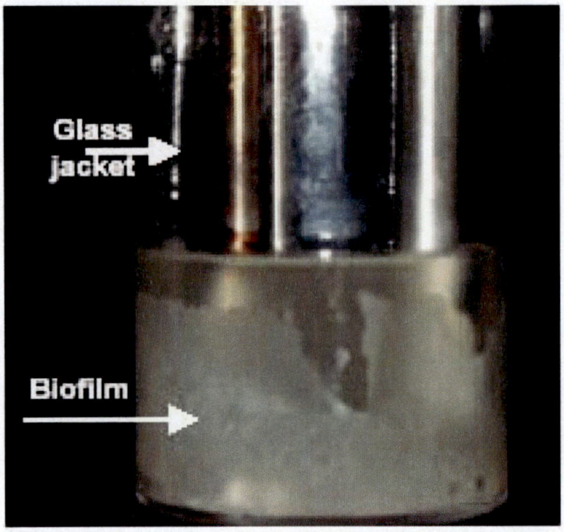

Figure 27. Example of the formation of a biofilm on the glass jacket of Lumisens 1B after 8 days of continuous culture of *E.coli* TBT3 in glucose medium at 30°C. Not published photographie.

9. Conclusions and Future Developments

The detection of TBT remains necessary with regard to its high toxicity on the food chain. National agencies as well as private compagnies need to monitor such toxic biocide.

The strain *Escherichia coli* TBT3 is currently the unique known bacteria able to detect specifically TBT and DBT either in sea water or in solid matrix like in contaminated sands. Nevertheless, further in-depth research will be necessary to conduct in order to understand the strain in a molecular point of view. In the present paper, we demonstrated that the *E.coli* TBT3 applications to biotools (bioassay or biosensor) need knowledges gathered during 4 years of research by our team. It is not finished and some futur developments are presented below.

The bioassay could be used as a screening method to identify samples containing TBT or DBT. Application on sediment could be possible for harbour or for the control of paint. As noted by Champ (2000, 2003) there is no quick, inexpensive and non destructive method for detecting the presence of TBT on a vessel. Ceulemans et al (1998) reported an abnormal high contamination of TBT in sediment from a shipyard used for maintenance of small boat (less than 25 m) suggesting the illegal use of antifouling paints. At the present time, an inspection method to control the paints based on X-ray fluorescence spectrometry (XRF) is propose by Senda et al (2003). Nevertheless the sampling of the paint is done by abrasive paper and hence, this method could be refused by the owner of the ship. The use of a bioassay with the bacteria *E.coli* TBT3 could be an alternative to discriminate the paint containing TBT as biocide. This work is under development in our laboratory and preliminary results show that the bacteria is induced by paint containing TBT, while no induction is observed when chlolothalonil and copper are used as active biocide.

The use of *E. coli* TBT3 in a biosensor as Lumisens 1B allowed a continuous measurement of the biocide but due to an important recovery time, only 2 or 3 measurements are possible per day. Lumisens 1B is versatile and some features are original, but the biosensor lack of simplicity for no-microbiologist. In this conditions, this first biosensor is a good device in the laboratory to study bioluminescent strains in a controlable and reproducible environment. We currently develop the range Lumisens 2, allowing an easy way to detect chemical compounds. This new biosensor is complementary to Lumisens 1 and it uses small and disposable cards in which, every bioluminescent strains can be immobilized. The cards are inserted in a new, transportable device that will allow an on line and automatic measurement of pollution in the field.

ACKNOWLEDGMENTS

This research was supported by the grant CER 2000-2006, Action n°15 (section I), research programme number 18035 (Ville de La Roche-sur-Yon, Conseil Général de Vendée, Conseil Régional des Pays de Loire, Ministère Français chargé de la Recherche).

We would like to thank Thomas Charrier for the figure 3.

REFERENCES

[1] Abalos M, Bayona JM, Compano R, Granados M, Leal C, Prat MD.1997. Analytical procedures for the determination of organotin compounds in sediment and biota. *J Chromatogr* A 788: 1-49

[2] AFNOR Commission de normalisation AFNOR T91 M: Eaux – Micropolluants organiques. Projet NF T90 –250 Dosage de composés organoétain dans les sédiments-Méthode par chromatographie en phase gazeuse (2003- 12-22)

[3] Atlas RM.1997. Handbook of microbiological media. Parks LC (ed), CRC Press, Boca Raton, New York, p 744

[4] Bancon- Montigny C, Lespes G, Potin- Gautier M. 2004. Organotin survey in Adour-Garonne basin . *Water Research* 38 : 933-946

[5] Bancon-Montigny C, Lespes G, Potin-Gautier M. 2000. Improved routine speciation of organotin compounds in environmental samples by pulsed flame photometric detection. *J Chromatogr A* . 896: 149-158

[6] Behra P, Lecarme-Théobald E, Bueno M, Ehrhardt JJ. 2003. *Journal of Colloid and Interface Science.* 263: 4-12

[7] Billard P, DuBow MS. 1998. Bioluminescence based assays for detection and characterization of bacteria and chemicals in clinical laboratories. *Clin Biochem*. 31: 1-14

[8] Biselli S, Bester K, Hühnerfuss H, Fent K. 2000. Concentration of the antifouling compounds Irgarol 1051 and the organotins in water and sediments of German North and Baltic Sea Marinas. *Marine Bulletin Pollution* 40 : 233-243

[9] Bitton G, Khafif T, Chataigner N, Bastide J, Coste CM. 1986. A direct INT-dehydreogenase assay (DIDHA) for assessing chemical toxicity. Toxicol Assess 1: 1-12

[10] Blum JL. 1997. *Bio-and Chemi-luminescent sensors*, World Scientific Publishing, Singapure, London.

[11] Bolton EK, Sayer GS, Nivens DE, Rochelle JM, Ripp S, Simpson ML. 2002. Integrated CMOS photodetectors and signal processing for very low level chemical sensing with the bioluminescent bioreporter integrated circuit. *Sens Actual B Chem.* 85: 179-185

[12] Brandt KK, Pedersen A, Sorensen J. 2002. Solid-phase contact assay that uses a lux-Marked *Nitrosomonas europea* reporter strain to estimate toxicity of bioavailable linear alkylbenzene sulfonate in soil. *Appl Environ Microbiol.* 68 (7) : 3502-3508

[13] Briscoe SF, Diorio C, DuBow MS. 1996. Luminescent biosensors for the detection of tributyltin and dimethyl sulfoxide and the elucidation of their mechanisms of toxicity. In: Moo-Young, M., Anderson, W.A., Chakrabarty, A.M. (Eds.), *Environmental biotechnology, principles and applications.* Kluwer Academic Press, Dordrecht, Boston, London. p. 645-655

[14] Bundy JG, Wardell JL, Campbell CD, Killham K, Paton GI. 1997. Application of bioluminescence-based microbial biosensors to the ecotoxicity assessment of organotins. *Let Appl Microbiol.* 25 : 353-358

[15] Ceulemans M, Slaets S, Adams A. 1998. Speciation of organotin in environmental sediment samples. *Talanta.* 46: 395-405

[16] Cha JM, Kim BW. 2002. Bioassay of environmental endocrine disruptors TBT and DMSO using recombinant *Escherichia coli.* Bioprocess Biosyst Eng 24: 405-410

[17] Champ MA. 2000. A review of organotin regulatory strategies, pending actions, related coast and benefits. *The Science of total Environment.* 258: 21-71

[18] Champ MA. 2003. Economic and environmental impacts on ports and harbours from the convention to ban harmful marine anti-fouling systems. *Marine Pollution Bulletin.* 46 : 935-940

[19] Cheung KC, Wong MH, Yung YK. 2003. Toxicity assessment of sediments containing tributyltin around Hong Kong harbour. *Toxicology letters.* 137: 121-131

[20] Choi SH, Gu MB. 2002. A portable toxicity biosensor using freeze-dried recombinant bioluminescent bacteria. *Biosens Bioelectron.* 17: 433-440

[21] Daunert S, Barrett G, Feliciano JS, Shetty RS, Shrestha SS, Smith-Spencer W. 2000. Genetically engineered whole –cell sensing systems: coupling biological recognition with reporter genes. *Chem Rev.* 100: 2705-38

[22] Durand MJ, Thouand G, Dancheva-Ivanova T, Vachon P, DuBow M. 2003. Specific detection of organotin compounds with recombinant luminescent bacteria. *Chemosphere.* 52: 103-111

[23] EEC N°2455/2001/EC, decision of the european parliament and of the Council of 20 November establishing the list of prioritary substances in the field of water policy and amending Directive 2000/60/EC, *Official Journal*, L331, 44, 15 December 2001

[24] Ellis RJ, Wright AG.1999. Optimal use of photomultipliers for chemiluminescence and bioluminescence applications. *Luminescence.*14 : 11-18

[25] Fent K. 1996. Ecotoxicology of organotin compounds. *Critical Reviews in Toxicology.* 26(1): 1-117.

[26] Gómez-Ariza JL, Morales E, Giraldez I, Sanchez-Rodas D, Velasco A. 2001. Sample treatment in chromatography-based speciation of organometallic pollutants. *J Chromatogr A.* 938: 211-224.

[27] Gu MB, Dhurjati PS, Van Dyk TK, LaRossa RA. 1996. A miniature bioreactor for sensing toxicity using recombinant bioluminesent *Escherichia coli* cells. *Biotechnology* progress. 12: 393-397

[28] Guzzo A, DuBow M.S, 1991. Construction of a stable, single copy luciferase gene fusions in *Escherichia coli, Arch. Microbiol.* 156 :444-448

[29] Heitzer A, Malachowsky K, Thonnard JE, Bienkowski PR, White DC, Sayler GS. 1994. Optical biosensor for environmental on line monitoring of naphtalene and salicylate bioavailability with an immobilized bioluminescent catabolic reporter bacterium. *Appl Environ Microbiol.* 60: 1487-1494

[30] Hoch M, Alonso-Azcarate J, Lischick M. 2003. Assessment of adsorption behavior of dibutyltin (DBT) to clay –rich sediments in comparison to highly toxic tributyltin (TBT). *Environ Pollut.* 123: 217-227

[31] Hoch M. 2001. Organotin compounds in the environment : an overview. *Appl Geochem.* 16, 719-743

[32] Holzman TF, Baldwin TO. 1983. reversible inhibition of the bacterial luciferase catalysed bioluminscence reaction by aldehyde substrate: kinetic mechanism and ligand effects. *Biochemistry.* 12: 2838- 2846

[33] Ikariyama Y, Nishiguchi S, Koyama, T, Kobatake E, Aizawa M, Tsuda M, Nakazawa T. 1997. Fiber-optic based biomonitoring of benzene derivatives by recombinant *E. coli* bearing luciferase gene fused TOL-plasmid immobilized on the fiber optic end. *Analytical Chemistry.* 69: 2600-2605

[34] Ivask A, François M, Kahru A, Dubourguier H.C, Virta M, Douay F. 2004. Recombinant luminescent bacterial sensors for the measurement of bioavailability of cadmium and lead in soils polluted by metal smelters. *Chemosphere.* 55: 147-146

[35] Jude F, Arpin C, Brachet-Castang C, Capdepuy M., Caumette P, Quentin C. 2004. TbtABM, a multidrug efflux pump associated with tributyltin resistance in *Pseudomonas stutzeri . FEMS Microbiology Letters.* 237:7-14.

[36] Kawai S, Kurokawa Y, Harino H, Fukushima M. 1998. Degradation of tributyltin by a bacteria strain isolated from polluted river water. Environ. Pollut. 102: 259-263

[37] Köhler S, Belkin S, Schmid RD. 2000. Reporter gene bioassays in environmental analysis. Fresen. *J. Anal. Chem.* 366: 769-779

[38] Kress-Rogers E. 1997. Biosensors and electronic noses for practical applications, in *Handbook of biosensors and electronic noses*, medicine, food and the environment. Erika Kress-Rogers (Ed.), CRC Press, Boca Raton. p 3-39

[39] Kubitscheck HE. 1970. *Introduction to research with continuous cultures.* Prentice-Hall, Inc. Englewood Cliffs, New Jersey.195 p.

[40] Meighen EA. 1994. *Genetics of bacterial bioluminescence.* Annual Review of Genetic. 28: 117-139

[41] Michel P, Averty B. 1999. Contamination of French coastal waters by organotin compounds : 1997 update. *Marine Pollution Bulletin.* 38: 268-275

[42] Midorikawa S, Arai T, Harino H, Ohji M, Cu ND, Miyazaki N. 2004. Concentration of organotin compounds in sediment and clams collected from coastal areas in Vietnam. *Environ. Pollut.* 131: 401-408

[43] Miller CE, Wuertz S, Cooney JJ, Pfister RM. 1995. Plasmids in tributyltin-resistant bacteria from fresh and estuarine waters. Journal of Industrial *Microbiology.* 14: 337-342

[44] Negri AP, Hales LT, Battershill C, Wolff C, Webster NS. 2004. TBT contamination identified in Antarctic marine sediments. *Marine Bulletin Pollution.* 48 : 1142-1144

[45] Nivens DE, McKnight TE, Moser SA, Osbourn SJ, Simpson ML, Sayler GS. 2004. Bioluminescent reporter integrated circuits : potentially small, rugged and inexpensive whole-cell biosensors for remote environmental monitoring. *J Appl Microbiol.* 96: 33-46

[46] Picart P, Bendriaa L, Daniel P, Horry H, Durand MJ, Jouvanneau L, Thouand G. 2004. New bioreactor for *in situ* simultaneous measurement of bioluminescence and cell density. *Rev. Sci. Instrum.* 75 (3) : 747-755

[47] Polyak B, Bassis E, Novodvorets A, Belkin S, Marks RS. 2001. Bioluminescent whole cell optical fiber sensor to genotoxicants: sustem optimisation. *Sens Actual B Chem.* 74: 18-26

[48] Quevauviller P, Morabito R. 2000. Evaluation of extraction recoveries for organometallic determinations in environmental matrices. Trends in analytical chemistry.19: 86-96.

[49] Roy U, Dubey SK, Bhosle S. 2004. Tributyltin chloride-utilizing bacteria from marine ecosystem of west coast of India. *Current Science.* 86(5): 702-705

[50] Ruiz JM, Bachelet P, Caumette P, Donard OFX. 1996. Three decades of tributyltin in the coastal environment with emphasis on Arcachon Bay, France. *Environ. Pollut.* 93 : 195-203

[51] Schaechter M, Maaloe O, Kjelgaard NO. 1958. Dependency on medium and temperature of cell size and chemical composition during balanced growth of *Salmonella typhimurium*, *J. Gen. Microbiol.* 19: 592-606

[52] Senda T, Miyata O, Kihara T, Yamada Y. 2003. Inspection method for identification of TBT-containing antifouling paints. *Biofouling.* 19: 231-237

[53] Shim WJ, Oh JR, Kahng SH, Shim JH, Lee SH. 1999. Horizontal distribution of butyltins in surface sediments from an enclosed bay system, Korea. *Environ Pollut.* 106: 351-357

[54] Shokri A, Sandén AM, G Larsson. 2002. Growth rate dependent changes in *Escherichia coli* membrane structure and protein leakage. Appl Microbiol Biotechnol. 58: 386-392

[55] Stanley PE. 1997. Commercially available luminometers and imaging devices for low-light level measurements and kits and reagents utilizing bioluminescence or chemiluminescence: survey update 5. *J. Biolumin. Chemilumin.* 12: 61-78

[56] Sternberg C, Eberl L, Kongsbak-Poulsen L, Molin S. 1997. Detection of bioluminescence from individual bacterial cells: a comparison of two different low-light imaging systems, *J. Biolumin. Chemilumin.* 12: 7-13

[57] Sturgeon RE, Wahlen R, Brandsch T, Fairman B, Wolf-Briche C, Alonso JI, Gonzalez PR, Encinar JR, Sanz-Medel A, Inagaki K, Takatsu A, Lalere B, Monperrus M, Zuloaga O, Krupp E, Amouroux D, Donard OF, Schimmel H, Sejeroe-Olsen B, Konieczka P, Schultze P, Taylor P, Hearn R, Mackay L, Myors R, Win T, Liebich A, Philipp R, Yang L, Willie S.2003. Determination of tributyltin in marine sediment: Comite Consultatif pour la Quantite de Matiere (CCQM) pilot study P-18 international intercomparison. *Anal. Bianal.Chem.* 376: 780-787

[58] Thouand G, Daniel P, Horry H, Picart P, Durand MJ, Killham K, Knox OGG, DuBow MS, Rousseau M. 2003a. Comparison of the spectral emission of *lux* recombinant and bioluminescent marine bacteria. *Luminescence.* 18: 145-155

[59] Thouand G, Durand MJ, Picart P, Daniel P, Massé J, DuBow MS. 2001. Detection of bacteria in the food industry with bioluminescent sensors. *Recent Res Develop Microbiol.* 5 : 79-93

[60] Thouand G, Horry H, Durand MJ, Picart P, Bendriaa L, Daniel P, DuBow MS. 2003b. Development of a biosensor for on-line detection of tributyltin with recombinant bioluminescent *Escherichia coli* strain. *Appl Microbiol Biotechnol.* 62: 218-225

[61] Ulitzur S, Hastings JW. 1979. Evidence for tetradecanal as natural aldehyde in bacterial bioluminescence. *Proc Natl Acad Sci USA.* 76: 265-267

[62] Ulitzur S. 1997. Established technologies and New Approaches in applying luminous bacteria for analytical purposes. *J Biolumin Chemilumi.*12: 179-192

[63] Wagner LW, Van Dyk TK. 1998. Cryopreservation and reawakening. In: LaRossa, R.A. (Ed.) *Bioluminescence methods and protocols.* Humana Press, Totowa, New Jersey. P 123-127.

[64] White S, Catterick T, Fairman B, Webb K. 1998. Speciation of organo-tin compounds using liquid chromatography-atmospheric pressure ionisation mass spectrometry and liquid chromatography-inductively coupled plasma mass spectrometry as complementary techniques. *J.Chromatogr.A.* 794: 211-218.

[65] Yotter RA., Wilson DM. 2003. A review of photodetectors for sensing light-emitting reporters in biological systems. *Sensors Journal, IEEE.* 3 (3): 288-303.

In: Trends in Agriculture and Soil Pollution Research
Editor: James P. Livingston, pp. 43-78
ISBN: 1-59454-325-9

Chapter 2

FATE OF PETROLEUM HYDROCARBONS IN THE SUBSURFACE ENVIRONMENT: ABIOTIC PROCESSES

Ishai Dror
Dept. of Environmental Sciences & Energy Research,
The Weizmann Institute of science Rehovot, Israel

ABSTRACT

Petroleum hydrocarbons (PHC) are used in nearly every aspect of life. These materials are utilized as the major source of energy for domestic, municipal and industrial purposes. For example PHC are used to heat homes, the fuel transportation, the power manufacturing processes and tools. Furthermore, PHC, are a basis for a variety of synthetic materials and distillates. Used as intended, PHC provide great benefits to society. However, when released to the subsurface environment as raw feedstocks or refined fuels or lubricants, a number of problems can result. For example the PHC, may contaminate land surfaces, the unsaturated zone and ground waters at numerous sites due to their intensive and versatile use. Some components of the PHC mixture are considered resistant and thus PHC represent a long-term source of environmental subsurface pollution.

Based on our own laboratory and field evidence and published studies of other research groups this chapter emphasizes (1) The effects of leaching on the abiotic processes controlling the fate of PHC in the soil environment. (2) The effect of soil amendments on PHC in situ attenuation, and (3) A new possible pathway for intrusion of PHC to groundwater. The experimental evidence of PHC behavior in porous media is linked to four major processes: retention, redistribution and attenuation in the subsurface, and dissolution/partition to the soil solution and groundwater.

1. Introduction

Petroleum hydrocarbons (PHC) are the world's major source of energy and as such they are exploited in nearly every aspect of life for domestic, industrial and municipal purposes. Used as intended these materials provide great benefits to society. However, when released to the subsurface environment, a number of problems can result.

The vast and versatile PHC usage, and transport that originate from a world oil consumption of 76.21 million bbl/day [World Fact Book 2003] cause numerous PHC polluted sites, and release of large amounts of PHC into the environment. Furthermore, almost every populated site and many unpopulated sites, are susceptible to PHC contamination. PHC may reach the soil surface as a result of a spill or may reach the soil together with other wastes.

Once in the soil the fate of hydrocarbon mixtures can be affected by the dynamics of moisture conditions in the subsurface as result of rainfall, irrigation, fluctuation of the ground water level, and any drying or flooding event or simple water flow. In this study the modification induced to the hydrocarbon behavior –by changes in water status in the porous media - will be considered as the effect of "leaching". During leaching the fate of petroleum contaminants in the unsaturated zone is affected by various processes simultaneously (e.g., retention, volatilization, dissolution, biodegradation), by the properties of the surrounding porous material and by the composition of the hydrocarbon mixture itself. A weathering phenomenon, affecting the physical and chemical properties of the mixture that remain in the subsurface and characterized by continuous changes in its composition, may lead to PHC attenuation. The extent of the processes influencing the fate of petroleum contaminants in soils is controlled by leaching parameters such as the amount of incoming water, the composition of soil solution and incoming water, the water infiltration rate, and by the properties of the porous media.

In a soil contaminated with a PHC, a four-phase system exists, three fluid (aqueous, non-aqueous-PHC, air) and a solid one. The relative phase saturation degrees (i.e., the volume of the phase per total pore volume) of water, PHC and gas are, in addition to the ambient boundary conditions (such as phase pressure or liquid inflow rates), governed by the soil pore geometry, the wetting order, and the capillary pressures between the phases. Mass transfer occurs among the various phases (PHC, air, aqueous and solid phases), and simultaneous biodegradation of the PHC may also influence the weathering processes. A scheme of the relations between the four phases is illustrated in figure 1.

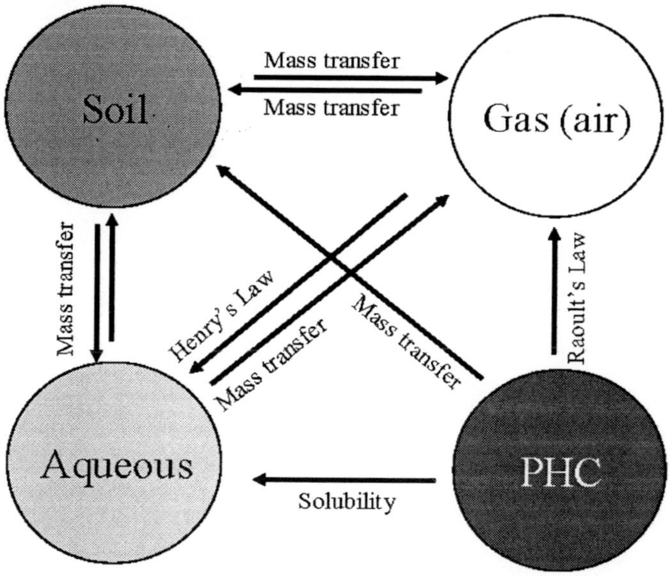

Figure 1. A scheme of possible partitioning relations between the four phases in porous medium: aqueous phase, water insoluble-PHC phase, gas or air phase and a solid phase.

Typically, PHC mixtures are composed of molecules that have low water solubilities and thus their aqueous concentrations are very limited. However, even minute concentrations of these contaminants can have a very significant effect on the taste and odor of the polluted water and might be toxic and/or carcinogenic. PHC represent a long-term source of water pollution because the contaminants will continue to enter the water phase to replace the amount that is degraded or transported away from the polluted site. Field observations show that non aqueous phase liquids (NAPLs) are often persistent for many years in sediments and in subterranean sites [Thorn and Aiken, 1998; Broholm et al., 2000; Loehr, et al., 2000]. Over the time in such contaminated sites the pollutants can potentially be degraded by microbial activity. This bioremediation will occur only when (i) the right environmental conditions prevail (e.g. pH, temperature, moisture, nutrients etc.), (ii) relevant microbial consortia are present and (iii) the contaminants are available for microbial attack. When these conditions exist the microorganisms in the subsurface may degrade the residual contamination. Natural intrinsic bioremediation, is generally a slow process but by optimization it may proceed faster. PHC are made up of variety of organic compounds and the amount of carbon available to potential degrading species is large. This results in a large microbial demand for inorganic nutrients, especially N and P. However, the concentration of these nutrients is frequently low or the rate of diffusion from the surrounding water or soil to the vicinity of the PHC, where the degradation is occurring, is too slow to meet the large demand. Remediation of petroleum-contaminated soils thus require adequate amounts of N and P to support microbial activity and these nutrients may have to be added to the soil to support or enhance the attenuation process. The nutrient supply for bioremediation enhancement may be added to the contaminated soil as sludge [e.g. Bossert et al. 1984; Stronguilo et al., 1994; Wilson et al. 1996; Jorgensen et al., 2000; Graber et al., 2001] or as a fertilizer [e.g. Bollag and Bollag, 1995; Sims and Sims, 1999; Romantschuk et al., 2000; Jorgensen et al., 2000]. Since the redistribution of the PHC in the soil profile is affected by the properties of the porous media

[Wurdemann et al., 1990] the application of different amendments to a contaminated site may result in different effects on the contaminant distribution in the soil profile as well as on its attenuation.

Usually PHC contamination of soil and groundwater is considered a result of a spill on the surface or from underground reservoirs and tubing near the soil surface, which is then transported further into deeper soil layers and the groundwater.

The following sections will first discuss a new potential pathway of PHC contamination of coastal aquifers. Then examination of the effects of leaching and soil amendments, under laboratory and field conditions, on several major processes will be discussed.

2. Intrusion of PHC to Coastal Aquifers

There is considerable evidence of major hazards to coastal aquifer water quality, due to the massive exploitation that has led to low water table levels, which in turn increase intrusion of the sea water and degradation of these aquifers. For example the European Environmental Agency [MOPTMA-MINER, 1994] notes that 58% of the 82 coastal hydrological units in Spain and the Balearic Islands show seawater intrusion as a direct result of over-exploitation of the freshwater resources. Similar seawater intrusion into aquifer has been reported in many locations around the world, including the eastern and western coastlines of the USA [e.g., Barbieri et al, 1986; Bond and Bredehoeft, 1987; Kashef, 1983; Magaritz and Luzier, 1985; Oude Essink, 2002; Siraz, 1998]. In all cases, the intrusion of seawater into coastal aquifers results in an increase of salt concentration in the initially fresh groundwater.

In addition to the high salt concentration of the seawater, in many cases as a result of anthropogenic activities, coastal seawater is also polluted by PHC (from petroleum products as well as from other industrial solvents) and other toxic chemicals. The contamination of the near-shore marine environment by petrogenic and pyrogenic hydrocarbons may result from sewage discharge, ship and industrial fuel emissions or spillage, and burning of petroleum following supertanker accidents and sinking [e.g. Pastor et al., 2001]. Petroleum hydrocarbon residues have been detected in both coastal seawater and near-shore sediments [e.g., Dachs et al., 1997; Budzinski et al., 1997; Notar et al., 2001; Pastor et al., 2001; Sole et al., 2001; Witt, 2002; Readman et al., 2002] and may become persistent contamination hazards for freshwater aquifers.

Over a century ago it was proved that the presence of electrolytes in aqueous solutions modifies the solubility of organic molecules in these solutions. Generally, most organic compounds are less soluble in a saline water than in fresh water, and the activity coefficient is inversely proportional to the salinity. This phenomenon, termed the "salting-out effect", was studied initially by Setschehov [1889]. He proposed the empirical equation $\log(S_0/S) = kC_s$ to describe this effect, where S_0 and S are the solubility of the organic compounds in water and in the salt solution, respectively, k is the salt effect constant, and C_s is the salt concentration in water.

The validity of the salting-out effect has been confirmed over the years [e.g., Osol and Kilpatrik, 1933; Morrison, 1944; Paul, 1952; Erlender and McGurie, 1968; Mackcay and Shiu, 1975; Sutton and Cader, 1975; Whitehouse, 1984; Xie et al., 1990; Dror et al., 2002]. In a critical review on the salting out effect, Xie et al. [1997] suggested 1.36 as a factor quantifying the decrease in solubility of aromatic and alkene hydrocarbons in seawater from corresponding data for freshwater.

In reality the concentration of organic pollutants under natural conditions may be much higher than expected only from theoretical considerations of solubility. This is because the total concentration of organic contaminants contained in natural seawater is given not only by the dissolved contaminant concentration but also by the presence of an emulsified phase or mechanically–dispersed organic droplets [Tkalich and Soon, 2002]. These emulsions and/or droplets can lead to a significant increase in the "carrying capacity" of the aqueous solution for organic contaminants [Dror et al., 2003a,b].

In a recently published work done in our lab [Dror et al., 2003a,b] it was found that while hydrocarbons migrate in response to chemical gradients between the source and the collector cells across an inert sand barrier, these gradients are significantly enhanced by the presence of salt in the source cell. Moreover, a "pulse-like" transport of hydrocarbons droplets was determined, indicating their role for in increasing concentrations of hydrocarbons in fresh water. The high concentrations of hydrocarbons found in the salt water, which were greater than the theoretical limits of solubility, indicate that the carrying capacity of seawater for hydrocarbons is controlled not only by the solubility, but also by factors such as the ability for droplets to form. Furthermore, the salting out phenomenon, which reduces the aqueous solubility of the organic compounds, will also act to force hydrocarbons across the interface between the two water phases with varying salinity.

Polluted marine environments are active open systems in which waves, tidal movement, streams and mixing occur continuously, leading to the formation of stable droplets, an increase in the carrying capacity of seawater for organic contaminants, and consequently the salting out of hydrocarbons. Under these conditions, the organic pollutants will be driven across an inert or active interface to the freshwater aquifer. Hence under such a scenario, contaminated seawater will act not only as a source but also as a driving force that magnifies transport of hydrocarbons into a freshwater aquifer.

3. Effect of Leaching on PHC Contaminants in the Subsufrace

To support the assumption that leaching can play a major role in determining the biotic and abiotic behavior of residual PHC in the soil environment we present experimental findings on the major processes affecting their attenuation and redistribution in the subsurface. Our data, as well as additional results from the literature on PHC retention, dissolution, volatilization, redistribution and weathering are reported and discussed.

General information on the experimental methodology used for determining entrapped PHC adsorption, desorption and mass transfer may be found in reviews by Abriola and Bradford [1998] and Pignatello [2000]. Details on laboratory and field experimental procedures used by our research group have been reported previously [Yaron et al., 1989; 1996, Galin et al., 1990, Fine and Yaron, 1993; Nye et al., 1994.; Gerstl et al.,1994; Jarsjo et al., 1997;Gustin and Prost, 1998, Dror et al., 2000a,b,c, 2001, 2002].

3.1. Retention

Physical entrapment is a form of non-adsorptive retention of non-aqueous-phase liquids (NAPL) in the subsurface. The mutually inmiscible fluids (aqueous and NAPL) each impede the transport of the other through the soil, until a maximum saturation is reached. PHC fluids can then become trapped as blobs or ganglia in pores in the unsaturated zone for an indefinite time [Schwille, 1994], and they act as a source of contamination which diminishes as the PHC volatilizes, dissolves in water, or degrades. The degree of soil saturation for a water-immiscible liquid can be expressed as the proportion of pore space occupied by a given liquid phase [van Dam, 1967; Schwille, 1981], or as the liquid volume held in unit weight of soil [Mercer and Cohen, 1990; Yaron et al. , 1998].

In a soil contaminated with a PHC three fluid phases (aqueous, water insoluble, and air) and a solid phase can be distinguish. The volume occupied by the aqueous phase is determined by the amount of incoming water and that of the water immiscible phase by the amount and composition of petroleum products spilled on the soil surface. The composition and properties of the solid phase, such as porosity of the soil subsurface will define the volume occupied by the air phase. This can be determined by the amount of free pores space occupied by the aqueous and PHC phases. Mass transfer occurs among the various phases (PHC, air, aqueous and solid phases), and simultaneous biodegradation of the PHC may also influence the weathering and attenuation processes. A scheme of the relations between the four phases is depecited in figure 1.

In a saturated soil, there is no air phase, the water is in contact with the pore walls and fills the small pores volume. The organic contaminants form entrapped ganglia which resemble discontinuous blobs within large pores [Mayer and Miller, 1992; Powers at al., 1994]. In unsaturated soils the centers of the large pores are filled with the gas phase, water is in contact with the solid phase and fills the small pores, but organic contaminants spread along the water-gas phase interface. Gwirtzman and Roberts [1991] showed that for a soil at a given PHC saturation the organic-water interfacial area will be much greater when the organic phase is distributed as a film than when it is entrapped as ganglia. The fractional wettability of the subsurface - where natural systems exhibit variations in solid surface properties - will determine the organic-water interfacial area [Bradford and Leij, 1997] and will influence the water flow field [Wang, 1998a].

During leaching the porous medium environment is in a dynamic condition: the incoming water continually changes the saturation status of the soil subsurface. Simultaneously an anisotropic soil profile develops, in which saturated and nonsaturated conditions alternate with changing depth and in which the mass transfer patterns between the organic liquid phase and the aqueous, air or solid phase become nonuniform along the soil profile.

Let us consider a simplified scenario, where the residual volatile organic contaminant mixture is found in the upper layer of a soil profile and three distinct situations can be defined:

1. A PHC layer rests on a soil profile saturated with water at the top and unsaturated at the bottom at the beginning of the leaching (Fig. 2a). If a low-density PHC mixture (LNAPL) is predominant, mass exchange occurs only at the interfaces between the immiscible liquid, and the atmosphere and the soil water, and no other direct interaction of the water-immiscible liquid with the air or the soil solid

phase is observed. A differential partitioning of the mixture components into the soil water occurs, after which the soluble fraction of PHC – depending on its properties - can volatilize into the soil air or be adsorbed on the solid fraction of the porous medium. In the case of a high-density organic contaminant (DNAPL) it is possible that, in time, the contaminant will move through the saturated soil layer and behave as in situation 2.

2. A PHC layer rests on an unsaturated soil profile (Fig. 2b). If the incoming leachate is not sufficient to bring the soil to saturation, we define a differential mass transfer between the organic (PHC) phase and the two other fluid phases: aqueous and air. Retention of selected PHC components on the soil solid phase, either as solute or in the gaseous phase, is also possible.
3. A PHC contaminated layer rests on a soil profile with an unsaturated upper layer and a saturated layer at the bottom (Fig. 2c). The incoming leach water promotes dissolution of the residual contaminant and reduces the volatilization of high-vapor-pressure components of the mixture. The transfer of contaminant into soil profile may continue only by diffusion at the interface between unsaturated and saturated soil layers. The retention of selected components in an adsorbed phase may occur in the unsaturated zone.

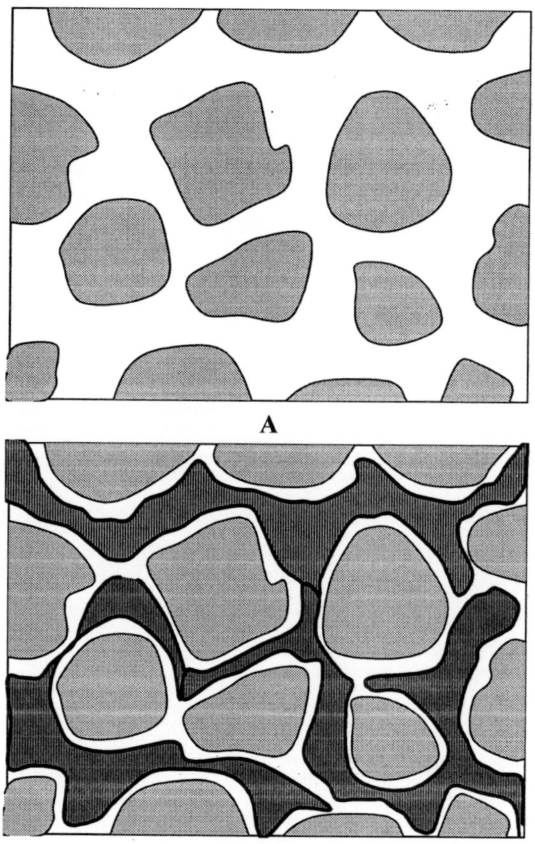

A

B

Figure 2 – Continued.

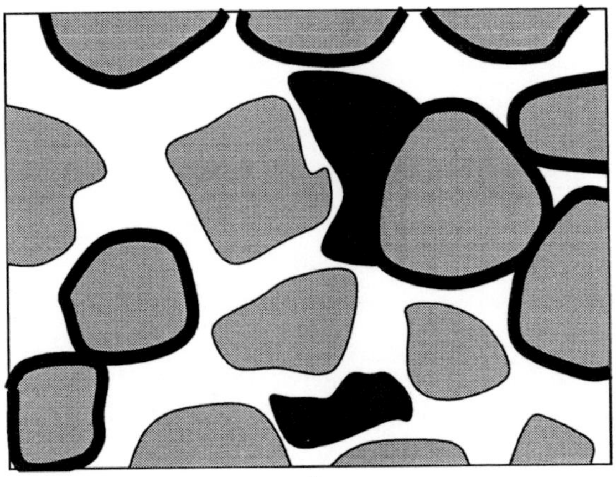

C

Solid Water PHC Air

Figure 2. Distribution scenarios of residual NAPL in the subsurface: A- for saturated zone; B unsaturated zones when the porous medium is water wet; and C in the saturated zone when the porous medium has fractional wettability .

From the above discussion we may conclude that the redistribution of petroleum products and their components in soils, leading to their attenuation in some parts of the profile and concentration in others, is ultimately governed by a mass exchange between the organic phase and the aqueous, gaseous and solid phases. Leaching indirectly controls both the redistribution and the abiotically induced attenuation of the contaminants in the subsurface.

The extent of PHC trapping is determined primarily by the physical properties of the vadose zone and by its moisture content. The soil retention capacities for kerosene [Fine and Yaron, 1993; Jarsjo et al., 1994] and for gasoline [Gustin and Prost, 1998] were quantitatively determined in a series of laboratory and outdoor experiments. An illustration of the effects of moisture content on retention capacities for kerosene and gasoline is shown in Fig. 3. It was observed that kerosene retention was inversely related to the moisture content and positively related to soil porosity, clay and organic matter contents [Fine and Yaron, 1993]. As soil moisture increased, PHC retention capacity decreased markedly to relatively low values in all the soils studied [Fine and Yaron, 1993; Jarsjo et al., 1994; Gustin and Prost, 1998]. It was postulated that the water was retained in small soil capillaries and thus reduced the effective soil pore volume. This would lead to significant reductions both in the physical entrapment of hydrocarbons in the small pores and capillaries and in their retention on mineral and organic matter surfaces [e.g., Chiou and Shoup, 1985; Pignatello, 1989; Mingelgrin and Prost, 1989]. At relative humidities of ~50% (most relevant to the soil environment) the following conditions persist: at least a monolayer of water is present on external surfaces of macro- and meso-pores, the micro- and meso-pores are filled with water; and the macropores are empty or only partially filled with water [Pignatello, 2000]. Low kerosene retention capacity (KRC) values were found for relatively high water contents [Fine and Yaron, 1993] and may be explained by blocking of the pore entrance with water, which prevents kerosene from filling up the available pore space in soils with relatively narrow pores. These results imply that, for the soil water contents expected to prevail in humid zones the differences among various types of soils will be small, because of the high soil water content expected there. In the arid and semiarid areas on the other hand, these differences will be significantly great in soils because of the small water content in the soil which allow large volume of free pores in the soil.

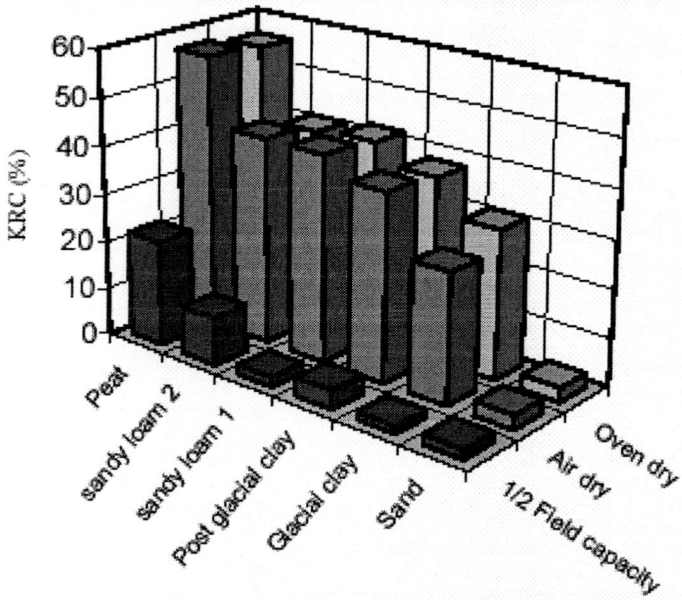

Figure 3. Possible soil moisture status when contamination with PHC begins.

Sorption of petroleum hydrocarbons from the vapor phase by the soil solid phase is also affected by the soil moisture status. Two distinct situations may exist.

1. *The spill occurs* immediately *after leaching, when the soil moisture content is high.* Under these conditions, when the soil is saturated or partially saturated with water, the adsorption sites of the solid phase are occupied by water molecules and the sorption of hydrocarbons is completely or partially prevented. The results of an experiment carried out to check the effect of initial soil moisture content on the adsorption kinetics of five hydrocarbons from a surrounding atmosphere onto a sandy clay soil are presented in Fig. 4. It was found that at 11% moisture content (W/W), which was equivalent to 70% of field capacity, hydrocarbon adsorption was almost negligible [Yaron et al., 1989]. This behavior was also observed for different hydrocarbon mixtures by Chiou and Shoup [1985], Barbex and Brown [1986], and Acher et al. [1989].
2. *The spill occurs a long time after leaching, when the soil is dry or partially hydrated.* An experiment performed by Nye et al. [1994] was designed to elucidate the sorption mechanism in the above conditions. The sorption of m-xylene and n-dodecane was studied on oven- and air-dried clay soil and on medium sand. m-Xylene reached sorption equilibrium faster on oven-dried

clay soil than on air-dried soil (Fig. 5). The total sorption capacity of the air-dried clay was nearly twice that of the oven-dried clay. The sorption capacity of the sand was relatively unaffected by its moisture content: the uptake rate of n-dodecane was equal in both the air-dried and oven-dried sand but the total sorbed amount in the air-dried clay was less than in the oven-dried clay. In the case of clay the water molecules compete for sorption sites, resulting in reduced sorption of n-decane compared with oven-dried clay. This behavior was partially explained on the basis of structurally related size-exclusion phenomena.

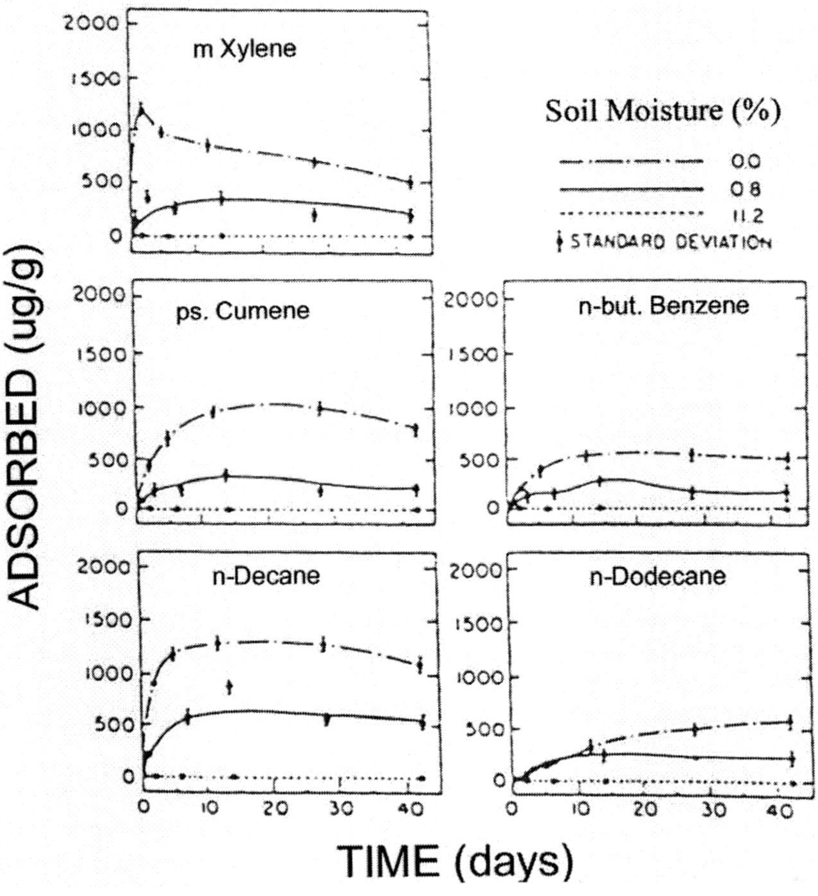

Figure 4. Effect of initial soil moisture content on kinetics of PHC adsorption from a surrounding atmosphere of a simulated kerosene by Gilat soil. [after Yaron et al. 1989]

The above findings confirm that in dehydrated systems the silicate layers of clay minerals are fully collapsed and the intercalation of non-polar organic liquids or gases is either absent or proceeds with difficulty, therefore, sorption occurs predominantly on external surfaces. [Mingelgrin and Prost, 1989].

Consideration of the time scale indicates that the soil moisture content immediately after leaching affects the retention capacity of the subsurface for petroleum hydrocarbon contaminants. A long time after leaching the soil moisture content in unsaturated or partially saturated soils achieves values that may affect their adsorption and redistribution in the subsurface.

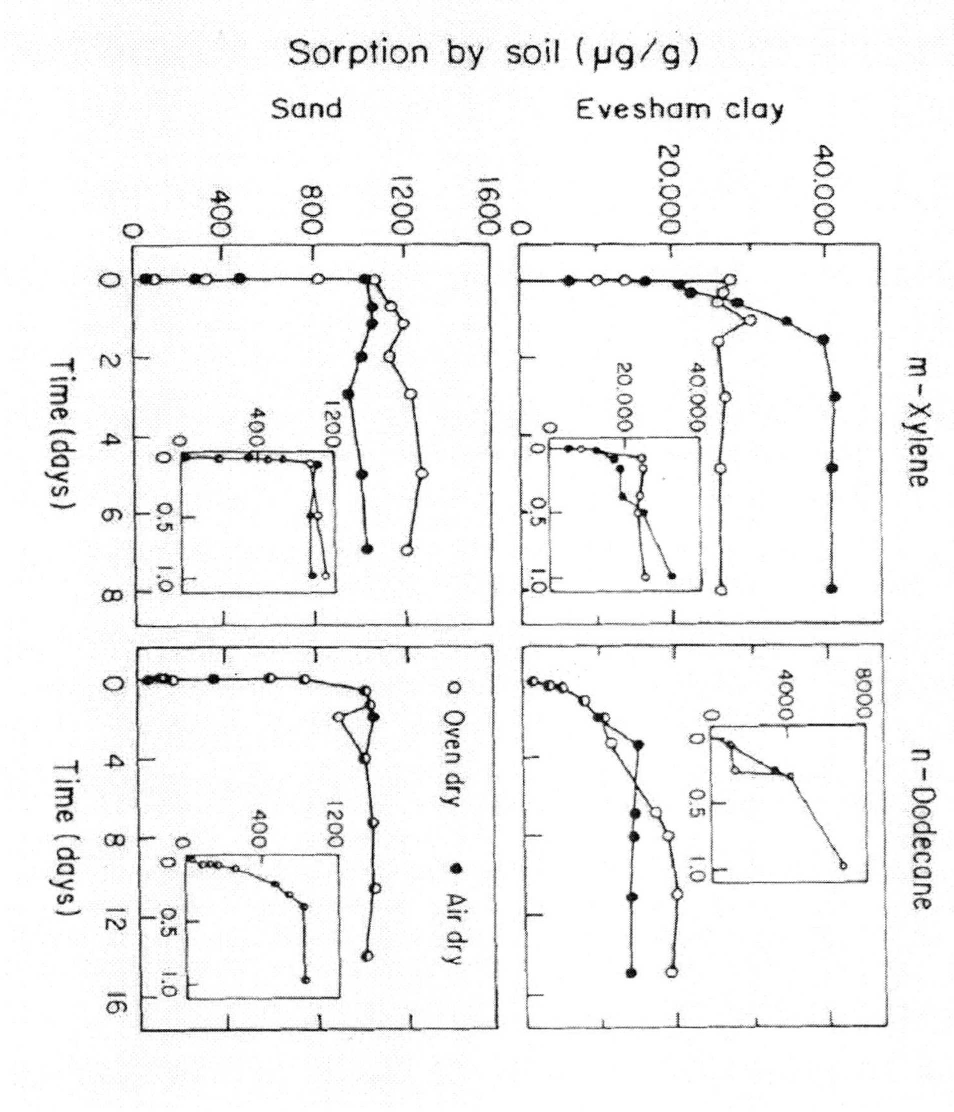

Figure 5. Kinetics of m-xylene and dodecane vapor sorption by medium sand and Evesham clay soil. [after Nye et al 1994]

3.2. Dissolution

The aqueous solubility of organic compounds is among the most important physical properties controlling their transport and fate in the subsurface. The dissolution process is defined as the transfer of chemicals from the non-aqueous phase to the aqueous phase in the contacting solution. The amount of NAPL that will dissolve in an aqueous phase depends on the solubility of the compounds, their mole fractions, the rate of transfer from nonaqueous to aqueous phase and the contact time between the two phases [Ptacek et al., 1987]. A bulk of experimental data on and models of PHC mixture dissolution have been published and quoted in review papers such as those published by Abriola and Bradford [1998], Lee et al. [1998] and Weber et al. [1998]. In the present study a discuss selected experiments in which dissolution of partially volatile PHC occurred as a result of leaching. Under equilibrium conditions a multi-component dissolution may be described in terms of a constant partition coefficient between the organic mixture and the aqueous solution. An example of kerosene–water partitioning at equilibrium, as reported by Dror et al. [2000a] is presented in Fig. 6. Kerosene – a petroleum product containing more than 100 hydrocarbons - was in contact with an aqueous electrolyte solution, 0.01 N NaCl, for 100 h at an ambient temperature of 22^0C. From the analysis of the chromatograms (Figs. 6a, b) we can see that aliphatic and branched aliphatic hydrocarbons in the range of C_9 to C_{16} make up the major group of components in neat kerosene, with only a minor set of aromatic compounds. In the aqueous electrolyte solution, on the other hand, aromatic compounds, especially the branched benzenes and naphthalenes, make up the majority of the compounds in the aqueous phase. The C_9 to C_{10} aliphatic and branched aliphatic components do not appear in the chromatogram of the aqueous kerosene. Aromatic components are several order of magnitude more soluble then aliphatic components, therefore, aromatic compounds, which were minor constituents of neat kerosene, dissolve in water solution much more than the major group of aliphatic compounds.

Under leaching - when flow conditions exist - the continuous release of components to the aqueous phase affects the composition of the mixture and, consequently, the partitioning coefficient between the organic and the aqueous phases. Augustjin [1993] used a miscible displacement technique to study the dissolution of polyaromatic hydrocarbons (PAH) from coal-tar-coated sand. The tar, which was homogeneously distributed on sand grains, was packed in columns and then leached with 0.01 N $CaCl_2$ solution at a pore water velocity of 15 cm/h. The elution profile of six PAHs from a column packed with sand-coated tar was described by Lee et al., [1998], it shows that the concentration of each component is initially constant and that it remains so until at least one of the constituents is depleted. The concentration of the remaining components reaches a new constant level in response to the changing molecular composition of the mixture.

Garg and Rixey [1999] reported the results of dissolution experiments in which mixtures of benzene, toluene, m-xylene and naphthalene (BTXN) were residually trapped in a relatively insoluble oil phase (tridecane). Leaching was done at a length:velocity ratio L/v of 0.0036 day and the dissolution was measured in the effluent, and differential dissolution of BTXN was found: aqueous concentrations followed equilibrium dissolution for reductions in concentration of approximately two orders of magnitude. For a longer time significant departures from dissolution were observed for all contaminants.

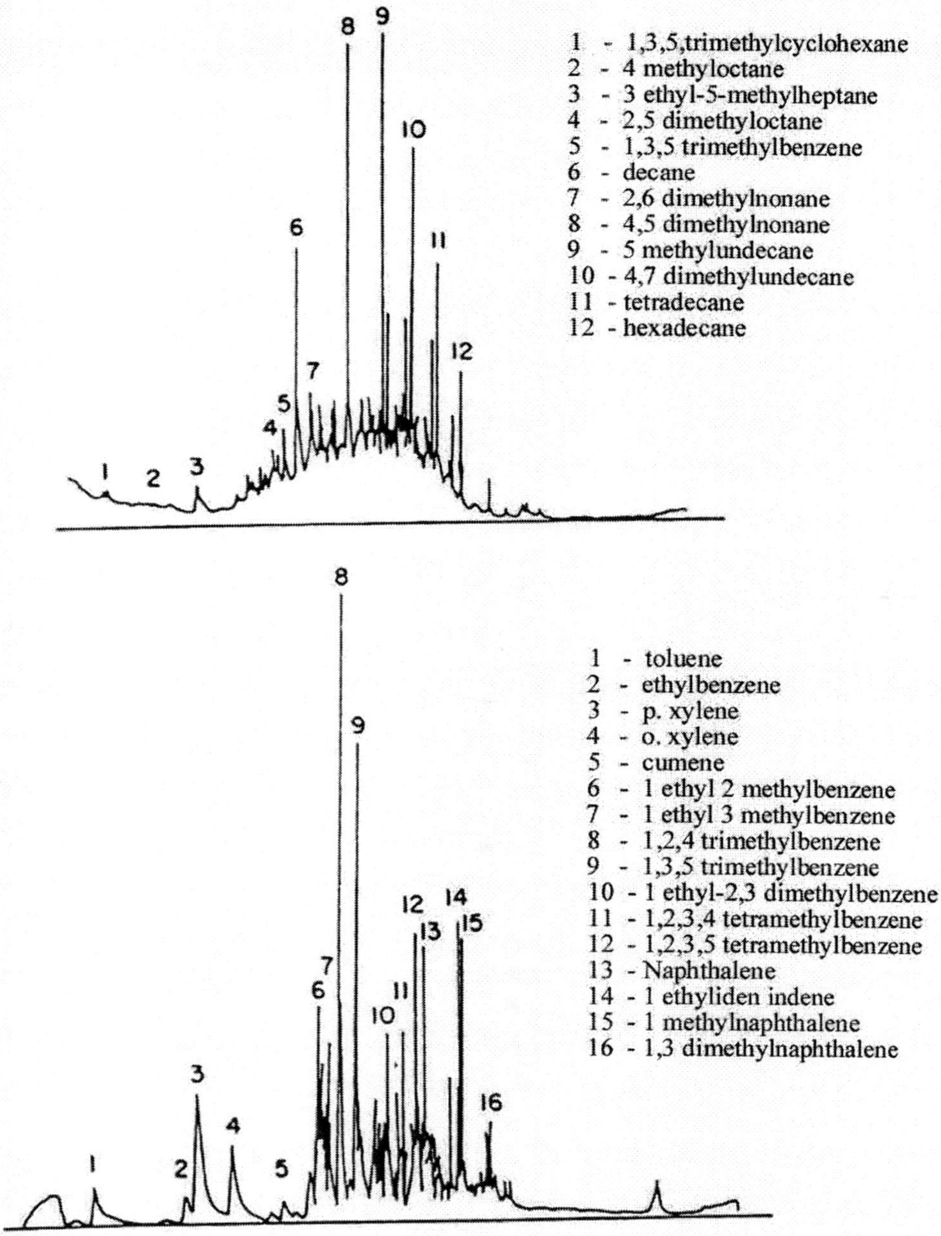

Figure 6. Gas chromatography of neat kerosene (a) and kerosene dissolved in 0.01N Nacl aqueous solution (b) . [after Yaron et al 1998].

The properties of the hydrocarbons affect their dissolution patterns too. Fortin et al. [1998] reported on the leaching of o-xylene (LNAPL) and o-chlorobenzene (DNAPL) trapped in a sand column. Major differences in the mobility and dissolution, between the two hydrocarbons were observed under continuous leaching. Under water-saturated conditions and downward flow, the trapped LNAPL was removed from the sand column by dissolution and mobilization of small droplets, whereas the DNAPL was removed by dissolution only.

The composition of the leaching water may influence PHC dissolution. Upon contact with soil the leaching water becomes a solution enriched in electrolytes and natural organics such as humic and fulvic substances.

Earlier studies - mainly looking at the effect of sea water - proved that the presence of electrolytes in aqueous solutions decreases the solubility of organic compounds in relation to their solubility in distilled water and a phenomenon called "salting-out".

Solubility in distilled water and in media of higher ionic strength is described by the Setschenow equation which empirically determines the magnitude of the "salting out" (see section 2). In a review on this topic, Xie et al. [1997] pointed out that the salting out effect is greater for organic solutes with larger molar volumes and that the Setschenow constants are similar for a sea water and a 0.5 M aqueous NaCl solution.

An experiment on the effects of the composition and concentration of an electrolyte solution on kerosene dissolution was set up in order to mimic a similar situation in a soil water environment [Dror et al., 2000a]: the kerosene dissolution in the electrolyte solution decreased as the NaCl concentration in water increased from 0.2 M to 2.0 M (Fig. 7) and no changes were observed when the electrolyte concentration was held constant but NaCl ions were replaced with $CaCl_2$ showing that ionic composition did not affect the dissolution in water of the PHC analyzed.

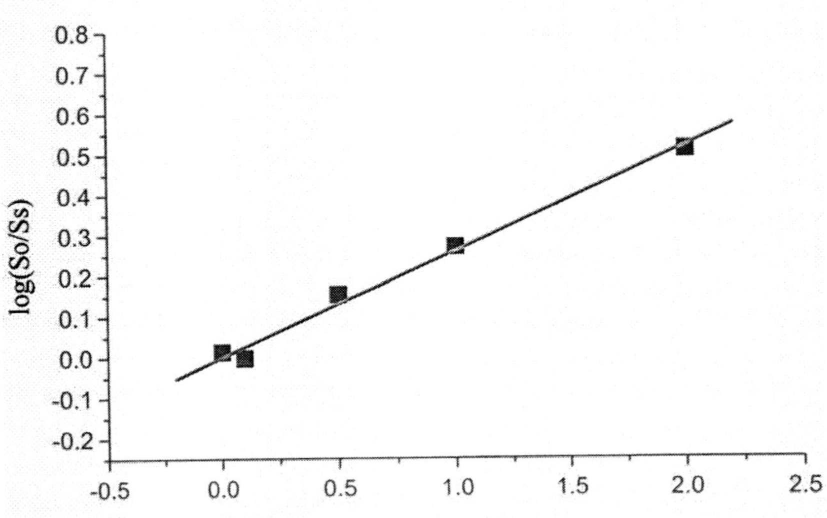

Figure 7. Dissolution of kerosene in aqueous solutions as affected by electrolyte concentration (after Dror et al 2000a)

Naturally occurring macromolecules in the soil medium are transported during the leachate flow through the subsurface. These macromolecules, including humic and fulvic substances, and negatively charged polysaccharides, may increase the dissolution of residual petroleum hydrocarbons. Whereas the solubility of the solute in distilled water may be regarded as an unambiguous physical constant, the apparent solubility and the physical state of a solute containing coexisting organic solutes are susceptible to modification by interacting with cosolvents. In principle, a cosolvent can enhance solute solubility, either by changing the solvency of the medium or by direct solute interaction, by adsorption or by partitioning [Chiou et al., 1986]. In a batch experiment on the effect of humic acid on kerosene dissolution in an aqueous solution, Dror [2000a] found a linear correlation between the amounts of humic acid and kerosene dissolved in the aqueous solution (Fig. 8). The enhancement of kerosene dissolution occurred even at low humic acid content in the aqueous solution. In view of the fact that humic substances are relatively high-molecular-weight species containing nonpolar organic moieties, Chiou et al. [1986] assumed that a partition-like interaction between a solute of very low solubility in aqueous solutions, and a "microscopic organic environment" of dissolved humic molecules would be a reasonable explanation for solute solubility enhancement.

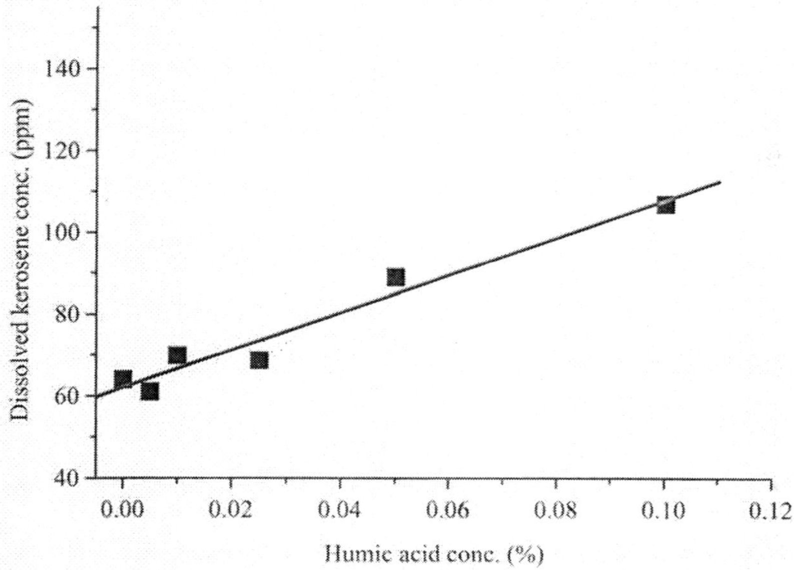

Figure 8. Dissolution of kerosene in aqueous solutions as affected by humic substances. [Dror et al 2000a]

3.3. Redistribution

The properties of the mixture components, and the hydraulic and physical properties and hydraulic status of the porous medium affect multiphase redistribution of PHC in the subsurface. Leaching of the soil subsurface before contamination might affect the vertical extent of the water-immiscible phase in the subsurface and, consequently, the depth of the entrapped zone.

The initial soil moisture content affects the penetration of PHC, both as immiscible liquids and as a vapor in gas phase. The rates and extents of liquid hydrocarbon infiltration in a sandy loam increased with increasing soil moisture content. Data presented in Fig. 9 describe the penetration of a mixture of five hydrocarbons (m-xylene, ps. cumene, n-but-benzene, n-decane and n-dodecane) in a sandy loam soil column. It may be observed that increase in the soil water content from oven dry (0.0%) to half field capacity (12%) led to an increase of up to 50% in penetration depth. However, the biggest increment was observed for the transition from an oven-dry to an air-dried soil (0.0 to 0.8%); a much smaller effect was observed when the soil moisture content was raised from 4 to 12%.

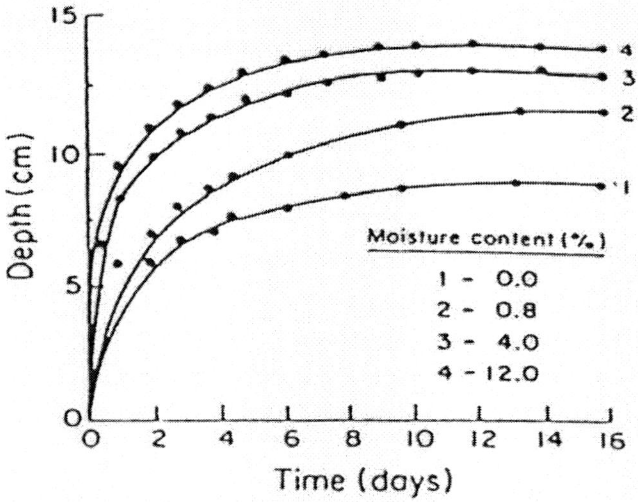

Figure 9. Penetration depth as a function of time, of simulated PHC in Bet Dagan soil column as affected by initial soil moisture content: oven dry 0.0% and air dry 0/8%. [after Acher et al 1989]

Gerstl at al. [1994] studied the effects of the relationship between soil type (texture and pore size distribution) and soil moisture content on kerosene conductivity. The kerosene conductivity were measured (relative to that in the respective oven-dry medium) for sand, loam and clay soil columns, as the initial moisture content varied from 0% to field capacity. Over this range of moisture contents the soil system is composed of two or three soil phases. It was observed that in sand, the kerosene conductivity remained nearly constant and in soils the kerosene conductivity changed as the initial soil moisture content increased. The kerosene conductivity of the clay soil increased as the soil moisture content increased from oven dry to 70% of field capacity. At 70% of field capacity the kerosene conductivity reached a maximum and at initial water content of field capacity the kerosene conductivity decreased to zero. In the loam soil the kerosene conductivity did not change when the soil was partially hydrated (air-dry) but increased as the moisture content reached field capacity. In the first stage of hydration (from oven-dry to air-dry) the water molecules partially_covered the sorption sites in the soil; water sorbed on the soil particles prevented interaction between the solid and hydrocarbon molecules and therefore a kerosene layer was not formed. In the oven-dry clay soil the kerosene infiltrating into the soil was adsorbed on the pore surfaces, creating a wall effect which retarded movement by formation of a viscous layer; however, since in the air-dry soil the pore walls were partially hydrated, the wall effect was reduced and the kerosene conductivity increased relative to that in oven-dry soil. The kerosene conductivity was not affected by partial hydration, in either the loam soil or the sand, both of which are characterized by larger pores than those of the clay soil.

Once the PHC transport in water-immiscible phases, and the water flow in the soil column stops, the contaminant becomes entrapped in the soil, according to its retention capacity, and the downward redistribution in the soil subsurface continues in the gaseous phase. In previous column studies it was shown that kerosene components moved faster and deeper in air-dry than in oven-dry columns [Acher et al., 1989]. This behavior is consistent with the adsorption pattern described above.

In additional soil column studies performed by Wilkins et al. [1995], it was found that during NAPL volatilization of three single components (styrene, toluene and tetrachlorethylene) contaminant vapor pressures deviated from local equilibrium values by 10-40% for pore velocities ranging from 0.25 to 1.5 cm/s.

In a 1-yr monitoring study reported by Johnson and Perrot [1991], on subsurface gasoline vapor concentration, it was found that concentrations at the sampling points near the center of the vapor plume remained relatively stable while those nearer to its periphery showed long-term increases. The principal reason for the relatively stable concentrations were that the diffusive transport of the vapors was very slow and the diffusive fluxes were small, primarily because of the high water content of the soil, which was approximately 90% of the saturation level. This finding confirms that the controlling factors in defining the conductivity for the water- immiscible liquid are the pore sizes and their distribution in the subsurface.

More realistic information was obtained from a recent field experiment on the effect of leaching on the redistribution of PHC components under field conditions (where natural attenuation of the contaminants occurs simultaneously with their downward transport [Dror et al. 2000b]. Kerosene was added to an air dry (2%) loamy clay soil and to a soil wetted to an initial moisture content equivalent to field capacity (17%). The field was then leached by sprinkler irrigation for 39 days, with a total of 500 mm water applied in 10 pulses of 50 mm each. The results presented in Fig. 10 show how the combined processes of natural attenuation, volatilization and/or leaching affect the redistribution of the PHC in the soil profile with passing time. The soil water content at the time of PHC application, and the amount of leaching also affect the redistribution dynamics. Application of kerosene to wet (field capacity) (Fig. 10a) and dry soil (Fig. 10b) showed the effect of the soil moisture content at the time of kerosene application on its redistribution under leaching. After 5 days there was a much higher concentration of kerosene ($9.4*10^6$ μg/kg) in the upper 10-cm layer of initially wet soil than in the corresponding layer of the initially dry soil ($5.3*10^6$ μg/kg). Infiltration of kerosene into the soil profile was lower in the prewetted field than in the dry one. This may be explained by the fact that during that time in the prewetted soil layers less free pore volume was available for the transport of kerosene. The high concentration of kerosene near the surface of the prewetted soil was reduced strongly (to $2.7*10^6$ μg/kg) in the next 14 days, volatilization being the major natural attenuation process; because the measurements indicated that there was no significant downward movement, we considered that the kerosene retained near the soil surface was highly affected by evaporation to the atmosphere. A larger fraction was lost from the prewetted soil treatment than from the dry treatment, since more kerosene was retained in the upper soil layer in the former case. This may be explained in terms of water partly filling the soil pores, causing strong reduction of the hydrocarbon retention on the soil solid phase.

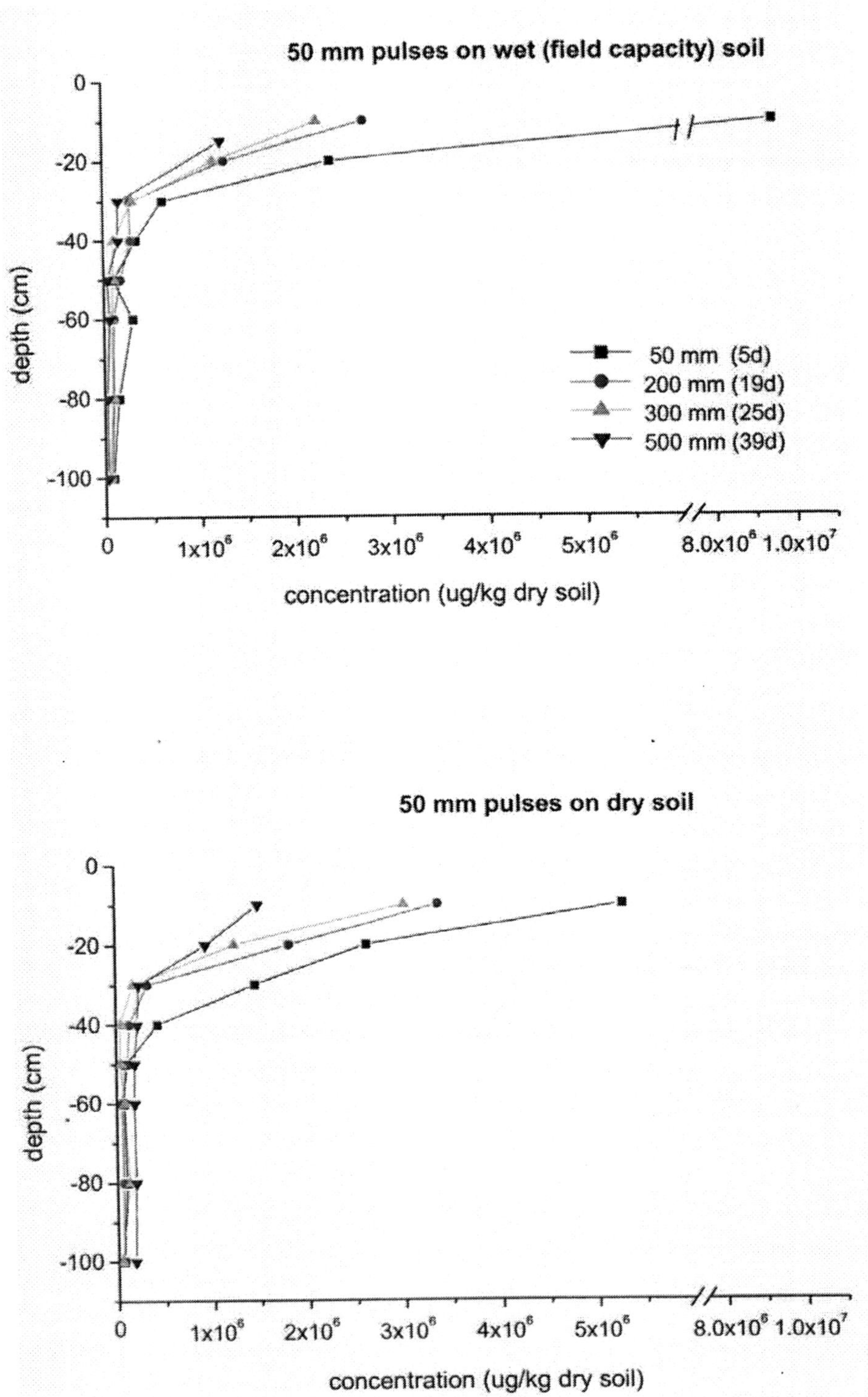

Figure 10. Effect of leaching on natural attenuation and kerosene redistribution under field conditions a – kerosene applied on wet (field capacity) soil; b- kerosene applied on air dry soil [Dror et al 2000b].

In a parallel set of experiments [Dror et al., 2000a] the effect of enrichment of PHC mixture with specific organic compounds on its distribution in soil was examined. The center of mass (CoM) redistribution of kerosene in the subsurface during leaching is presented in Fig. 11 for kerosene enriched with xylene, 1,3,5-trimethylbenzene and naphthalene (Fig. 11a) and for kerosene enriched with xylene, 1,3,5-trimethylbenzene and dichlorobenzene (Fig. 11b). The CoM redistribution for the total hydrocarbon was found almost the same for these two treatments, while single components of the mixture move to the subsurface at different rates. Hence, the addition of solvents to kerosene has little or no effect on the redistribution of the total hydrocarbon mixture in the field. Single components of that mixture might move to deeper soil layer as shown for naphthalene.

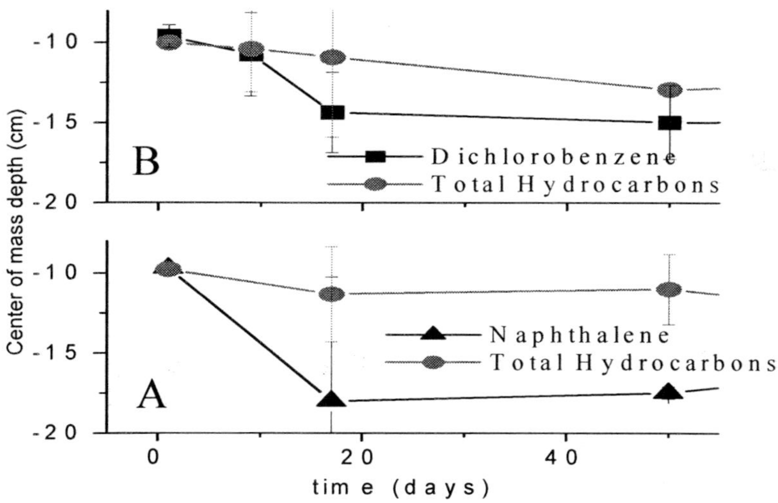

Figure 11. Center of mass redistribution in subsurface during leaching of neat kerosene and kerosene enriched with dichlorobenzene (B) and naphthalene (A).

During leaching the CoM of naphthalene is found at a deeper depth then that of the total kerosene and the CoM of dichlorobenzene is similar to that of the total kerosene thusshowing the differential transport of single components.

As discussed before the properties of the kerosene hydrocarbon mixture and the environmental conditions in the soil profile affect the redistribution of the kerosene in the field soil. The distribution of kerosene in the soil profile during the experiment is shown in Fig. 12. The kerosene redistribution differed for soils amended with sewage sludge, NPK fertilizers and the untreated soil. The kerosene concentration in the treated profile is much lower than for the untreated soil. However for the plots treated with sewage sludge the concentration in the layer depth of 10 cm and deeper, after 17 days from kerosene application is higher for the treated soil. This was explained by the fact that the addition of sewage sludge to the soil surface enriched the leaching water with natural organic matter and hence increase the solubility of the kerosene components. These dissolved hydrocarbons can redistribute to a deeper layer with the leaching water. The chromatogram of the kerosene treated with sewage sludge left at the soil surface after 50 days support this explanation. In this chromatogram - Fig. 13c the light and more soluble compounds disappeared almost completely which can be explained by enhance dissolution and transport of these compounds in addition to their degradation. Moreover the hydrocarbon concentration for the sewage sludge treatment after 50 days show that the concentration in the surface is lower than in the 10-20 and 20-30 cm depth. This also supports the transport as a solute explanation. After 100 days the degradation become the most important process and hence the transport of kerosene is not as obvious as after 50 days from kerosene application.

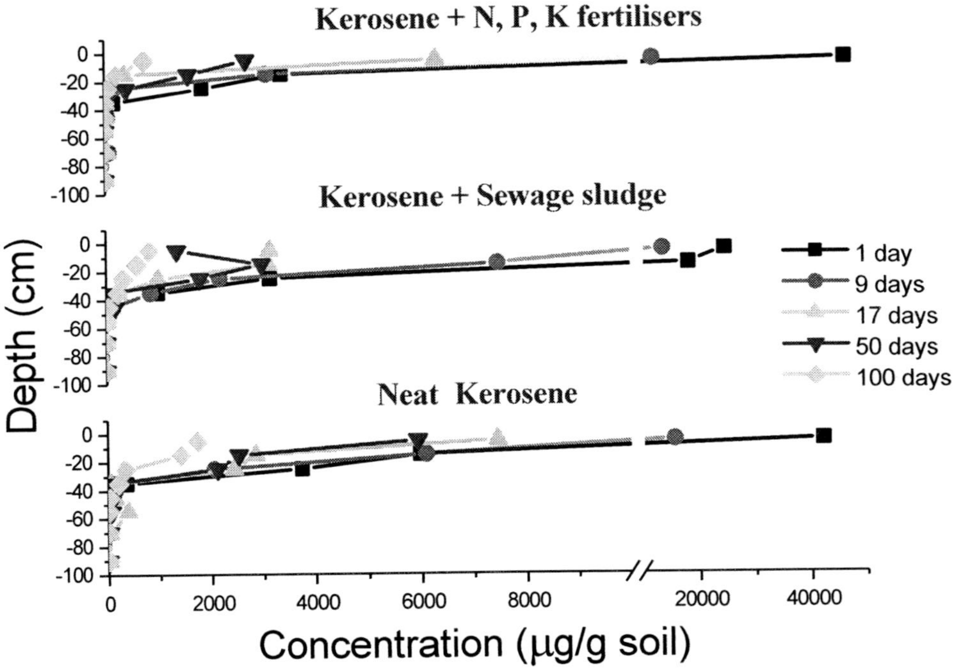

Figure 12. Effect of amendments on the redistribution with depth of kerosene in the field soil after 100 days and 650 mm leaching

For the addition of NPK fertilizer treatment the kerosene hydrocarbons is retained near the soil surface throughout the experiment. This behavior was explained by reduced solubility of the kerosene hydrocarbons due to addition of electrolytes as fertilizer. This in turn led to less transport of kerosene with the leaching water. Moreover the dissolved kerosene would probably degrade by the enhanced microbiological activity induced by the large amount of nutrients available.

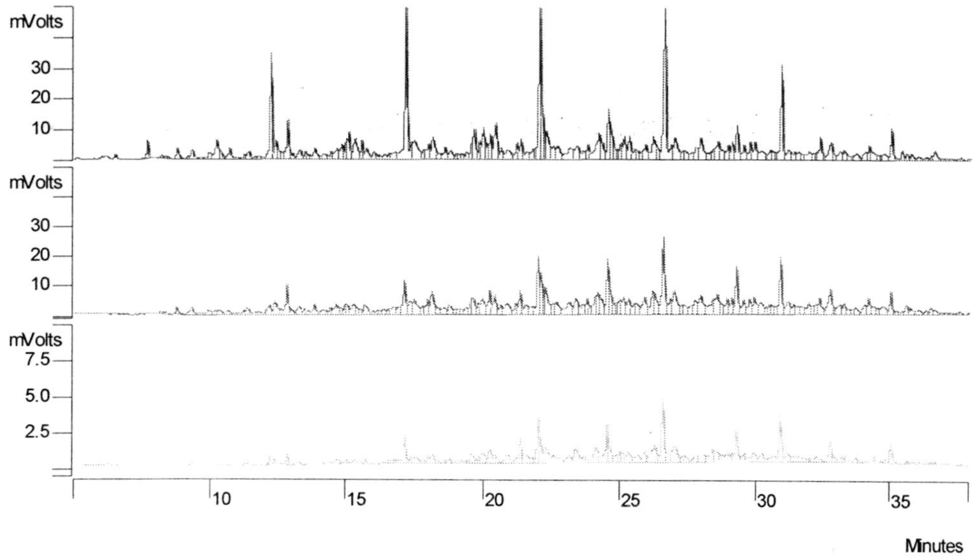

Figure 13. Kerosene weathering in field (0-10cm) 50 days after application under different soil treatments. A -neat kerosene; B - kerosene + NPK fertilizers; C- kerosene + sewage sludge; identified components: 1. n-decane; 2. n-undecane; 3. n-dodecane; 4. n-tridecane; 5. n-tetradecane;

3.4. Attenuation

The main attenuation processes of entrapped PHC in the soil porous medium are volatilization and degradation (mainly biological). These processes may occur in the organic, aqueous and adsorbed phases, and both are directly affected by the soil moisture status and indirectly by the leaching pattern.

PHC attenuation in the soil environment through volatilization is related to the vapor pressure of each of its components. Therefore, during volatilization, changes in the composition of the initial solution occur and the entrapped contaminant undergoes weathering. An additional factor influencing the PHC weathering at any particular time is the rate of vapor movement through the soil pores to the soil surface and thus to the atmosphere. An increase in the soil moisture content results in a reduction in the free pore space available for vapor movement and, consequently, to a decrease in ability of the PHC to evaporate out subsurface. Henry's Law describes the ratio of gas phase concentration to liquid concentration for volatile compounds under equilibrium conditions, but equilibrium as defined by Henry's Law does not always exist in dynamic systems such as porous media during leaching. Therefore, experimental evidence is required to quantify this process. The experimental results of Karimi et al. [1987] showed the effect of soil water content on benzene volatilization flux. An increase in the soil water content following leaching reduced the available pore space for volatilization (Fig. 14), and by virtue of the solubility of benzene in an aqueous phase, an increase in the soil water content led to an increase in the total amount of soluble benzene in the soil solution, and a consequent reduction in the amount of contaminant available for volatilization. Karimi et al [1987] suggested that the effect of soil water content on benzene volatilization flux through soil could be attributed to its effect on the air-filled pore space, which, in turn, is the major soil parameter controlling benzene volatilization through soil. Gidda et al. [1999] showed by a column experiments with three Canadian soils, that higher soil water content inhibited the volatilization of gasoline. In that study water was shown to impact both the diffusive and the wicking movement of entrapped gasoline to the atmosphere. As a result, a decrease of PHC attenuation occurred because of reduced volatilization. This behavior would be expected only for rigid porous medium (e.g., sand or sandy soils) or when the soil moisture content does not change from its initial value or increases during volatilization. It would be different if an initially wet nonrigid soil (e.g., clayey) exhibited a trend toward dryness during volatilization. Rogers and Ong [2000], as a result of a controlled laboratory venting experiment, concluded that under saturated conditions the properties of a rigid porous medium may affect the mass removal of benzene vapors when there are void channels in the saturated zone. They related this conclusion to the fact that a larger-grained porous medium has bigger air channels, which favor a greater vapor mass transfer to the atmosphere and therefore greater attenuation. In an outdoor experiment, Fine and Yaron [1993] measured the volatilization of kerosene in fine clayey soil during 18 days following PHC application on soils with initial moisture contents of 6.8% (air dry) and 27% (field capacity). Fig. 15 shows the kerosene concentration attenuation for the experimental soil during free and enhanced volatilization. The removal of kerosene from the dry-free combination (Fig. 15 A 1) after 18 days was no more than 35% of the KRC, whereas the comparable loss from the moist-free (Fig. 15 A 2) combination was twice as much, throughout the soil column. However, the relative basis is some what misleading. The two free venting treatments, actually removed very similar total amounts of kerosene. They differed on a relative basis, because their initial KRC's were different. This was explained by the fact that during the 18 days of measurements the moisture contents in the clayey soil decrease because of the formation of cracks and the consequent restoration of the air permeability to its prewetting value. The authors commented that this was the only feasible explanation, because the air permeability in the wet soil treatment remained lower than that in the dry soil treatment. When the same treatments were submitted to venting, enhanced volatilization was observed that led to a diminution of the entrapped PHC concentration.

The weathering of PHC through volatilization is influenced by the properties of the soils and of each component of the mixture which quantify its tendency to be removed preferentially. The mixture composition depends on the properties of the components in the gas and liquid phases. Jarsjo et al. [1994] reported on the weathering at 27^0C of kerosene entrapped in a sand, sandy loam soil, and peat, 7 days after application. Using a published method [Martinez de la Cruz 1990] Jarsjo et al. [1994] determined the temporal changes in the concentrations of the remaining kerosene components during volatilization, as functions of the relative numbers of moles of the components (Fig. 16). The components $C_9 - C_{11}$ were removed preferentially during the whole process and their selectivity reached zero as the molar fraction of these components in the liquid approached zero. As the composition of the remaining liquid changed the C_{12} was also removed preferentially, whereas the selectivities of the C_{14} and C_{15} components remained below zero, at least until volatilization had progressed to an extent corresponding to 0.8 molar fraction of the initial amount. The effect of soil type on the weathering of kerosene may be observed in the gas chromatograms (G.C.s) of contaminants recovered from a sand, sandy loam and peat soil (Fig. 17).

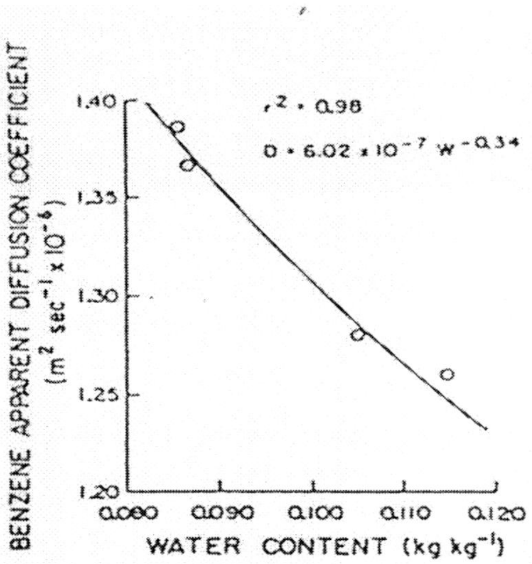

Figure 14. Effect of soil water content (wt/wt) on the specific benzene volatilization flux through a 2.54 cm soil cover with a bulk density 1.4 kgm^{-3} [after Karimi et al., 1987]

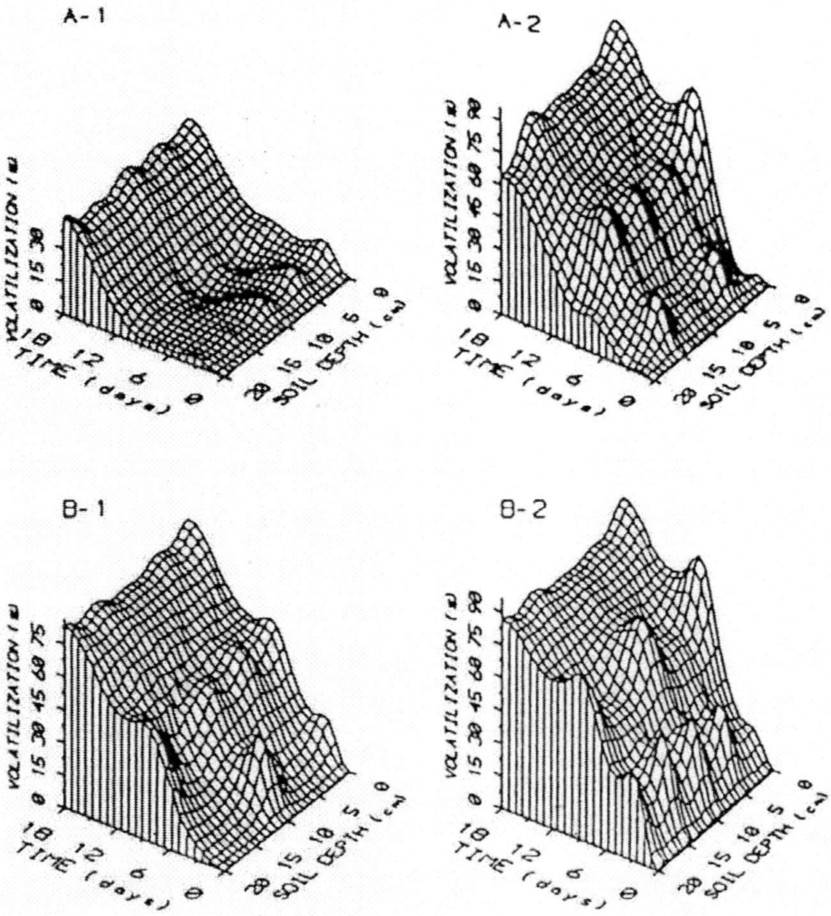

Figure 15. Changes in the kerosene concentration of the Nahsholim vertisol as affected by free (A) and enhanced venting (B) volatilization from initially dry (1) and moist (2) soils. [after Fine and Yaron 1993].

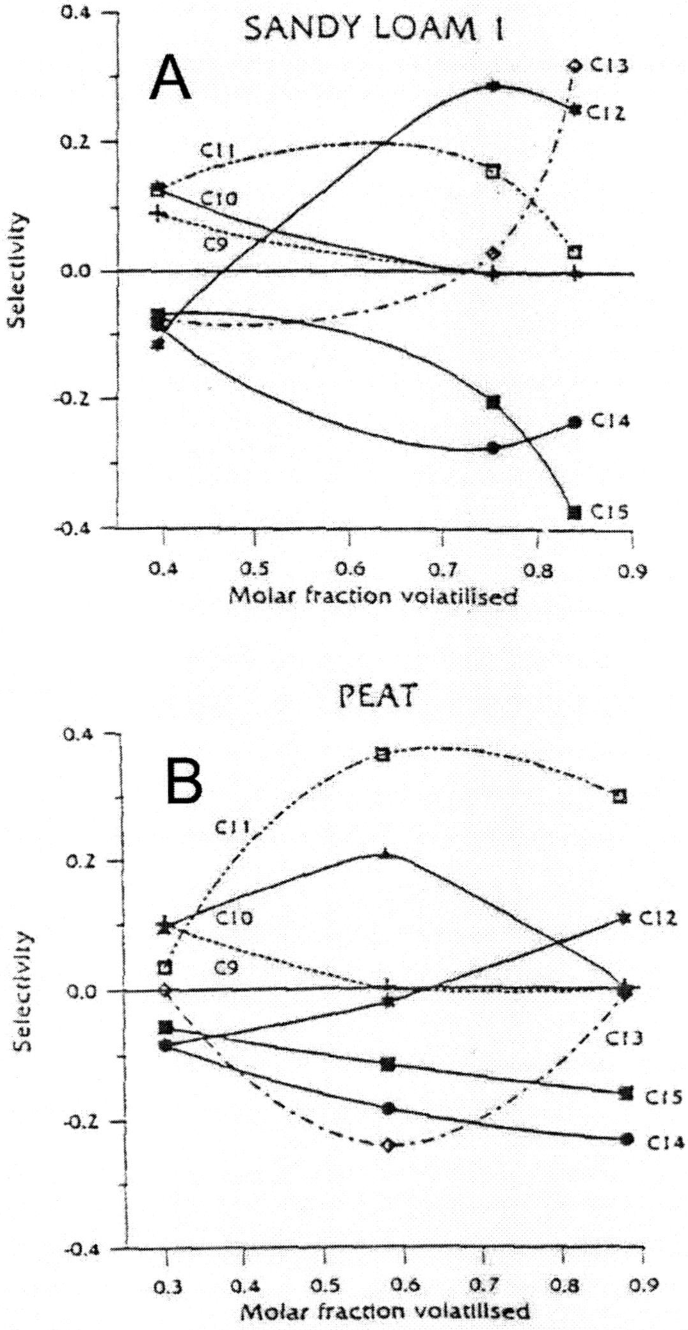

Figure 16. Selectivity of C_9-C_{15} during volatilization from sandy laom (a) and peat (b). [after Jarsjo et al 1994]

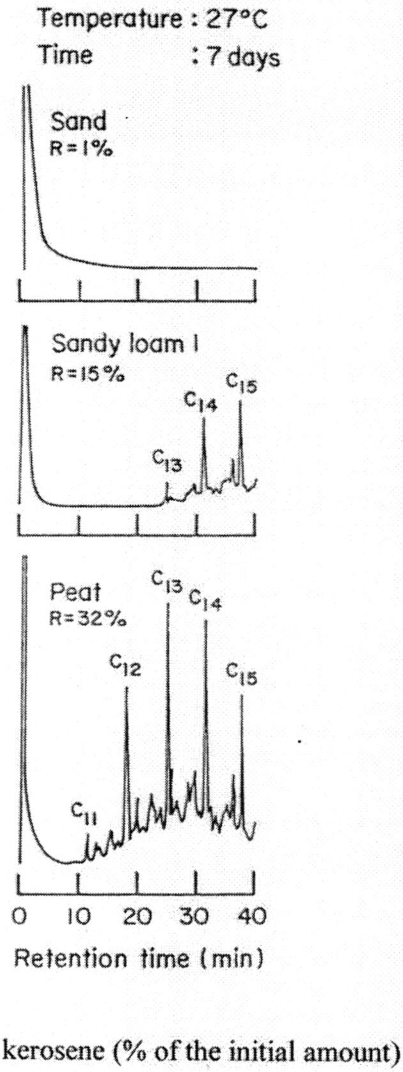

Figure 17. Gas chromatograms of kerosene recovered from sand, sandy laom and peat after volatilization at 27^0C and 7 days. [after Jarsjo et al, 1994]

In situ data on the cumulative effect of volatilization, redistribution in depth and biodegradation, as reflected in temporal changes of PHC concentration in the subsurface, may be obtained through field experiments in which environmental conditions are favorable for both mass transfer and bioactivity. Dror et al. [2000a, b, 2001] report on two field experiments in which temporal changes in the concentration of kerosene as a result of field leaching were measured to a depth of 100 cm. The dynamics of kerosene attenuation for 180 days after contaminant application and leaching with 1,000 mm of irrigation and rainwater are presented in Fig. 18. Kerosene attenuation rates in the 0-20cm upper soil layer and in the whole soil profile (0-100 cm) were compared. The attenuation process occurred primarily in the soil upper layer and was slightly affected by leaching and redistribution with depth down the soil profile. One hundred and eighty days after the kerosene "spill", its residual concentration had decreased from an average of 20,000 to 1,900 µg/g dry soil in the upper layer and to an average of 1,600 µg/g dry soil in the soil profile (0-100 cm). This attenuation pattern could be divided in to two parts. In the first 50 days after application enhanced volatilization predominated, and the initial increase of the n-alkane group in the subsurface, relative to the total PHC composition, confirms the volatilization of low-vapor-pressure compounds. From 50 days onward slow attenuation of kerosene was observed, with a persistent concentration of hydrocarbons of about 1,900 µg/g dry soil. The attenuation pattern may be explained by the fact that in the first period the light fraction of kerosene, including aromatic compounds such as alkylated benzene, was drastically reduced, and in the second period biological degradation become the most significant attenuation process. This behavior is in accordance with existing knowledge [e.g., Douglas et al., 1992], i.e., that the rate of biodegradation follows the descending order n-alkanes > aromatics > branched alkanes.

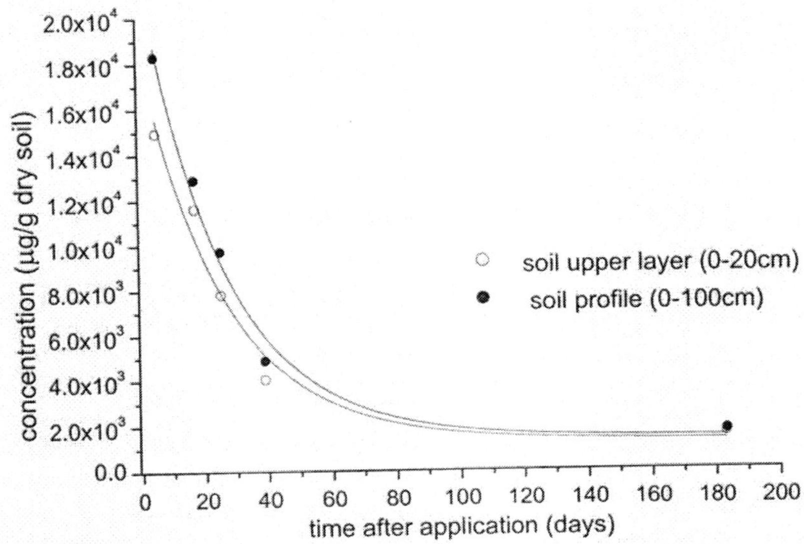

Figure 18. Natural attenuation of kerosene in soil upper layer (white dots) and in the whole soil profile (black dots) during leaching with 1000 mm [after Dror et al 2000b]

The properties of PHC components, such as solubility in aqueous solution and vapor pressure, as well as their interactions within the mixture affect their attenuation in the subsurface, which is also affected by leaching. The persistence of selected components of a kerosene entrapped in a soil during 39 days after addition was determined by Dror et al. [2000b] in an unirrigated field and in a field leached with 500 mm of water; the results are presented in Fig. 19 and present the dynamics of n-alkanes in the leached soil (based on GC area peaks). As a general feature a decrease in concentration was observed in comparison with the amount applied, but the kerosene attenuation in the leached field was less than that in the dry field. This is explained in terms of the amount of free pore space available for mass transfer of the volatile compounds, which was greater in the dry soil than in the leached field. The concentration of all the compounds decreased in comparison with the amounts applied, but different behavior patterns were observed for each of them. The concentrations of the lighter compounds, xylene, n-nonane, trimethylbenzene, n-decane and C_9-alkylated cyclic hydrocarbons in the 20-30cm layer were the same as or slightly greater than those in the 0-10cm layer in the leached field. This was explained in terms of a combination of volatilization of the light compounds from the surface to the atmosphere and transport in solution in the water phase. The heavier, less soluble and less volatile, compounds were found in the 60-80cm layer in the same relative concentrations as the light components, which indicates that PHC transport to the deeper layers was in the form of microemulsions or by preferential flow. In the dry field there was a constant decrease in concentration of all the compounds with increasing depth and, in general, lower concentrations than in leached soils. For the four heavy compounds, n-dodecane, C_{12} alkene, n-tridecane and tetradecane the residual concentration in the dry field was similar to or greater than that in the leached field, which may be attributed to smaller effect of volatilization of the heavy compounds and their biodegradation. The preferential biodegradation of the heavy components in the leached field, because of more favorable environmental condition may account for the higher concentrations in the dry field.

In additional work [Dror et al., 2000a] the attenuation pattern of the total petroleum hydrocarbons content in the soil profile of the neat kerosene and neat kerosene treated with amendments (NPK fertilizer or sewage sludge) during 100 days of field experiment is presented in Fig. 20. Both amended and non-amended soils show attenuation of kerosene with time and leaching. After 100 days of field experiment the total kerosene hydrocarbon concentration was reduced to less than 10% of the amount applied for all cases. In all cases most of the hydrocarbons were lost during the first 20 days. However the rate of kerosene attenuation was faster for the amended soils. The difference in the rate of attenuation is seen especially between 10 to 50 days from the begging of the experiment. For all the treatments a minimum contaminant concentration level was asymptotically approached. However, for the plots where amendments were added, this level was reached much faster.

Generally the biodegradation-weathering pattern of PHC is different from the physical-chemical degradation pattern. In biologically enhanced degradation, selected chemicals will be affected more then other compounds with similar physical properties [e.g. Park et al., 1990; Wang et al., 1998b] whereas for physical-chemical degradation compounds generally degrade in accordance with their properties.

The weathering pattern of kerosene in the field may be observed in Fig. 13. Chromatogram 13a represents the kerosene hydrocarbons in the 0-10 cm field soil after 50 days when no amendment was added to the soil. Chromatogram 13b shows the kerosene hydrocarbons in the 0-10 cm soil layer when amended with NPK fertilizer - and chromatogram 13c displays the kerosene hydrocarbons in the (0-10cm) to which sewage sludge was added. By comparing chromatograms a – c we can see the weathering effect of the amendments both on the composition and concentration of the remaining hydrocarbons after 50 days in the 0-10cm field soil layer. The treated soils (b and c) shows much smaller peak areas for all the compounds seen in the chromatogram indicating that weathering of kerosene is enhanced in the treated soils. The ratio of the total kerosene hydrocarbon concentrations in the 0-10 cm layer is more than 4:2:1 comparing no treatment to addition of fertilizer and sludge respectively. This may be explained by enhanced biological activity in the treated soil plots due to the supply of nutrients which support enhance microbial growth and hence consuming of hydrocarbons as a substrates.

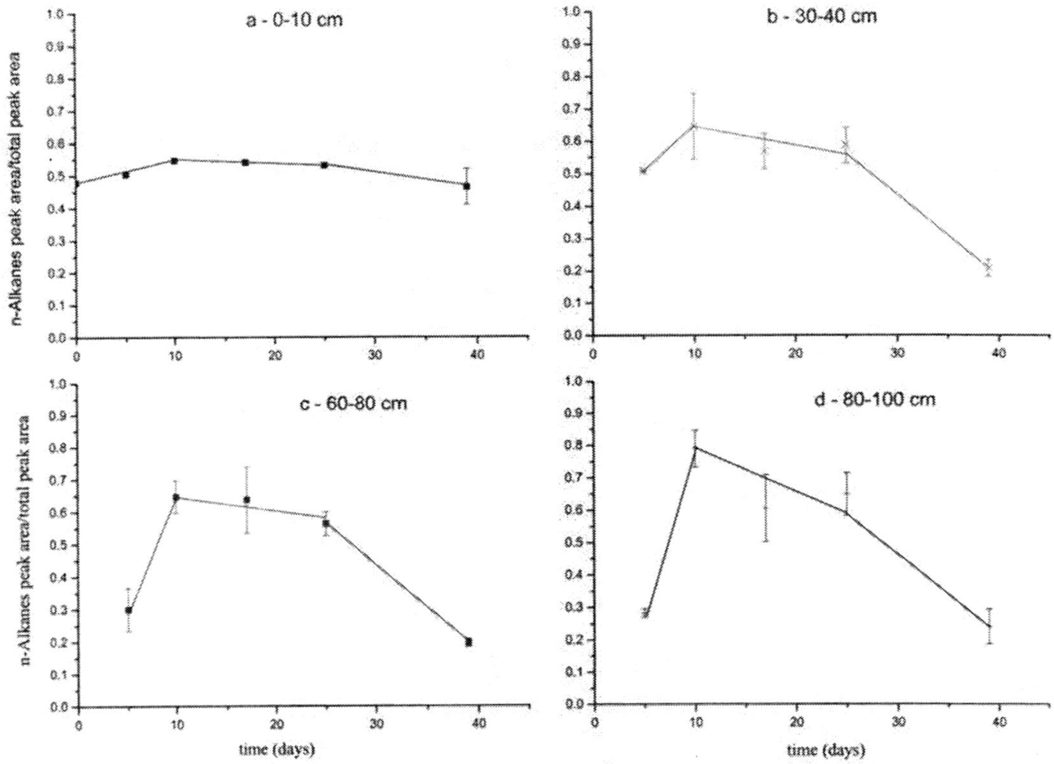

Figure 19. Ratio of n-alkanes to total hydrocarbons in soil profile during 39 days and 500 mm water leaching [after Dror et al 2000a].

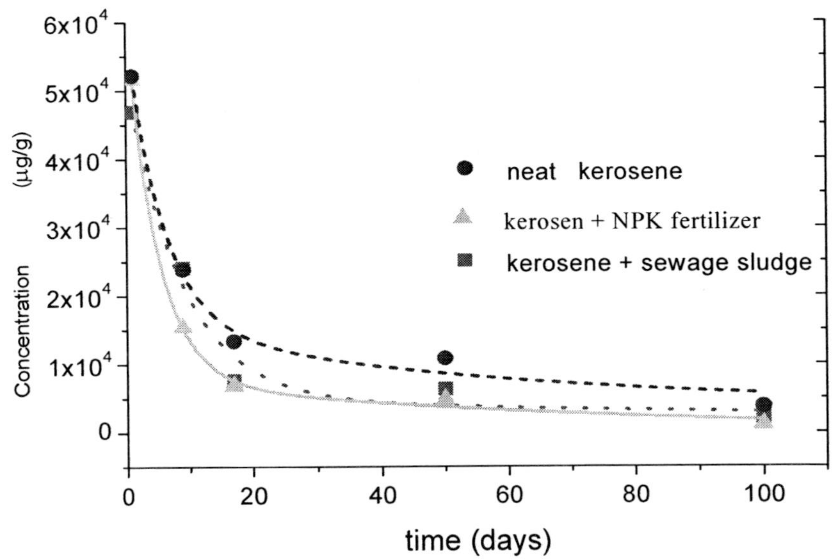

Figure 20. Effect of amendments on natural attenuation of kerosene in the soil profile during 100 days and leaching with 650 mm [after Dror et al 2000c].

The effect of weathering on the composition differs in the two amended soils. In the soil treated with sewage sludge (13c) the hydrocarbons found in the chromatogram up to a retention time (r.t.) of 20 minutes disappear almost completely whereas for the fertilizer treatment (13b) the weathering does not affect all the compounds leaving several in r.t. 0-20 minutes. The compounds usually found in the beginning of the chromatograms are the more soluble and volatile compounds. These compounds are expected to dissolve and volatilize more readily than the compounds eluting at the end of the chromatogram. The addition of sewage sludge to the soil surface enriched the leaching water with dissolved organic matter. Such an increase in the dissolved organic matter was shown to increase the solubility of kerosene hydrocarbons [Dror et al., 2000c]. This suggests that the enhanced degradation of these compounds is caused by greater dissolution of these compounds by the dissolved organic matter from the sewage sludge and the degradation of the dissolved compounds (and transport to deeper layer - see explanation in the redistribution part. The weathering picture for the fertilizer treated soil shows preferred degradation of the n – alkanes all over the chromatogram range which are know to be preferred substrates for biodegradation [e.g. Lee and Levy 1989; Douglas et al., 1992] and are present in larger amounts than other compounds in the kerosene mixture. This may suggest that the enhanced degradation caused by the optimization of the conditions for biological growth, which then degrade the hydrocarbons available in the soil. For the plots treated with either amendment the n-alkane concentration in the soil upper layer are reduced much faster than in the untreated soil. For example the ratio of n-dodecane (peak 3) to branched aliphatic (peak 3a) is 2.55 for the untreated soil and 0.98 and 1.14 for the treated soils, respectively.

The ratio of n-alkanes (peaks 1-5) to the other hydrocarbons such as branched aliphatics, aromatic, etc. (unnumbered peaks) indicates enhanced biodegradation as the biodegradation occurs preferentially for the n-alkanes [e.g. Lee and Levy 1989; Douglas et al., 1992].

4. Summary and Further Implications

Petroleum hydrocarbons are very important source of energy that power the world but they arealso one of the most abundant pollutants of subsurface environments. Therefore we must understand thier behavior in the subsurface in order to protect and remediate natural resources. In this work we discussed a recently recognized mechanism which suggests the potential for significant intrusion of organic contaminants from polluted seawater to freshwater aquifers. Initial work on this subject suggests the existence of a "salt pump" mechanism: while migration of organic compounds is due to chemical gradients, it is significantly enhanced by the presence of salt. Then, the existence of droplets that cautfc se increased carrying capacity of aqueous solutions (CCAS) for organic compounds was described.

Polluted marine environments are active open systems in which waves, tidal movement, streams and mixing occur continuously, leading to the formation of stable droplets, an increase in the carrying capacity of seawater for organic contaminants, and consequently the salting out of hydrocarbons. Under these conditions, the organic pollutants will be driven across an inert or active interface to the freshwater aquifer. Hence under such a scenario, contaminated seawater will act not only as a source but also as a driving force that magnifies transport of hydrocarbons into a freshwater aquifer.

The environmental conditions are very important for the fate of PHC in the subsurface. In this work we studied the effect of leaching on PHC in the subsurface under field conditions. Once the residual PHC reach the soil surface, their concentration is influenced by the soil moisture content. The transport and attenuation of the PHC in the unsaturated zone is affected by the amount and rate of leaching. Since PHC is a mixture of volatile and non-volatile hydrocarbons whose components differ by several orders of magnitude in their vapor pressure and water solubility, their fates in the soil environment will be diverse too. This will influence the temporal concentrations of the PHC and their distributions with depth, as vapors, solutes or water-immiscible solutions. The fate of PHC in the soil medium is influenced by both biotic and abiotic processes.

The relevant microbial consortia for bioremediation exists in the field and the addition of amendments to the soil enhances degradation of PHC in the field soil. Soil amendments may enhance the rate of PHC degradation and reduces the residual amount left compared to the untreated soil. The addition of sewage sludge may enhance the redistribution with depth of PHC. Although the attenuation of kerosene for soil amended with sewage sludge and N,P,K fertilizer was found to be similar, the distribution of the kerosene in the soil profile was very different. The kerosene treated with fertilizer was retained near the soil surface and the kerosene treated with sewage sludge was distributed to the deeper soil layers. This was explained by the properties of the leaching water. In the sewage sludge treatment the leaching water was enriched with organic matter which increase kerosene dissolution whereas for the NPK fertilizer treatment the leaching water was enriched with electrolytes which reduce the dissolution of kerosene hydrocarbons.

The abiotic processes controlling soil-PHC interactions were divided to the following patterns:- if the amount of PHC in the soil upper layer is smaller then the soil retention capacity, then 1. retention by the solid phase occurs via entrapment into the soil air-filled pores; 2. the highly volatile fraction evaporates into the gas phase in the porous medium, from where it may be lost to the atmosphere, sorbed onto the soil solid phase or dissolved in the soil water or ground water; 3. differing aqueous solubilities for the disparate components and variable quality of the soil water will result in differential dissolution and transport as solutes; 4. less volatile components are transported as nonaqueous phase through the porous medium.

if the amount of PHC in the soil upper layer is greater than the soil retention capacity, the hydrocarbon mixture will travel into the vadose zone as a separate nonaqueous phase, and will reach the ground water or be redistributed along the unsaturated zone, in accordance with the patterns described above.

Leaching the soil before or after the PHC reach it affects the distance to which the abiotic processes governing PHC behavior extend into the porous media; these processes include retention, redistribution and natural attenuation.

The history of sites polluted by petroleum hydrocarbons as depicted by various surveys reveals major changes in their composition as result of weathering of the initial product. Understanding the long-term fate of PHC-polluted sites necessitates the development of new experimental techniques, able to mimic the natural leaching effect for periods of tens or even hundreds of years. Only parameters obtained from especially designed experiments will be able to validate models for predicting the long-term behavior of petroleum hydrocarbons in the soil subsurface.

5. REFERENCES

[1] Abriola, L.M., and Bradford, S.A. (1998) Experimental investigations of the entrapment and persistence of organic liquid contaminants in the subsurface environment. *Environ. Health Perspectives* 106, 1083-1095.

[2] Acher, A.J., Boderie, P., and Yaron, B. (1989). Soil pollution by petroleum products, I. Multiphase migration of kerosene components in soil columns. *J. Contam.Hydrol.*4, 333-345

[3] Augustijn, D.C.M. (1993). Chemodynamics of complex waste mixtures: applications to contamination of soils and aquifer media. Ph.D Diss, University of Florida, Gainesville.

[4] Barbex, O.C, and Brown, K.W., 1986. Movement of xylene through unsaturated soils following simulated spills. *Water Air Soil Pollut.* 29, 321-333.

[5] Barbieri, G., Barrocu G., and Ranieri G., (1986) Hydrogeological and geophysical investigations for the evaluation of salt intrusion phenomena in Sardinia, *Proc.9th SWIM-Delft,* 659-670.

[6] Bollag, J.M. and Bollag, W.B. (1995) Soil Contamination and Feasibility of biological remediation. *Bioremediation: Science and Applications.* pp. 1-11. SSSA, Madison, WI

[7] Bond L.D, and Bredehoeft J.D., (1987) Origins of sea water intrusion in a coastal aquifer: a case study of the Pajaro Valley, *Journal of Hydrology,* 92, 363-388.

[8] Bossert, I., Kachel, W.M., and Bartha, R. (1984), Fate of hydrocarbons during oily sluge disposal in soil. *Applied and Environmental Microbiology* 763-767.

[9] Bradford, S.A., and Leij, F.J., (1997). Estimating interfacial areas for multi fluid soil systems. *J.Contam. Hydrol.* 27:83-105.

[10] Broholm,K., Nilsson, B., Sidle, R. C., and Arvin, E. (2000) Transport and biodegradation of cerosote compounds in clayey till, a field experiment. *J.Contam.Hydrol.* 41, pp. 239-260

[11] Budzinki H., Jones I., Bellocq J., Pierard C., and Garrigues P., (1997), Evaluation of sediment contamination by polycyclic aromatic hydrocarbons in the Gironde estuary, *Marine Chemistry,* 58, 85-97.

[12] Chiou, C.T. and Shoup, T. D. (1985). Soil sorption of organic vapors and effects of humidity on sorptive mechanism and capacity. *Environ.Sci.Technol.* 19, 1196-1200

[13] Chiou, C.T., Malcolm, R. L., Brinton, T. I., and Kile, D. E. (1986). Water solubility enhancement of some organic pollutants and pesticides by dissolved humic and fulvic acids. *Environ.Sci.Technol.* 20, 502-508

[14] Dachs J., Bayona J.M., and Albaiges J., (1997), Spatial Distribution, vertical profiles and budget of organochlorine compounds in Western Mediterranean seawater, *Marine Chemistry,* 57, 313-32.

[15] Douglas, G.S., McCarthy, D.T., Dahlen, D.T., Seavey, J.A., Steinhauer, W.G., Prince, R.C., and Elmendorf, D.L., (1992), *The use of hydrocarbon analyses for environmental assessment and remediation.* In P.T. Kostecki and E. J. Calabrese, (eds). Contaminated Soils: Diesel Fuel Contamination, Lewis Pub. An Imprint of CRC Press, Boca Raton, FL, pp. 1-21.

[16] Dror, I. Gerstl, Z., Prost,R., and Yaron, B (2000a). Behavior of neat and enriched volatile petroleum hydrocarbons mixture in the subsurface during leaching Leaching. *Land Contamination and Reclamation,* Richmond UK 8:341-348

[17] Dror,I. Gerstl, Z., Rubin, H. Braester, C. and Yaron, B. (2000b) Persistence of petroleum hydrocarbon mixture in the soil profile during leaching: a field experiment. In *Persistnt, bioaccumulative and toxic chemicals I: fate and exposure*, R.L.Lipnick, D. C.G. Muir, and J. L. M. Hermens. Eds. American Chemical Society Book Department Symposium Series # 772. 70-84

[18] Dror,I. Gerstl, Z., and Yaron, B. (2000c). In Situ Effect of Soil Amendments on the Dynamics of Kerosene Attenuation Leaching. *Land Contamination and Reclamation*, Richmond UK 8: 349-356.

[19] Dror, I., Gerstl, Z., and Yaron, B. (2001) A field experiment on redistribution and dissipation of residual kerosene in soil subsurface during leaching. *J. of Cont. Hydrology* 48: 305-323.

[20] Dror I., Prost R., and Yaron B., (2002), Abiotic behavior of entrapped petroleum products in the subsurface during leaching. *Chemosphere,* 49, 1375-1388.

[21] Dror, I., Amitay T., Yaron B., and Berkowitz B., (2003a) Salt-pump mechanism for intrusion of marine contaminants into coastal aquifers, *Science*, 300, 950.

[22] Dror, I., Yaron B., and Berkowitz B., (2003b) Response to comment on "Salt-pump mechanism for intrusion of marine contaminants into coastal aquifers", *Science*, 302, 784c.

[23] Erlender S.R., and McGurie J.P., (1968), Solubilities of hydrocarbons in concentrated salt solutions and their correlation with the clathrate water theory, ion-ion associations and structures of polymers, *Journal of Macromolecular Science and Chemistry*, 2, 859-875.

[24] Fine, P. and Yaron, B. (1993), outdoor experiments on enhanced volatilization by venting of kerosene component from soil. *J.Contam.Hydrol.*12, 355-374.

[25] Fortin, J., Jury, W. A., and Anderson, M. A. (1998). Organic chemicals in the environment dissolution of trapped nonaqueous phase liquids in sand columns. *J.Environ. Qual.* 27, 38-45.

[26] Galin, T. McDowell, C. and Yaron, B. (1990) The effect of volatilization on the mass flow of a nonaqueous pollutant liquid mixture in an inert porous medium: experiments with kerosene. *J Soil Sci.* 41: 631-641.

[27] Garg, S., and Rixey, W. G. 1999. The dissolution of benzene, toluene, m-xylene and naphthalene from a residually trapped non-aqueous phase liquid under mass transfer limited conditions. *J.Contam.Hydrol.* 36, . 313-331.

[28] Gerstl, Z., Galin, Ts., and Yaron, B. (1994). Mass flow of a volatile organic liquid mixture in soils. *J.Environ. Qual.* 23 487-493.

[29] Graber,E. R., Dror, I., Berkovitch, F., and Rosner, M. (2001) Sewage sludge amendment and pesticide transport in a field trail. *Chemosphere* 44: 805-811

[30] Gidda, T., Stiver, W. H., and Zytner, G. R. (1999). Passive volatilization behaviour of gasoline in unsaturated soils. *J. Contam. Hydrol.* 39 137-159.

[31] Gustin, M.G. and Prost, R., (1998). Transport of hydrocarbons in an unsaturated porous media *INRA Report* 28 pp.

[32] Gwirtzman, H., and Roberts, P.V., (1991). Pore scale analysis of two immisiscible fluids in porous media. *Water Resour. Res.* 37:1165-1176.

[33] Jarsjo, J., Destouni, G., and Yaron, B. (1994), Retention and volatilisation of kerosene: Laboratory experiments on glacial and post-glacial soils. *J. Contam. Hydrol.* 17 167-185.

[34] Jarsjo, J., Destouni, G., and Yaron, B. (1997a). On the relation between viscosity and hydraulic conductivity for volatile organic liquid mixtures in soils. *J.Contam.Hydrol.* 25, . 113-127.

[35] Jarsjo, J., Prost, R., and Yaron, B. (1997b). An experimental approach for the characterisation of rigid porous media and unsaturated conductivity relations. *J. Porous Materials* 4, 199-209

[36] Jorgensen, K.S., Puustinen, J., and Suortti, A. M. (2000) Bioremediation of petroleum hydrocarbon contaminated soil by composting in biopiles. *Environmental Pollution* 107, pp. 245-254.

[37] Karimi, A.A., Fermer, W. J., and Cliath, M. M. (1987), Vapor-phase diffusion of benzene in soil. *J.Environ. Qual.* 16, . 38-43.

[38] Kashef A.I., (1983), Salt-water intrusion in the Nile Delta, *Ground Water*, , 21, 160-167.

[39] Lee, K., and Levy, E. M., (1989), *Biodegradation of petroleum in the marine environment and its enhancement.* In J.O. Nriagu and J.S.S. Lakshminarayana, eds., Aquatic Toxicology and Water Quality Management, Vol 22, John Wiley & Sons, Inc., NY pp. 217-243.

[40] Lee, L.S., N. Priddy, and Augustijn, D.C.M. (1998), Estimating mass transfer of polyaromatic hydrocarbons from coal tar contaminated soil. In: *Soil and Aquifer Pollution* Rubine, H., Narkis, N. and Carberry, J. Eds. pp. 91-107 Berlin Springer – Verlag.

[41] Loehr, R.C., Webster, M.T., and Smith, J. R. (2000) Fate of treated and weathered hydrocarbons in soil - long term changes. *Practice Periodical of Hazardous, Toxic, and Radioactive Waste Management* 4, pp. 53-59

[42] Mackay, D., and Shiu, W.Y., (1975) Determination of solubility of hydrocarbons in aqueous sodium-chloride solutions, *Canadian Journal of Chemical Engineering,* 53, 239-242.

[43] Magaritz M., and Luzier J.E, (1985), Water-rock interactions and seawater-freshwater mixing effects in the coastal dunes aquifer, Coos Bay, Oregon, *Geochimica Cosmochima Acta,* 49, 2515-2525.

[44] Martinez de la Cruz, J (1990), Convective drying of solids wetted with multicomponent solvent mixtures. Int. Report Dep. Chem. Eng. R. Inst. Technol., Stockholm.

[45] Mayer, A.S., and Miller, C.T., (1992), The influence of porous medium characteristics and measurement on pore scale distribution of residual nonaqueous phase liquids. *J. Contam. Hydrol.* 11, 189-213.

[46] Mercer, J.W. and Cohen, R. M. (1990), A review of immiscible fluids in the subsurface: properties, models, characterization and remediation. *J. Contam. Hydrol.* 6, 107-163.

[47] Mingelgrin, U. and Prost, R., (1989),Surface interactions of toxic organic chemicals with minerals. In ., *Toxic Organic Chemicals in Porous Media.* Gerstl, Z., Chen, U., Mingelgrin, U. and Yaron, B. Eds. pp. 91-135 Berlin Springer Verlag.

[48] MOPTMA - MINER. (1994) Libro blanco de las aguas subterráneas. ministerio de obras públicas, transportes y medio ambiente (moptma) y ministerio de industria y energía. madrid, españa.

[49] Morrison T.J., (1944), The salting out effect. *Transactions of the Faraday Society*, 40, 43-48.

[50] Notar M., Lescovesek H., and Faganeli J., , (2001), Composition, distribution and sources of polycyclic aromatic hydrocarbons in sediments of the Gulf of Trieste, Northern Adriatic Sea, *Marine Pollution Bulletin*42, 36-44.

[51] Nye, P.H., Yaron, B., Galin, Ts., and Gerstl, Z. (1994), Volatilization of multicomponent liquid through dry soil: testing a model. *Soil Sci.Soc.Am.J.* 58, . 269-277

[52] Osol, A., and Kilpatrik, M., (1933), The "salting-out" and "salting-in" of weak acids. I. The activity coefficients of molecules of ortho, meta and para chlorobenzoic acids in aqueous salt solutions, *Journal of the American Chemical Society*, , 55, 4430-4444.

[53] Oude-Essink, G.H.P., (2002), Salinisation of the Wieringermeerpolder, The Netherlands, *Proc.17th Salt Water Intrusion Meeting, Delft,*.

[54] Park, K.S., Sims, R.C., Dupont, R.R., Doucette, W.J., and Matthews, J.E. (1990) Fate of PAH compounds in two soil types: influence of volatilization, abiotic loss and biological activity. *Environ.Toxicol.Chem.* 9, pp. 187-195

[55] Pastor D., Sanchez J., Porte C., and Albaiges J., (2001), The Aegean Sea oil spill in the Galicia coast (NW Spain) 1. Distribution and fate of crude oil and combustion products in subtidal sediments, *Marine Pollution Bulletin,* 42, 895-904.

[56] Paul M.A., (1952), The solubilities of naphthalene and biphenyl in aqueous solutions of electrolytes, *Journal of the American Chemical Society*, 74, 5274-5277.

[57] Pignatello,J.J., (1989), Sorption dynamics of organic compounds in soils and sediments. In: Reaction and Movement of Organic Chemicals in Soils. 45-80 S.S.S.A.. Madison WI (Special pub.)

[58] Pignatello,J.J., (2000), The measurement and interpretation of sorption and desorption rates for organic compounds in soil media. *Adv. Agron.* 69:1-153.

[59] Powers, S.E., Abriola, L. M., Dunkin, J. S., and Weber, W. J. Jr. (1994), Phenomenological models for transient NAPL water masstransfer processes. *J.Contam.Hydrol.*16, 1-33.

[60] Ptacek, C.J., Cherry, J.A., and Gillham, R.W. (1987). Mobility of dissolved petroleum derives hydrocarbons in sand aquifers. In Oil in freshwater: Chemistry, Biology, countermeasure technology Vandermeulen, J. H. and Hrudey, S. E. Eds. pp. 195-214 NY Pergamon Press.

[61] Readman J.W., Fillman G., Tolosa I., Bartocci J. Villeneuve, Catinni C, and Mee L.D., (2002), Petroleum and PAH contamination of The Black Sea, *Marine Pollution Bulletin,* 44, 48–62.

[62] Rogers, S.H., and Ong, S.K., (2000), Influence of porous media , air flow rate and channel spacing on benzene NAPL removal during air sparging *Environ. Sci. and Technol.* 34:764-770.

[63] Romantschuk,M., Sarand, I., Petanen, T., Peltola, R., Jonsson-Vihanne, M., Koivula, T., Yrjala, K., and Haahtela, K. (2000) Means of improve the effect of in situ bioremediation of contaminated soil: an overview of novel approaches. *Environmental Pollution* 107, pp. 179-185.

[64] Schwille,F., (1981), Groundwater pollution in porous media by fluids immiscible with water. *Sci. Total Environ.* 21,173-185.

[65] Schwille, F., (1994), Migration of organic fluids immiscible with water in the unsaturated zone. In Pollutants in Porous Media. Yaron, B. , Dagan, G. and Goldshmid, J. Eds. pp. 27-47 Berlin, Springer-Verlag.

[66] Setschehov, J. (1889), Uber die Konstitution der salzlosungen auf ground ihres verhaltens zu kohlensaure, *Z. Physik. Chem.*, 4, 117.

[67] Sims, R.C. and Sims, J. L. (1999) Landfarming of petroleum contaminated soils. *Bioremediation of contaminated soils* (eds Adriano,D.C., Bollag,J.M., Frankenberger,W.T.Jr., & Sims,R.C.), pp. 767-795. ASA, CSSA, SSSA, Madison, Wisconsin, USA.

[68] Siraz, L., (1998), Salinisation mechanism in the coastal groundwaters of Palghar Taluka, Maharashtra, India, in: Weaver T.R. and Lawrence C.R (eds) *IAH, IGC Groundwater Sustainable Solutions,* 347-352..

[69] Sole, M., Porte, C., and Albaiges, J., (2001), Hydrocarbons, PCBs and DDT in the NW Mediterranean deep-sea fish, *Deep-Sea Research Part 1,* 48, 495-513.

[70] Stronguilo, M.L., Vaquero, M.T., Comellas, L., and Borto-Puig, F. (1994) The fate of petroleum aliphatic hydrocarbons in seawage sludge amended soils. *Chemosphere* 29, 273-281.

[71] Sutton, C., and Calder, J.A., (1975), Solubility of alkylbenzenes in distilled water and seawater at 25.0^0C, *Journal of Chemical Engineering Data*, , 20, 320-322.

[72] Thorn, K.A. and Aiken, G.R. (1998), Biodegradation of crude oil into nonvolatile organic acids in a contaminated aquifer near Bemidji, Minnesota. *Org.Geochem.* 29, pp. 909-931.

[73] Tkalich, P., and Soon, C.E., (2002), Vertical mixing of oil droplets by breaking waves, *Marine Pollution Bulletin*, 44, 1219-1229.

[74] van Dam, J., (1967), The migration of hydrocarbons in water–bearing stratum. In: The joint problems of the oil and water industries.. Hepple, P. Ed,. pp. 55-96.Inst of Petroleum, London, UK.

[75] Wang, Z., Fingas, M., Blenkinsopp, S., Sergy, G., Landriault, M., Sigouin, L., and Lambert, P. (1998a). Study of 25 year old Nipis Oil Spill: Persistence of oil residues and comparisons between surface and subsurface sediments. *Environ Sci Technol* 32, 2222-2232

[76] Wang,Z., Fingas, M., Blenkinsopp, S., Sergy, G., Landriault, M., Sigouin, L., Foght, J., Semple, K., and Westlake, D. W. S. (1998b) Comparision of oil composition changes due to biodegradation and physical weathering in different oils. *Journal of Chromatography A* 809, pp. 89-107.

[77] Weber, W.J.Jr., Mukherji. S, and Peters, C.A. (1998), Aqueous dissolution of constituents of composite non aqueous phase liquid contaminants. In: *Soil and Aquifer Pollution* Rubine, H., Narkis, N. and Carberry, J. Eds. pp. 123-135 Berlin Springer – Verlag.

[78] Whitehouse, B.G. (1984), The wffects of temperature and salinity on aqueous solubility of polynuclear aromatic hydrocarbons, *Marine Chemistry,* 14, 319-332.

[79] Wilkins, M.D., Abriola, L. M., and Pennell, K. D. (1995), An experimental investigation of rate limited nonaqueous phase volatilization in unsaturated porous media: steady state mass transfer. *Water Resour. Res.* 31, 2159-2172.

[80] Wilson,S.C., Duckham, C., and Jones, K. C. (1996) Quantification of volatile organic compound losses from sluge amended soil: a pilot scale study. *Chemosphere* 33, 1261-1272.

[81] Witt G., (2002), Occurrence and transport of polycyclic aromatic hydrocarbons in the water bodies of the Baltic Sea. *Marine Chemistry,* 79, 49-66.

[82] World Factbook (2003), www.cia.gov/cia/publications/factbook/rankorder/2174rank.html

[83] Wurdemann, S., and Alexander, M. (1999) *Biodegradation and Bioremediation*, second edn. Academic press, San Diego, California.

[84] Xie, W.H., Su, J. Z., and Xie, X. M. (1990), Studies on the activity coefficient of benzene and derivatives in aqueous salt solutions. *Thermochimica Acta* 169, 271-286

[85] Xie, W.H., Shiu, W. Y., and Mackay, D. (1997), A review of the effect of salts on the solubility of organic compounds in seawater. *Mar. Environ. Res.* 44, 429-444.

[86] Yaron, B, (1989), On the behavior of petroleum hydrocarbons in the unsaturated zone: abiotic aspects. In: *Toxic Organic Chemicals in Porous Media* Gerstl, Z., Chen, Y., Mingelgrin, U and. Yaron, B.Eds. pp. 211-230 Berlin, Springer-Verlag. ;.

[87] Yaron, B., Calvet, R. and Prost, R. (1996), *Soil Pollution Processes and Dynamics*. Springer-Verlag Berlin. 320 pp.

[88] Yaron,B., Dror,I., Graber,E., Jarsjo, J., Fine,P. and Gerstl, Z. (1998), Behavior of volatile organic mixtures in the soil environment. . In *Soil and Aquifer Pollution,* Ruine, H., Narkis, N. and Carberry, J. Eds pp.37-56 Berlin, Springer –Verlag.

In: Trends in Agriculture and Soil Pollution Research
Editor: James P. Livingston, pp. 79-93 ISBN: 1-59454-325-9

Chapter 3

EXPOSURE ASSESSMENT TO DIOXINS OF PEOPLE LIVING IN THE SEVESO (ITALY) AREA

Elena Fattore and Roberto Fanelli
Department of Environmental Health Sciences, "Mario Negri" Institute for Pharmacological Research, Via Eritrea 62, 20157 Milan, Italy

ABSTRACT

Exposure of the general population to some persistent organic pollutants, such as polychlorinated dibenzo-*p*-dioxins (PCDDs), furans (PCDFs), also generally named dioxins, and biphenyls (PCBs), is close to the exposure levels that have been shown to induce neurobehavioral changes and disturbances to the reproductive system in experimental animals. In this context it is important to investigate specific situations where, for different reasons, subpopulations can be exposed to levels of these pollutants mentioned above to the "background" contamination. This is the case of the area close to Seveso (north Italy) where, because of a chemical plant accident occurred in 1976, the residue levels of the most toxic dioxin congener, the 2,3,7,8-tetrachlorodibenzo-*p*-dioxin (TCDD), in soil exceed the limit established by the Italian law for residential areas.

This chapter illustrates an exposure assessment to PCDDs and PCDFs of adult people living in this area by using environmental concentrations experimentally measured and from the literature. The purpose was to investigate the excessive of exposure due to a specific source as contaminated soil in comparison to a reference location not involved in the 1976 accident. The main exposure pathways, such as ingestion and dermal contact of soil particles, inhalation, food, vegetables grown in local orchards, and locally reared animal were taken into account in three different scenarios, characterized by different environmental concentrations and different exposure pathway parameters. The total daily intake of these compounds, expressed as WHO-TEQ, has been estimated for the three different scenarios, to be compared to the Tolerable Daily Intake (TDI) of 2 pg WHO-TEQ/kg body weight recently adopted by the *Scientific Committee on Food* (SCF) of the European Commission (EC).

INTRODUCTION

Polychlorinated dibenzo-*p*-dioxins (PCDDs), dibenzofurans (PCDFs) and biphenyls (PCBs) are environmental pollutants mainly released into the environment during some decades in the past and accumulated into the food chain because of their high environmental persistence and bioaccumulation potential. Nowadays these compounds are ubiquitously distributed in the environment, including fatty tissues of animals and humans [1,2] and the general population is exposed to these pollutants mainly through the diet. Among these compounds certain congeners are highly toxic to all animal species and seem to exert their biological activity through the same mechanism of action, which is mediated by the aryl hydrocarbon (*Ah*) receptor [3]. These compounds are always present in the environment, as well in the living organisms, as complex mixtures of different congeners whose relative concentrations might be different as consequence of selective bioaccumulation processes. In order to assess the toxicological risk and to set reference values for regulation purposes for these compounds, the toxic equivalency (TEQ) concept has been developed for humans and wild life [4,5]. In the TEQ concept, the toxic potency of each congener is evaluated in comparison to that of 2,3,7,8-tetrachlorodibenzo-*p*-dioxin (TCDD), which is the most toxic congener, and a toxic equivalency factor (TEF) is assigned to each congener. The sum of chemical concentrations multiplied by the corresponding TEF values gives the TEQ concentration for an environmental mixture:

$$TEQ = \sum_i Ci \times TEFi$$

Where:

- TEQ: concentration expressed in Equivalent Toxicity;
- C*i*: concentration of the *i* congener;
- TEF*i*: Toxic Equivalency Factor of the *i* congener.

This approach, which makes it feasible to obtain a single toxic measure of a complex exposure situation, is based on the assumption that binding to the *Ah*-receptor is a necessary first step for all important biological effects and that these compounds act in an additive manner via the same mechanism of action.

Different TEF systems have been proposed and used in the past years [5,6] (Table 1). The latest has been developed from the European Centre for Environmental Health of the World Health Organization (WHO-ECEH) and from the International Programme on Chemical Safety (IPCS) [6], and it is the one utilized in the present study (Table 1).

Table 1. Toxic Equivalency Factors, adopted from the International system (I-TEF) and from the World Health Organization (WHO-TEF) for human risk assessment and other mammals for the 2,3,7,8-chlorine substituted congeners of the polychlorinated dibenzo-*p*-dioxins (PCDDs) and dibenzofurans (PCDFs)

Congener	I-TEQ	WHO-TEQ
Polychlorinated dibenzo-p-dioxins (PCDDs)		
2,3,7,8-tetrachlorodibenzo-p-dioxin	1	1
1,2,3,7,8-pentachlorodibenzo-p-dioxin	0.5	1
1,2,3,4,7,8-exachlorodibenzo-p-dioxin	0.1	0.1
1,2,3,6,7,8-hexachlorodibenzo-p-dioxin	0.1	0.1
1,2,3,7,8,9-hexachlorodibenzo-p-dioxin	0.1	0.1
1,2,3,4,6,7,8-heptachlorodibenzo-p-dioxin	0.01	0.01
1,2,3,4,6,7,8,9-octachlorodibenzo-p-dioxin	0.001	0.0001
Polychlorinated dibenzofurans (PCDFs)		
2,3,7,8-tetrachlorodibenzofuran	0.1	0.1
1,2,3,7,8-pentachlorodibenzofuran	0.05	0.05
2,3,4,7,8-pentachlorodibenzofuran	0.5	0.5
1,2,3,4,7,8-hexachlorodibenzofuran	0.1	0.1
1,2,3,6,7,8-hexachlorodibenzofuran	0.1	0.1
1,2,3,7,8,9-hexachlorodibenzofuran	0.1	0.1
2,3,4,6,7,8-hexachlorodibenzofuran	0.1	0.1
1,2,3,4,6,7,8-heptachlorodibenzofuran	0.01	0.01
1,2,3,4,7,8,9-heptachlorodibenzofuran	0.01	0.01
1,2,3,4,6,7,8,9-octachlorodibezofuran	0.001	0.0001

During these last decades, a lot of new scientific evidence on toxicological effects due to these pollutants came up, suggesting that some adverse effects can occur at much lower doses than earlier expected. In particular, environmental levels to which general population is exposed, mainly trough the diet, seem to be close to the exposure doses which have been shown to induce neurobehavioral changes and disturbances to the reproductive system in experimental animals [7-12].

On July 1976 an explosion in the 2,4,5-trichlorophenol production reactor occurred at the ICMESA (*Industrie Chimiche Meda Società*) chemical plant, causing the emission of a vapour cloud containing an undefined amount of TCDD. The toxic cloud spread and settled along a linear path of 6 km SE from the plant, causing a heavy contamination of the soil in Seveso and other surrounding municipalities, about 30 km around Milan. The area was subsequently divided into three risk zones with decreasing levels of contamination in soil: the zones A, B, and R [13,14]. Meanwhile the zone A (87.3 ha, 706 inhabitants) was evacuated and the first 20 cm of soil were scarified, the zone B (269 ha, 4613 inhabitants) and zone R

(1430 ha, 30774 inhabitants), were reclaimed by mixing the top soil with the underlying layers, so as to lower the TCDD concentration to levels that were supposed to be safe.

Results of a recent monitoring program to investigate the environmental levels of polychlorinated dibenzo-*p*-dioxins (PCDDs) and dibenzofurans (PCDFs) in the Seveso area [14-16] showed that residues of TCDD, arising from the accident, are still present in the top soil of some residential areas. In particular, in the ex zone B, mean PCDD and PCDF concentrations, expressed as TEQs, were in the range 1.5-250 ng WHO-TEQ/kg soil, dry weight (dw). These values are between 10-100 times the concentrations detected in general urban soil [17] and exceed the limit of 10 ng WHO-TEQ/kg soil dw established by the Italian law for residential areas. Moreover the fingerprint of the accident was clear from the high contribution of the TCDD congener to the total WHO-TEQ content (about 80%), whereas such contribution, in soil non specifically impacted, is usually less than 20% [17]. Thus the case of the Seveso area was an unique study case to assess the relevance of soil related exposure routes to the total PCDD and PCDF daily intake.

This chapter illustrates an investigation to assess the chronic total PCDD and PCDF daily intake of adult people living in the Seveso area in comparison to people living in an area where the PCDD and PCDF soil contamination was within the regulatory limit of 10 ng WHO-TEQ/kg soil. Since the high "background" of PCDDs and PCDFs due to the intake through the diet, both soil related and non soil related exposures pathways were taken into consideration. Even if the TCDD represented the main contributor to the total WHO-TEQ, the assessment was carried out for all the 2,3,7,8-chlorine PCDD and PCDF substituted congeners. The purpose was to estimate the excess of exposure due a specific source such as contaminated soil and to assess the total PCDD and PCDF daily intake in relationship to the Tolerable Daily Intake (TDI) of 2 pg WHO-TEQ/kg body weight (bw) estimated from the Scientific Committee on Food (SCF) of the European Commission (EC) [18].

Material and Methods

Assessment of indirect exposure via environment was carried out following the methodology described in the US Environmental Protection Agency (EPA) draft dioxin reassessment [19]. Three different scenarios, characterized by different environmental concentrations (Table 2) and different exposure pathway parameters (e.g. soil ingested, air inhalated, *etc.*) and behaviour of the hypothetical exposed persons (Table 3) were identified. In the first one, "zone B central scenario", the soil contamination was assumed to be the median value (50_{th} percentile) of the concentration values detected in zone B and residents were assumed to live in houses with a garden, carry out gardening, and grow some vegetables included in their normal diet. In the second one, "zone B high-end scenario", soil contamination was assumed to be the 95_{th} percentile of the concentration values detected in zone B, whereas inhabitants were assumed to perform farming activity and consume home-raised vegetable and animal food products (chicken meat and eggs). Finally, the third selection, "control central scenario", was defined as the first scenario for the exposure pathway parameters (Table 3), whereas the soil WHO-TEQ concentration was assumed to be that of the regulatory limit value. The following exposure pathways were investigated: a) ingestion of contaminated soil particles, b) dermal contact with contaminated soil particles, c) inhalation, d) water ingestion, e) general diet, f) vegetables grown in local yards and g) locally reared chickens and eggs.

Table 2. Environmental concentration values of the three different scenarios utilized for the exposure assessment

Zone B			
Environmental Media	Central Scenario	High End Scenario	Control Central Scenario
Soil (ngWHO-TEQ/kg d.w.)	30.6 [a]	107 [b]	10 [c]
Air (pgWHO-TEQ/m3)	0.378 [d]	0.378 [d]	0.263 [e]
Water (pgWHO-TEQ/L)	0.00056 [f]	0.00056 [f]	0.00056 [f]

[a,b] 50_{th} and 95_{th} percentiles of soil concentration values in Zone B, respectively. [c] Regulatory limit concentration value for residential areas. [d,e] Average annual concentration in air in zone B and Milan, respectively [14]. [f] Concentration value utilized from the US EPA for exposure assessment of the general population in the United States.

Table 3. Exposure pathways parameters for both the central scenarios (zone B central scenario and control central scenario) and the high end scenario

Exposure Pathway	Exposure Parameter		Central Scenarios	High End Scenario
Ingestion of soil particles	*children (1-6 years)*	CR (mg/day):	100	600
		ED (y):	6	6
		BW (kg):	16	16
	Adults:	CR (mg/day):	50	50
		ED (y):	24	24
		BW (kg):	60	60
Dermal contact of soil particles	*Outdoor:*	SA (cm^2):	5700	3600
		CR (mg/cm^2-event):	0.03	0.1
		EF (events/y):	100	350
		ED (y):	30	30
		BW (kg):	60	60
	Indoor:	SA (cm^2):	1000	1000
		CR (mg/cm^2-event):	0.005	0.005
		EF (events/y):	365	365
		ED (y):	30	30
		BW (kg):	60	60
Inhalation		CR (m^3/day):	13	20
		ED(y):	70	70
		BW (kg):	60	60
Water Ingestion		CR L/day:	1.4	2
		ED y:	70	70
		BW kg:	60	60

The general equation utilized to estimate the exposure was the following:

1) $$LADD = \frac{C \times CR \times CF \times ED}{BW \times LT}$$

Where:

- $LADD$ = lifetime average daily dose
- C = concentration exposure media (WHO-TEQ concentration in soil, air, food, etc.)·
- CR = contact rate of the environmental media (the rate of soil contact, of food ingested, air inhalated etc.)
- CF = contact fraction (between 0 and 1) reducing the contact rate in the specific exposure pathway
- ED = exposure duration: this parameter has been assumed 30 years for the exposure pathways directly related to soil contamination (soil ingestion, soil dermal contact, and vegetable and animal products locally derived), and 70 years for diet, air inhalation, and water ingestion.
- BW = body weight (60 kg was assumed to be the average body weight for female population)
- LT = average lifetime period (70 years)

Ingestion of Soil Contaminated Particles

This pathways was divided in two different exposure periods, the childhood period (1–6 years) and the adult period. Rigorously a such division should be done for all the exposure pathways, nevertheless soil ingestion is the only exposure pathways which is significant only during childhood, and negligible during the adult age. The specific equation utilized was the following:

$$2)\quad LADD_{si} = \frac{C_s \times CR \times ED \times A}{LT \times BW}$$

Where:

- $LADD_{si}$ = lifetime average daily dose for soil ingestion (pg WHO-TEQ/kg bw-day)
- C_s = PCDD and PCDF soil concentration (pg WHO-TEQ/kg soil dw)
- CR = contact rate (kg soil ingested/day)
- ED = exposure duration (years)
- LT = lifetime (years)
- BW = body weight (kg)
- A = absorption fraction (0.3)

Except for the different soil concentrations the central scenario and high end scenario differ for the amount of soil ingested during the childhood period (Table 3).

Dermal Contact of Soil Contaminated Particles

The individual equation utilized was the following:

$$3)\quad LADD_{dc} = \frac{C_s \times CR \times SA \times EF \times ED \times A}{LT \times BW}$$

Where:

- $LADD_{dc}$ = lifetime average daily dose due to soil dermal contact (pg WHO-TEQ/kg bw-day);
- C_s = dioxin concentration in soil particles (pg WHO-TEQ/kg soil);
- CR = contact rate (kg/cm^2-evento)
- SA = surface area of the exposed body (cm^2)
- EF = event frequency (event/day)
- ED = exposure duration (30 years)
- LT = lifetime (70 years)
- BW = body weight (30 kg)
- A = absorption factor (0.03)

This exposure pathway was divided in the indoor and outdoor component. The indoor component was not assumed to differ in the key parameters in the three scenarios (Table 3) and assumed a CR of 1, the SA of 1000 cm^2 (only hands), and a CR of 0.005 mg/cm^2-event. The outdoor component was modelled in different way in the central scenarios and high end scenario: in both the central scenarios the event frequency and the contact rates were assumed to be less than in the high end scenario, meanwhile the exposed body surface is assumed to be larger in the central scenario (5700 cm^2) than in the high end scenario (3600 cm^2) (Table 3).

Inhalation

The LADD due to inhalation was estimated multiplying the average annual concentration of PCDDs and PCDFs (Table 2) experimentally detected in Seveso zone B and in Milan [14] and using as CR 13 and 20 m^3/day of inhalated air for the central and high end scenarios, respectively, as indicated in the EPA methodology [19].

Water Ingestion

Since no experimental data were available for PCDDs and PCDFs levels in drinkable water in the area under investigation, data reported in the scientific literature were utilized [19,20] (Table 2), making no difference in the concentration levels between the scenarios. The CR were assumed to 1.4 and 2 L/day for the central and high end scenario, respectively, as suggested in the EPA methodology (Table 3).

General Diet

Although some evaluations on the dietary daily intake of PCDDs and PCDFs for the European and Italian general population are available [21], we decided to repeat such estimation to use more specific data about the dietary habit of people living in the North Italy [22,23]. For this exposure pathway no distinction has been done for the three different scenarios. The data about PCDD and PCDF concentrations in the individual foodstuffs were those reported in the *Opinion of the SCF on the Risk Assessment of Dioxins and Dioxin-like PCBs in Food* of the European Commission [24]. Since the PCDD and PCDF concentrations were expressed as I-TEQ were transformed in WHO-TEQ using an empirical transformation factor derived for biological samples. Data on food consumption arose from two different investigations in order to be able to detail the different foodstuffs within a certain food category. This was important because PCDD and PCDF concentrations were expressed on lipid base (except for cereals, fruits and vegetables) so that concentrations data should be transformed on whole product base in order to quantify the amount of PCDDs and PCDFs ingested. Nutritional composition tables of foodstuffs were used to carry out such transformation [25].

Home Grown Vegetables

The consumption of home grown vegetables was assumed to be the same for the three scenarios and it was calculated on the base of their seasonal nature and assuming that in the period when a given vegetable was available from home no other was supplied from the market.

Concentrations of PCDDs and PCDFs in below vegetables were estimated starting from soil concentration, respectively and assuming that the outer portions of the vegetables were those impacted. The equation utilized was the following:

$$4)\quad C_{bgv} = \frac{C_s \times RCF \times VG_{bg}}{K_{ds}}$$

Where:

- C_{bgv} = concentration in below ground vegetables (pg WHO-TEQ/g fresh weight)
- C_s = concentration in soil (pg WHO-TEQ/g soil)
- RCF = root concentration factor (log RCF = 0.77 log (K_{ow})- 1.52)
- K_{ow} = octanol-water partition coefficient
- VG_{bg} = empirical correction factor converting the dioxin concentration of outer portion into the concentration of the whole plant
- K_{ds} = soil-water partition coefficient (L/kg) = (K_{oc} x OC)
- K_{oc} = contaminant organic partition coefficient
- OC = fraction of organic carbon in soil (0.0364)

Concentrations in the home grown above vegetables were calculated starting from concentrations of PCDDs and PCDFs in air, since the translocation from roots has been assumed negligible. The following equation was utilised:

$$5)\quad C_{abv} = C_{vpa} + C_{ppa} = \frac{B_{vpa} \times C_v \times VG_{ag}}{d_a} + \frac{F_p}{1000 \times K_w \times Y_j}$$

Where:

- C_{abv} = concentration in above ground vegetables (pg WHO-TEQ /g d.w)
- C_{vpa} = contribution due to vapor phase absorption or air-born contaminants (pg WHO-TEQ /g)
- C_{ppa} = contribution due to dry and wet deposition of contaminated particulate (pg WHO-TEQ /g)
- B_{vpa} = mass-based air to leaf biotransfer factor
- C_v = concentration of contaminant in vapor phase (pg WHO-TEQ/m3)
- VG_{ag} = empirical correction factor
- d_a = density of air g/m3, 1190
- F_p = wet and dry contaminant deposition rate onto plant surface (pg/m2-y)
- K_w = first order weathering dissipation constant (1/y)
- Y_j = dry matter yield of crop (kg/m2)

The assumption of no translocation from roots is probably not properly correct, since soil-plant transfer has been shown for plants belonging to the genus Cucurbita for some vegetables such as zucchini and pumpkin [26]. Arnoldi and coworkers [16] investigated PCDD and PCDF concentrations in some zucchini fruits grown in zones A, B and R of Seveso and in control area. These experimental data were included in the present evaluation and utilized to estimate the daily intake due to the ingestion of home grown zucchini for the three scenarios.

Concentrations in locally reared chickens and eggs were also estimated starting from PCDD and PCDF concentrations in soil and vegetables in zone B. A scenario where chickens were allowed to freely scratch, consuming their feed, which was spread out on the grounds, and soil and vegetation incidentally, was assumed. The equation utilised was the following:

$$6)\quad C_{fat} = (BCF \times DF_s \times B_s \times AC_s) + (BCF \times DF_g \times AC_g) + (BCF \times DF_f \times AC_f)$$

Where:

- C_{fat} = dioxin concentration in chicken/egg fat (pg WHO-TEQ/g)
- BCF = bioconcentration factor
- DF_s = fraction of chicken diet that is soil (10%)
- B_s = bioavailability of contaminant on the soil vehicle relative to the vegetation vehicle

- AC_s = average contaminant concentration in soil (pg/g)
- DF_g = fraction of chicken diet that is incidental vegetation (5%)
- AC_g = average contaminant concentration in vegetation (pg WHO-TEQ/g).

Contamination of PCDDs and PCDFs in feed was assumed to be negligible, so that all the contamination in the chickens and eggs fat arose from PCDDs and PCDFs in soil and vegetation. The amount of fat in chickens and eggs was assumed to be 14 and 8%, respectively. Lifetime average daily dose (LADD) for this exposure pathway was taken into account for the high end scenario only and it was assumed that all the eggs and chickens consumed were of own production.

RESULTS

Table 4 shows the resulting LADD values, as pg WHO-TEQ/kg bw per day, for the different exposure pathways taken into consideration in this study. Those directly related to soil contamination (i.e., soil ingestion and soil dermal contact) are of minor importance since they accounted together about 1% and 6% for the two central scenarios and the high end scenario, respectively. This is mainly due to the application of the absorption fractions, which strongly reduces the LADDs for these two exposure pathways in all the scenarios. The differences in the resulting LADD values for soil ingestion and soil dermal contact, reflect rather well the original difference in soil concentration.

Table 4. Lifetime Average Daily Doses (LADDs) for PCDDs and PCDFs expressed as pg WHO-TEQ/kg bw, due to different exposure pathways, for Seveso zone B residents (central and high end scenarios) and for residents of a surrounding reference area (control central scenario)

	Zone B		
Exposure Pathways	Central Scenario	High End Scenario	Control Central Scenario
Soil ingestion	0.02	0.23	0.01
Soil dermal contact	0.001	0.02	0.0005
Inhalation	0.08	0.13	0.05
Water ingestion	1×10^{-5}	2×10^{-5}	1×10^{-5}
Diet	1.18	1.15	1.18
Vegetables (home derived)	0.16	0.4	0.08
Chicken and eggs (home derived)	—	2.3	—
Total	**1.44**	**4.2**	**1.32**

Even though PCDD and PCDF air concentrations in Seveso and surroundings (Table 2) were relatively high if compared to the values reported in the scientific literature for urban air [14], inhalation pathway does not seem to be an important route as it represents between 3 and 6% of the total WHO-TEQ intakes for all three the scenarios. As expected, water ingestion is not a significant pathway (<0.001% of the total daily intake), whereas the general diet represents 82, 89 and 27% of the total WHO-TEQ intake for the zone B and control central scenarios and for the high end scenario, respectively. The lower relative contribution of the general diet in the high end scenario is due to the high contribution due to eggs and chickens grown in the contaminated soil. The estimated daily intake from the general diet resulted 1.18 WHO pg/kg bw day for both the zone B and the control central scenarios, and WHO 1.15 pg/kg bw day for the zone B high end scenario. This slightly lower value for this scenario is due to the subtraction from the general diet of the contribution due to eggs chickens locally derived. The total average food consumption estimated in the present evaluation was 1.1 kg/person day, for the three scenarios. The relative contribution of the different foodstuff categories to the total food consumption, as well as to the total daily intake of PCDDs and PCDFs is shown in Figure 1a and 1b respectively. Cereals and cereal products are those most contributing to the total WHO-TEQ daily intake (22%), followed from dairy products (19%) and fish and sea food (18%) (Figure 1b). The highest contribution of cereal and cereal products is due to the large amount of this foodstuff eaten in the Mediterranean diet (Figure 1a).

Results of the estimated concentrations in vegetables and animal products grown in contaminated soil corresponding to the three different scenarios are shown in Figure 2. There are not differences in the concentrations in the above vegetables in the three scenarios because it was assumed that they were only impacted from the atmospheric deposition (translocation from roots has been assumed negligible) which was assumed the same for the two scenarios of Seveso and very similar for the central control scenario (Table 2). The concentrations values estimated in above vegetable resulted quite similar to the concentration values experimental detected in zucchini grown in the zone B of Seveso [16].

With regards to below vegetables and animals grown in an area with a soil contamination corresponding to that of the three scenarios, the concentrations of PCDDs and PCDFs estimated rather high (Figure 2), in particular for the high-end scenario, which shows values corresponding to about 9 and 5 pg WHO-TEQ/g whole product in chicken and eggs respectively.

The LADD values due to the consume of home raised below and above vegetables (Table 4) represent 11, 6 and 10% the of the total LADD in the zone B and control central scenarios and in zone B high end scenario, respectively. This exposure pathway is the most important, after the general diet, for the two central scenarios, and the third for the high end scenario. Consume of chickens and eggs grown in the contaminated area is definitely the highest contributing exposure pathway to the total PCDD and PCDF intake, causing a substantial increase of the LADD which exceed the TDI of 2 pg WHO-TEQ/kg bw estimated by the EC SCF (Table 4).

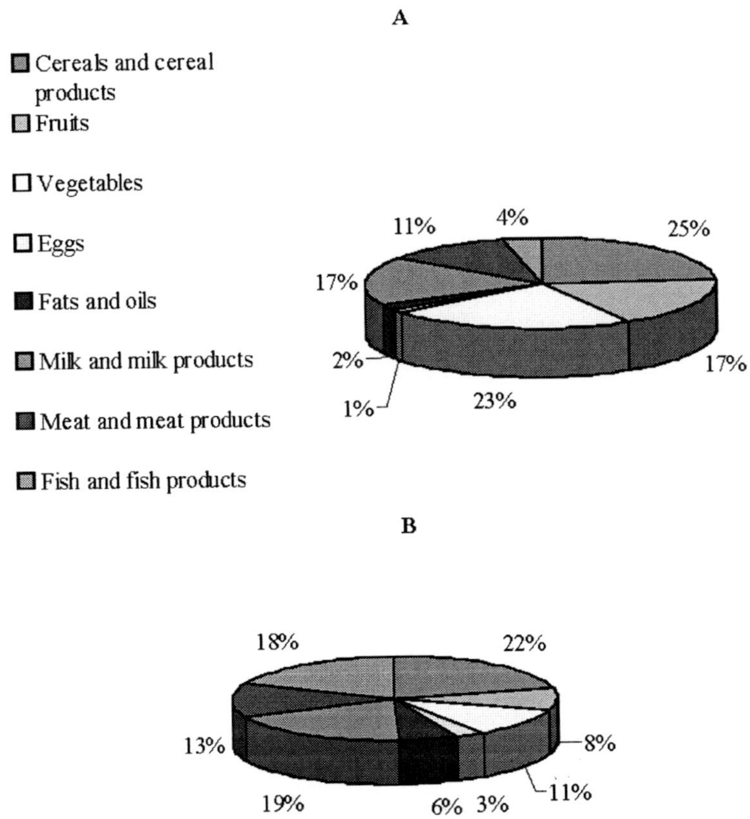

Figure 1. Average contribution (%) for the central scenarios of A) different foodstuffs to the total average daily consumption of food, and B) different foodstuffs to the total PCDDs and PCDFs dietary daily intake.

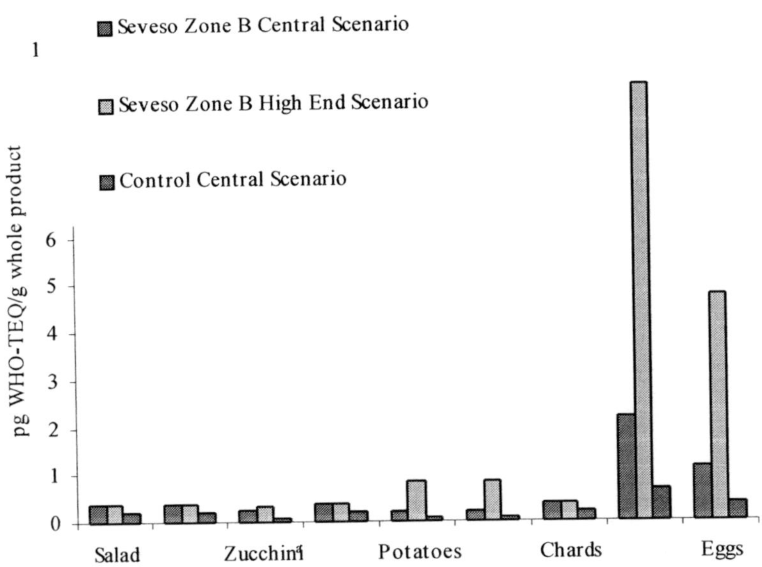

Figure 2. Estimated WHO-TEQ concentrations in vegetable and animal products grown in areas with soil contamination corresponding to that of the three different scenarios.
[a] Experimental values

DISCUSSION

The present study shows a theoretical exposure assessment starting from environmental concentrations, some of them experimental detected in the area under investigation, some of them derived from the scientific literature. This assessment assumes that environmental PCDD and PCDF concentrations remain constant overtime and an exposure duration of 30 years for the exposure pathways directly related to the soil contamination and 70 years for exposure pathways representing the "background" (general diet, inhalation and water ingestion). Adoption of the three different scenarios allowed researchers to compare the exposure in the contaminated area with that in a non-contaminated area, and to estimatc the exposure for a sort of "worst case" which is anyway possible. The resulting LADDs for the two central scenarios suggest that in term of exposure, a soil contamination of 30 or of 10 ng WHO TEQ/kg soil does not make any substantially difference, if dermal contact and ingestion of soil particles are the only exposure pathways taken into consideration. A soil concentration of about 100 ng WHO TEQ/kg (high end scenario) causes just a small increase to the total LADD.

This is due to high background of dioxin intake from the general diet. The mean average dietary intakes for adult people, estimated from data on eight European countries, was in the range of 0.4-1.5 pg I-TEQ/kg bw day for PCDDs and PCDFs and in the range of 0.8-1.8 PCB-TEQ/kg bw day for the "dioxinlike" PCBs [21]. Thus, the results of the dietary intake of PCDDs and PCDFs for the population in Nord Italy reported in this chapter is within the range estimated for the European population. Nevertheless, in this evaluation the contribution of the "dioxinlike" PCBs was not taken into account since experimental data regarding these compounds were not available for the area under investigation. In a previous study on PCB exposure trough the general diet in the Italian population [27], the daily intake due to the "dioxinlike" PCBs resulted 0.6 pg PCB-TEQ/kg bw day. If such contribution were added to the dietary intake of PCDDs and PCDFs estimated in this study, the daily intake through the diet would be close to the TDI in all three the scenarios.

Substantial differences in the daily intake in the three scenarios appear when food is directly produced in the area with contaminated soil. This is due to the high estimated concentration of home grown animal products. Considering a fat content of 8 % and 14 % for eggs and poultry, respectively, the WHO-TEQ concentrations in these two products would exceed the maximum limits of 2 and 3 pg WHO-TEQ/fat basis recently established from European Commission for poultry meat and eggs, respectively, in all three the scenarios [28]. Of course, the results of egg and chicken concentrations shown in this study are obtained by a theoretical model, so that they represent a rough estimate of the real concentration which would need to be experimentally validated. Moreover, the model assumed a scenario of free grazing of chickens in the contaminated area which is probably not so common anymore in these areas. Because of the uncertainties mentioned above, these results suggest that a survey to investigate if, and how much home-grown products are utilized in the area is important, in order to assess if there are any sub-group of population which may experience high exposure.

The exceeding of the TDI estimated from the SCF in the high-end scenario is not anyway surprising since the risk assessment reports from the European Commission estimate that considerable proportion of the European population exceed the TDI. These are groups of population that can suffer higher exposures for different reasons such as particular dietary habits, or for occupational exposures or, as in this case, for the presence of local contamination.

The estimation of TDI has been carried out starting from animal studies which showed the occurrence of some subtle neurobehavioural changes and effects on reproductive system at low doses of exposure [9-12]. These effects in the animals appeared at certain levels of dioxin concentration in the body (body burden) which are ten times higher than those obtained in humans who are exposed to a daily intake equivalent to the TDI and for a prolonged time period of 20-30 years. Thus the slightly exceeding of the TDI in subgroups of population does not mean that these effects necessarily occur but that the margin of safety for these effects is eroded.

ACKNOWLEDGEMENTS

We thank the members of the technical commission on risk assessment of Lombardy for the Environment Foundation, Milan (Italy) for the fruitful discussion and for funding. EF is supported from the European Commission Bonetox project, QLK4-CT-2002-02528.

REFERENCES

[1] WHO/IPCS-World Health Organization/International Programme on Chemical Safety (1989) Polychlorinated Dibenzo-*para*-dioxins and Dibenzofurans. *Environmental Health Criteria* 88.

[2] WHO/IPCS-World Health Organization/International Programme on Chemical Safety (1993) Polychlorinated Biphenyls and Terphenyls (second editions). *Environmental Health Criteria* 140.

[3] Hankinson O. *Annu. Rev. Pharmacol. Toxicol.* 1995, 7-340.

[4] Ahlborg U. G., Becking G. C., Birnbaum L. S., Brouwer A., Derk H. J. G. M., Feely M., Golor G., Hanberg A. Larsen, J. C., Liem A. K. D., Safe S.H., Schlatter C., Wærn F., Younnes M., and Yrjänheikki E. *Chemosphere* 1996 26, 1049-1067.

[5] NATO CCMS (North Atlantic Treaty Organization. Committee on Challenges of Modern Society). Report N° 176, August 1988.

[6] van den Berg M., Birnbaum L., Bosveld A. T.C., Brunström B., Cook P., Feeley M., Giesy J. P., Hanberg A.;, Hasegawa R., Kennedy S. W., Kubiak T., Larsen J. C., van Leeuwen F. X. R., Liem A. K. D., Nolt C., Peterson R. E., Poellinger L., Safe, S., Schrenke D., Tillitt D., Tysklind M., Younes M., Wærn F., and Zacharewski T. *Environ. Health. Perspect.* 1998, 106, 775-792.

[7] Schantz S.L., Bowman R. E. *Neurotoxicol. Teratol.* 1989, 11, 13-19.

[8] Rier S.H., Martin D.C., Bowman R.E., Dmowski W., Pbecker J.L. *Fundam. Appl. Toxicol.* 1993, 21, 433-441.

[9] Gray L.E. Jr., Ostby J. S., Kelce W.R. *Toxicol. Appl. Pharmacol.* 1997, 146, 11-20.

[10] Mably T.A., Bjerke D.L., Moore R.W. *Toxicol. Appl. Pharmacol.* 1992, 114, 118-126.

[11] Faqi A.S., Dalsenter P.R., Merker H.-J., Chahoud I. *Toxicol. Appl. Pharmacol.* 1998,150, 383-392.

[12] Ohsako S., Miyabara Y., Nishimura N., Kurosawa S., Sakaue M., Ishimura R., Sato M., Takeda K., Aoki Y., Sone H., Tohyama C., Yonemoto J. *Toxicol. Sci.* 2001, 60, 132-143.

[13] Fara, G. M. The ICMESA accident. First intervention for the protection of man and environment. In *Chemistry, Man and Environment the Seveso Accident 20 Years on: Monitoring, Epidemiology and Remediation*; Ballarin-Denti, A.; Bertazzi, P.A.; Facchetti, S.; Fanelli, R; Mocarelli, P., Eds, Elsevier, Amsterdam, 1999; pp. 3-16.

[14] Fattore E., Di Guardo A., Mariani G., Guzzi A., Benfenati E., Fanelli R. *Environ. Sci. Technol.* 2003, 37, 1503-1508.

[15] Umlauf, G., Barbiere M., Facchetti S. *Organohalogen Compd.* 1999, 44, 311-314.

[16] Arnoldi, A. Boschetti, M. Mariani, G., Guzzi, A., Facchetti S., Fanelli, R. *Organohalogen Compd.* 1999, 44, 329-332.

[17] Duarte-Davidson R., Sewart A., Alcock R.E., Cousins, I.T., Jones K.C. *Environ. Sci. Technol.* 1997, 31, 1-11.

[18] European Commission, Health & Consumer Protection Directorate-General. *Opinion of the Scientific Committee on Food on the Risk Assessment of Dioxins and Dioxin-Like PCBs in Food.* Update based on new scientific information available since the adoption of SCF opinion of 22nd November 2000. Adopted on 30 May 2001.

[19] EPA (Environmental Protection Agency) Draft Dioxin Reassessment. Draft Exposure and Human Health Reassessment of 2,3,7,8-Tetrachlorodibenzo-*p*-Dioxin (TCDD) and Related Compounds: http://cfpub.epa.gov/ncea/cfm/part1and2.cfm?ActType=default.

[20] Jobb B., Uza M., Hunsinger R., Roberts K., Tosine H., Clement R., Bobbie B., LeBel G., Williams D., Lau B. *Chemosphere* 1990, 20, 1553-1558.

[21] Reports on tasks for scientific cooperation 7 June 2000, *Assessment of dietary intake of dioxins and related PCBs by the population of EU Member States.*

[22] La lettera della nutrizione, An Italian news letter on nutrition: *Come mangiano gli italiani.* Anno III Nov/Dic, 1991.

[23] Turrini A., Lombardi-Boccia D. *Nutr. Res.* 2002, 22, 1151-1162.

[24] European Commission, Health & Consumer Protection Directorate-General. Scientific Committee on Food. *Opinion of the SCF on the Risk Assessment of Dioxins and Dioxin-Like PCBs in Food.* Adopted on 22 November 2000.

[25] Istituto Nazionale della Nutrizione, Tabelle di composizione degli alimenti (con aggiornamenti al 31/12/94) a cura di E. Carnovale- L. Marletta- F. Ciuccio.

[26] Hülster A., Müller J. F., Marschner H. *Environ. Sci. Technol.* 1994, 28, 1110-1115.

[27] Zuccato E., Calvarese S., Mariani G., Mangiapan S., Grasso P., Guzzi A. and Fanelli R. *Chemosphere* 1999, 38, 2753-2765.

[28] Council Regulation (EC) N° 2375/2001 of 29 November 2001 amending Commission Regulation (EC) N° 466/2001 setting maximum levels for certain contaminants in foodstuffs.

In: Trends in Agriculture and Soil Pollution Research
Editor: James P. Livingston, pp. 95-119
ISBN: 1-59454-325-9

Chapter 4

CHEMICAL DEGRADATION OF PHENYLUREA HERBICIDES IN THE SOIL

Elio Coppola, Stefano Salvestrini, Giuseppe Santoro and Sante Capasso*

Dipartimento di Scienze Ambientali, Seconda Università di Napoli, via Vivaldi 43, I-81100 Caserta, Italy

ABSTRACT

Based on the critical review of information from the literature and new data from experimental work, this paper analyses the chemical degradation of phenylureas in homogeneous water phase and in the water/soil system.

Kinetic evidence suggests that the formation of phenylisocyanate, the initial product in the phenylurea hydrolysis in water phase, occurs *via* an intermediate zwitterion. Depending on pH and buffer concentrations, the zwitterion can be produced through three parallel routes. Bifunctional acid-base buffers such as HCO_3^-/CO_3^{2-}, $H_2PO_4^-/HPO_4^{2-}$ and CH_3COOH/CH_3COO^- are very efficient catalysts. At high buffer concentration, as well as at pH < 3 or > 12, the uncatalysed breakdown of the zwitterion is the rate-determining step. Consequently, at high buffer concentration the reaction rate reaches a limiting value that is independent of the type and concentration of the buffer.

The sorption isotherms for phenylureas on soil were non-linear and analysis of data by Langmuir model using Scatchard-like plots gave the maximum amount of adsorptive that can be adsorbed (q_{max}) and the microscopic equilibrium binding constants. The q_{max} value increases linearly with the percent of organic carbon (% OC) in the soil.

A kinetic equation has been derived which permits to evaluate the microscopic rate constant (k) for the hydrolysis of phenylurea herbicides in a soil/water mixture from the time-dependence of phenylurea concentration in the water phase. This equation is based on the assumption that the reaction occurs only in the water phase and that the herbicide is reversibly adsorbed by the soil with a higher rate. By this equation the soil catalytic capacity on hydrolysis of phenylureas was evaluated. Our results showed that soil organic

* Dipartimento di Scienze Ambientali, Seconda Università di Napoli, via Vivaldi 43, I-81100 Caserta, Italy; Fax: +39 0823 274605; E-mail: sante.capasso@unina2.it

matter (SOM) is an efficient catalyst of the hydrolysis. The k values increase with the % OC in the soil and with the fraction of soil in the reacting mixture. As a consequence of the double effect of the SOM, on both the sorption equilibrium and on k, the disappearance rate for the total amount of phenylureas in soil/water systems is independent on the amount of SOM. For diuron $t_{1/2}$ was about 200 days.

Keywords: phenylurea herbicide, degradation, sorption, hydrolysis, soil catalysis

INTRODUCTION

Phenylureas are important herbicides used world-wide for the selective control of germinating grass and broad-leaved weeds in many crops and for total weed control in non-cultivated areas. They work by shutting down the photosynthetic (food producing) process in susceptible plants by binding to specific sites within the plant's chloroplasts. Inhibition of photosynthesis could result in a slow starvation of the plant; however, the plant experiences a more rapid death that is believed to be due to the production of secondary toxic substances. Because of the massive amounts employed, the environmental fate of phenylureas has been the subject of several studies focussed on degradation and on interactions of these compounds, and of their degradation products, with the soil [1-5]. In Italy more than 16 tons of the phenylurea derivative *diuron* are used every year [6].

The length of the carbonyl bond in ureas is greater than that recorded for other carbonyl compounds (the general order is ureas > amides > esters > carbolic acids > ketons > aldeydes) [7]. The greater length in ureas can be expected to make these compounds less susceptible to the electrophilic addition to the double bond and to increase the relative weight of the Lewis resonance structure with a positive charge on the carbon atom and a negative charge on the oxygen atom, and a single bond between the two atoms. These structural features might be expected to promote degradation processes involving a zwitterionic intermediate. Moreover it is known that the π system of the >N-CO-N< fragment is scarcely deformed by the steric interaction between the lateral chains and the phenyl ring [8]. The molecular structure of *diuron* (see Table 1), as obtained by X-ray diffraction, is reported in Fig. 1 [9].

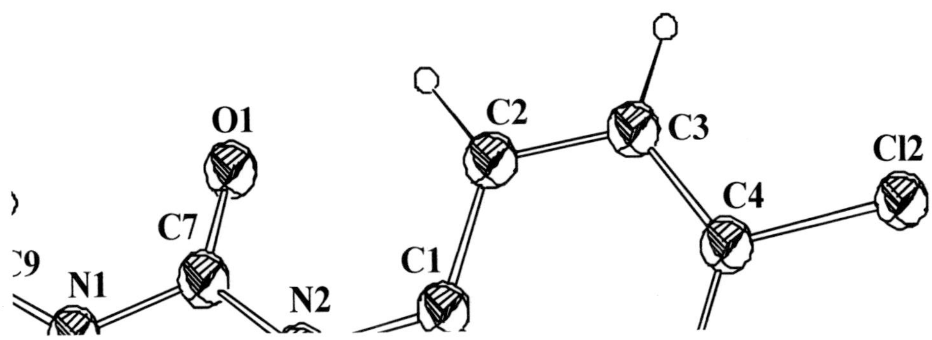

Figure 1. Molecular structure of the phenylurea diuron

The phenylureas are relatively unstable in soil, with half-lives from 1 month to 1 year [10]. Berger [11] has reported the half-lives of some phenylurea herbicides in native and sterile soil and analysed the relationship between reactivity and structure. Micro-organisms capable of phenylurea derivative rapid degradation have been isolated and their degradative capacity when inoculated into soil has been investigated [12].

Under aerobic condition, microbial degradation of these herbicides is believed to occur by successive demethylation of side chains, followed by cleavage of the ureidic group to give aniline derivatives [13-15]. Laboratory studies have demonstrated that aniline derivatives can irreversibly bind to humic acids, a major fraction of soil organic matter [16-18].

As regards abiotic hydrolysis of the ureidic group, it has been postulated that, at least in some cases, this is a multi-stage reaction involving first the formation of isocyanate and ammonia derivatives (i), followed by hydrolysis to the carbamic acid derivative (ii) and then scission to carbon dioxide and the aniline derivative (iii) [19-21].

R-NH-CO-N <	$\rightarrow$	R-NHCO + N <	(i)
R-NHCO + H_2O	$\rightarrow$	R-$NHCO_2H$	(ii)
R-$NHCO_2^-$ + H^+	$\rightarrow$	R-NH_2 + CO_2	(iii)

The reaction (i) is rate-limiting for the overall process; therefore, the isocianate and carbamic acid derivatives form a small proportion of the reacting mixtures.

However, despite the extensive work done in many laboratories, up to few years ago no pathway had been proposed for the first step of the phenylurea hydrolysis that fully satisfies the kinetic behaviour. It had been proposed [22] that in acidic media the reaction proceeds through the protonation of the substrate followed by a rate-determining attack by water, giving a tetrahedral intermediate. This intermediate then decomposes to an amine and a phenylcarbamic acid that quickly decarboxylates under acidic conditions to form the corresponding aniline. In basic media it has been proposed [23] that the hydrolysis of phenylureas occurs *via* an intermediate hydroxide ion complex whose breakdown creates the final products. On the other hand, earlier works have produced evidence of the formation of zwitterionic intermediates in the conversion of ureas into isocianates in aqueous solutions [20, 21]. In the formation of these intermediates water molecules act as proton transfer agent to either the unprotonated or the N-protonated substrate.

A herbicide added to water solutions in contact with soil phases distributes in the two phases, according to its partition coefficient. This is essentially a physical process and, as such, it usually has a low free energy of activation and therefore is relatively fast. On the contrary, the reaction of hydrolysis of ureas is a relatively slow process. This difference facilitates the measurement of kinetic parameters of the two processes. However, the kinetic constants cannot be derived straightforwardly from the time dependence of the concentration of the herbicide in the water and/or soil phase, but requires a specific equation. As the reaction proceeds in one phase, more product is released by the other phase in which the reaction is slower, or completely absent, as a consequence of the effect of the different media.

The aim of the work reported here was: a) to derive, from kinetic experiments, the mechanism of the hydrolysis of phenylureas and the catalytic capacity of chemicals added to the reaction solution; b) to gain information on the sorption of phenylureas by soils; c) to analyse the catalytic properties of soil components on the chemical hydrolysis of phenylureas. A better understanding of the reactivity and of sorption properties of these xenobiotic molecules is essential for more accurate predictions of their fate after introduction in the environment.

The study was carried out on 3-(3,4-dichlorophenyl)-1,1-dimethylurea (**I**, diuron), selected as a representative of the substituted phenylurea herbicide class, and on five soil samples collected in the Campania Region (Southern Italy) with different physico-chemical properties, mainly their organic matter content. Kinetic experiments in homogeneous water phase were also carried out using the phenylureas 3-(4-isopropylphenyl)-1,1-dimethylurea (**II**, isoproturon), 1-butyl-3-(3,4-dichlorophenyl)-1-methylurea (**III**, neburon) and 3-(3-chloro-p-tolyl)-1,1-dimethylurea (**IV**, chlorotoluron) as substrates.

EXPERIMENTAL

Materials

Diuron (**I**), isoproturon (**II**,), neburon (**III**), chlorotoluron (**IV**) and 3,4-dichloroaniline, were supplied by Dr. Ehrenstorfer GmbH (Table 1).

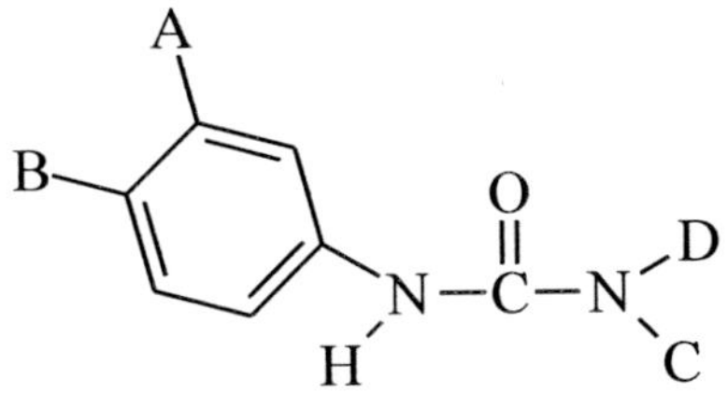

Table 1. Formula of the Phenylureas Investigated

Compound	A	B	C	D
3-(3,4-dichlorophenyl)-1,1-dimethylurea (diuron, **I**)	Cl	Cl	CH_3	CH_3
3-(isopropylphenyl)-1,1-dimethylurea (isoproturon, **II**)	H	$CH(CH_3)_2$	H	CH_3
1-butyl-3-(3,4-dichlorophenyl)-1,1-dimethylurea (neburon, **III**)	Cl	Cl	CH_3	$(CH_2)_3CH_3$
3-(3-chlorop-tolyl)-1,1-dimethylurea (chlorotoluron, **IV**)	Cl	CH_3	H	CH_3

The aniline derivative 3-isopropylaniline and 3-chloro-p-tolylaniline were obtained from Riedel-Dehaën, 3,4-dichloroaniline was from Dr. Ehrenstorfer GmbH. 3,4-dichlorophenyl isocyanate was from Fluka and all the other chemicals were from Carlo Erba.

Soil samples were collected at a depth of 0-20 cm (Ap horizon) in the following sites of the Campania Region (Southern Italy): Dugenta, Pignataro Maggiore, Dragoni, Mondragone and Castelvolturno.

Methods

Soil Analysis

The following analyses of the soil samples (Table 2) were carried out according to the Italian standard methods for soil analysis [24]: particle-size analysis was performed by a sedimentation procedure; organic carbon (OC) content was determined by the Walkley-Black method; organic nitrogen (ON) content was determined by the Kjeldahl method; soil pH were determined in 1:2.5 (w/v) soil/H_2O suspension; and the cation-exchange capacity (CEC) was measured by the $BaCl_2$-triethanolamine (TEA) method; carbonate content ($CaCO_3$) was determined with a Dietrich-Frühling calcimeter.

Table 2. Properties of the *Campania* investigated soils

Soil Collected in:	Sand	Silt	Clay	ON	OC	pH*	CEC	$CaCO_3$
	(%)	(%)	(%)	(%)	(%)		(cmol[+] Kg-^{1})	(%)
Castelvolturno	12.36	62.16	25.48	0.06	0.44	7.7	25.1	14.37
Mondragone	52.19	38.72	9.09	0.07	0.54	7.7	8.6	4.20
Dragoni	56.34	31.36	12.30	0.10	0.72	6.0	14.6	1.22
Dugenta	58.54	35.68	5.78	0.15	1.05	7.0	15.7	0.81
Pignataro Maggiore	19.47	48.44	32.09	0.26	2.00	7.7	37.7	2.03

*(1:2.5 w/vol)

HPLC Analyses

HPLC analyses were performed on a chromatographic Waters system, consisting of a 515 HPLC Pump and a 2487 dual λ Absorbance Detector, equipped with a C_{18} reversed-phase column (3.9 x 150 mm). Phenylureas and/or their degradation product were eluted by a linear gradient 35 – 42 % of acetonitrile in water in 15 min, flow rate 1 mL / min, $\lambda = 248$ nm.

Kinetic Measurements in Homogeneous Water Phase

Water solutions of the selected phenylurea (0.10 mM) at the desired pH and buffer concentrations, ionic strength 1M (KCl), were filtered through a 0.45 μm membrane filter and then stored in a thermostated bath kept in the dark, unless otherwise stated, at 60.0 ± 0.1 °C. At preselected times, aliquots were removed and analysed by HPLC. The rate constants were calculated by least-squares analysis, assuming the reaction to be first-order in the starting phenylurea. The fitting of the experimental data was satisfactory for all samples. The buffers were used in the concentration range of 10 – 500 mM and at the fraction (α) of the free base in the buffer of 0.2, 0.4, 0.6 and 0.8 for CH_3COOH/CH_3COO^-, pH 4.0 – 5.2 and $TrisH^+$/Tris, pH 6.9 – 8.1; at $\alpha = 0.2$, 0.6 and 0.8 for HCO_3^-/CO_3^{2-}, pH 8.9 - 10.1 and $H_2PO_4^-/HPO_4^{2-}$ pH 6.0 – 7.2. The buffer 4-methyl-morpholineH^+/4-methyl-morpholine was used only at $\alpha = 0.6$, pH 7.4. Dilute KOH and HCl were used in the pH ranges 1.0 – 3.0 and 11.0 – 13.0, respectively.

The temperature effect on the reaction rate was analysed by performing kinetic runs in the ranges 30 – 70 °C, 0.5 M phosphate buffer, 0.6 fraction of the free base in the buffer, and calculating the pre-exponential factor and activation energy from the linear regression of ln(k)

versus 1/T by least-squares method. The entropy and enthalpy of activation were calculated by the standard formulae derived from the absolute theory of reaction rates [25].

Characterisation of the Reaction Products

The compounds of the degradation reaction of **I**, **II**, **III** and **IV** in water solutions were identified by comparison with HPLC traces and UV spectra of pure samples. The compound obtained from **I** by the spontaneous degradation in water-methanol mixtures and by the reaction of 3,4-dichlorophenyl isocianate with pure ethanol was identified by ^{1}H- and ^{13}C-nmr and mass spectra.

pK$_a$ Values

The equilibrium constants for the acid-base reactions of the Aryl-NH-CO-N< group to Aryl-N$^{(-)}$-CO-N< and Aryl-NH-CO-NH^{+}< of **I**, **II**, **III** and **IV** were calculated by the computer program SPARC developed by Carreira [26, 27]. This program, based on structure activity relationships and perturbed molecular orbital theory, enables the calculation of the ionization pK$_a$s for a large number of organic compounds from the molecular structures.

Sorption Isotherms

Sorption experiments of *diuron* on soils were conducted using the batch equilibration method in 15-mL screw-cap vials at 30 °C. Three mL of *diuron* solution at the selected concentration in 1 mM sodium phosphate buffer, pH value of the soil, were added to 1.0 g of soil. The samples were stored in a thermostated bath and periodically shaken gently. At selected times aliquots were centrifuged at 10,000 rpm for 15 min, filtered through Filtek filters (RC 0.45 μm, Chemtek), that had been found to retain undetectable amounts of *diuron*, and analysed by HPLC. After 1 day, the equilibrium concentrations were reached. Analyses carried out after longer times (2-3 days) did not give significantly different results.

Kinetic Measurement in Soil/Water Systems

25 mL of an aqueous solution of *diuron*, 15 mg/L, were added to 5, 10, 17.5 or 25 g of sterilised soil samples collected in Castelvolturno (% OC = 0.44, Table 2). The mixtures were stored in a thermostat at 40.0 ± 0.1 °C in the dark and gently shaken at intervals. At pre-selected times, aliquots were removed and the water phases, separated by centrifugation (15 min, 10,000 rpm) filtered, and then analysed by HPLC. During the reaction course, the pH value of the soil-water mixture was periodically tested and no significant change was observed. The soil samples were sterilised by autoclaving three times at 121 °C for 30 min. All the water solutions had been sterilised before use by filtration on 0.2 μm membrane filter.

Soil samples with different % OC were used to test the influence on the reaction rate of the organic matter content. In these experiments 25 g of soil per 25 mL of *diuron* solution were used.

In a few experiments at the end of the kinetic runs, the sorbed phenylurea was extracted with methanol. The soil was shaken for 2 h with methanol (2 mL methanol / 1 g soil), the mixture was centrifuged at 5000 g for 20 min. and the supernatant was collected. This extraction was repeated three times. The combined aliquots of methanol were concentrated on a rotovapor and analysed by HPLC.

RESULTS

Degradation Reaction in Homogeneous Water Phase

In homogeneous water phase, under all the experimental conditions, phenylureas underwent spontaneous irreversible hydrolysis at an appreciable rate giving detectable amounts of a single aromatic product, which was identified as the corresponding aniline derivative by HPLC co-injection of pure compounds and mass spectrometry techniques. The concentration of the phenylurea plus that of the aniline derivative was constant during the kinetic runs, indicating the absence of reaction intermediates in an appreciable amount. Figure 2 reports typical time-concentration curves. Moodie and co-workers [28] reported a half-life value of 20 s for the uncatalysed hydrolysis of phenylisocianate in aqueous solution at 25 °C, and, moreover, showed that the reaction is buffer-catalysed. Fast degradation to aniline derivatives may explain why we could detect no phenylisocyanate derivatives in water solution by HPLC. However, an indication that the reaction proceeds through phenylisocianates was given by the analysis of the reaction products of **I** obtained in water-ethanol mixture (50 % ethanol). In this medium HPLC analysis revealed, besides the aniline derivative, another product, in 60 % relative yield after 15 days, which was identified by ^{1}H- and ^{13}C-NMR, and by mass spectra, as ethyl 3,4-dichlorophenylcarbamate: ^{1}H-NMR δ1.30 (3 H, t, CH_3), 4.22 (2 H, q, CH_2), 7.19 (1 H, dd, 6-H), 7.46 (1 H, d, 6-H), 7.59 (1 H, d, 6-H), 7.93 (s, 1 H, NH); ^{13}C –NMR δ14 (CH_3), 61 (CH_2), 119 (ortho CH), 120 (ortho CH), 124.5 (para CCl), 130 (meta CH), 132 (meta CCl), 138.5 (CN); 153 (CO). Electron ionization mass spectrum, recorded on the quadrupole instrument Trio 2000, Fisons, gave m/z 233 (M^+,100%).

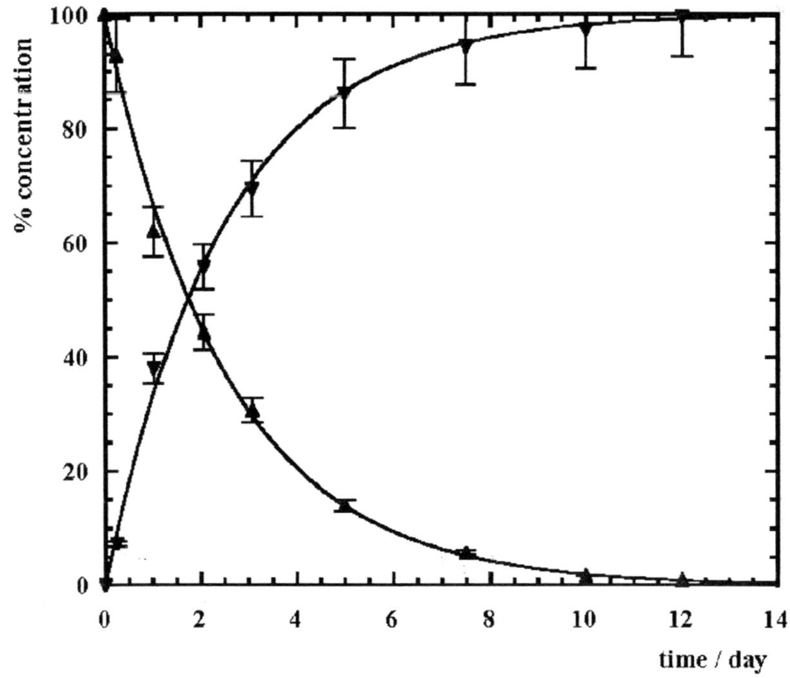

Figure 2. Typical time profiles for the chemical degradation of *diuron*: T = 70 C°, pH 7.2, 0.38 M phosphate buffer, ionic strength 1 M (KCl); *diuron* (▲), 3,4 dichloroaniline (▼)

This compound was the exclusive product of the reaction of 3,4-dichlorophenylisocyanate with pure ethanol.

pH Dependence of the Reaction Rate

Figure 3 shows a plot of k_o *versus* pH for the phenylurea *diuron*, where k_o is the observed first-order rate constant extrapolated to zero buffer concentration.

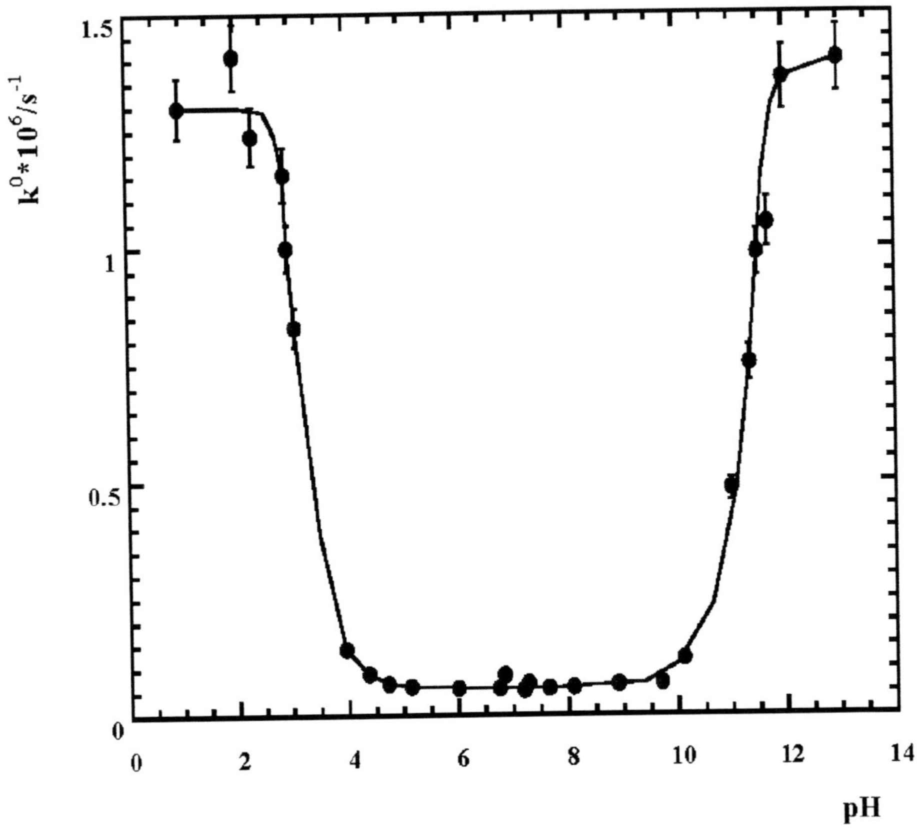

Figure 3. pH rate profile for the degradation reaction of *diuron* at T = 60 °C and μ = 1 M (KCl). k_o is the first–order rate constant at zero buffer concentration.

In the pH range 5 – 9, the rate constant is substantially pH-independent, while at lower and higher pH values the slope of the curve markedly increases, suggesting an efficient catalysis by hydronium ion at low pH and by hydroxide ion at high pH. At extreme pH values (< 2 or > 12), the k_o value is again constant. The value of the pH at the inflection point ($d^2k/dpH_2 = 0$) observed at basic pH (pH = 11.5 ± 0.3) was close to, although not coincident with, the pK_a value computed for the deprotonation equilibrium of **I** at the same temperature as for the kinetic experiments and zero ionic strength (pK_a = 10.3). Similar pH-rate constant plots were observed for the other phenylureas studied in this work.

Buffer Catalysis

The observed pseudo-first-order rate constant (k_{obs}) determined at constant pH reveals a marked dependence on the buffer concentrations. For all the buffers tested the kinetic constant first increases rapidly at low buffer tested concentrations and then gradually levels off at higher concentrations, reaching a limit value that is independent of the type of buffer and the fraction of free base in the buffer. The k_{obs} – *buffer concentration* plots for **I** at the fraction 0.6 of free base in the buffers are shown in Fig. 4.

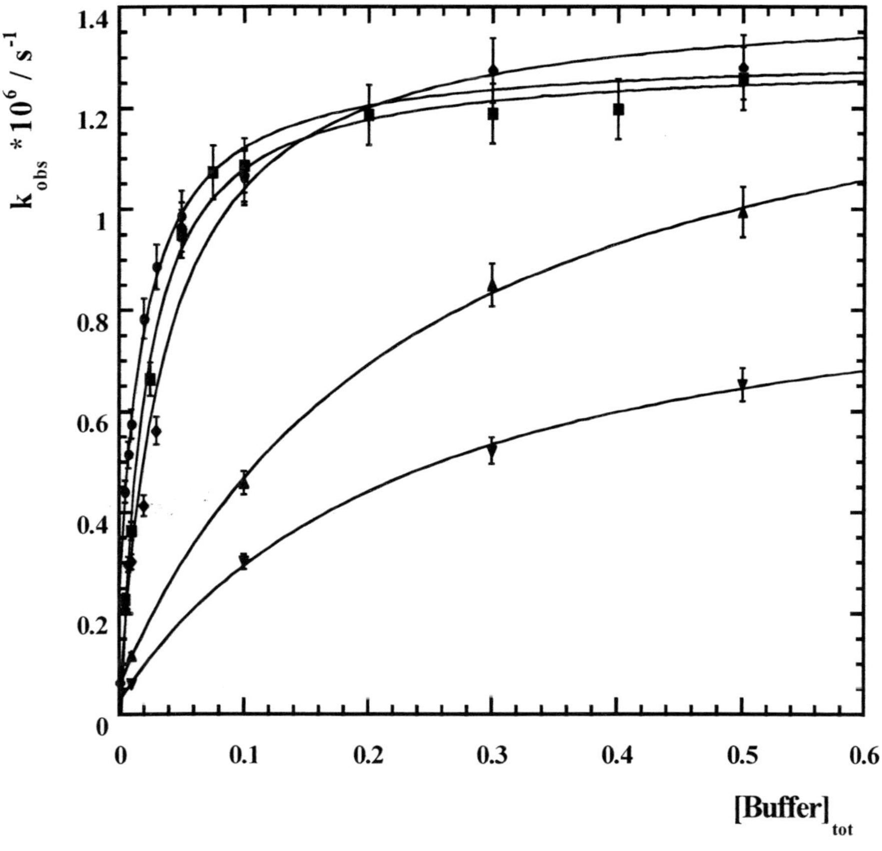

Figure 4. Dependence of the observed first-order rate constant (kobs) for the degradation reaction of diuron on the total buffer concentration at the fraction 0.6 of free base, T = 60 °C and μ = 1 M: (•) = HCO3-/CO32-, (▲) = TrisH+/Tris, (▼) = 4-methyl-morpholineH+/4-methyl-morpholine; (■) = H2PO4-/ HPO42-, (♦) = CH3COOH/CH3COO ¯.

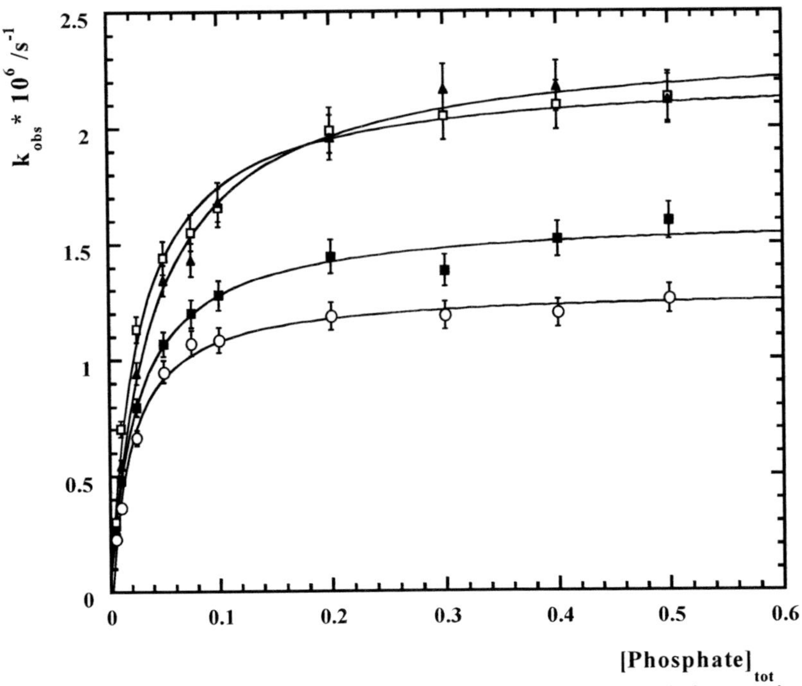

Figure 5. Dependence of the observed first-order rate constant (k_{obs}) for the degradation reaction of *diuron* (**I**) (○), *isoproturon* (**II**) (▲), *neburon* (**III**) (□) and *chlorotoluron* (**IV**) (■) on the total phosphate buffer concentration at the fraction 0.6 of free base, at T = 60 °C and μ = 1 M

Figure 5 shows the dependence of k_{obs} on the phosphate concentration for the hydrolysis of **I**, **II**, **III** and **IV**. The shape of the curves is the same for all the phenylureas tested. The dependence of the slopes of the k_{obs}-*buffer concentration* curves for **I** at zero buffer concentrations (k^o_{cat}), which measure the catalytic efficiency of the buffers, are shown in Fig. 6.

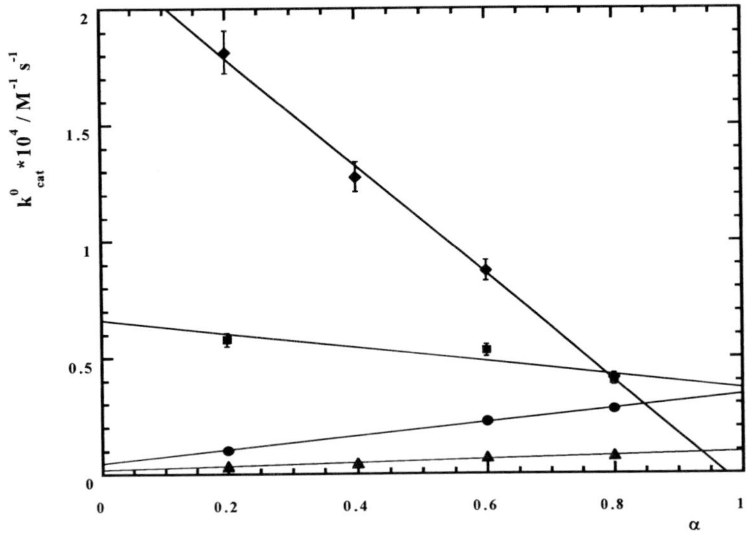

Figure 6. Plot of the catalytic constant of the buffers at zero buffer concentration (k^o_{cat}) against the fraction (α) of the free base in the buffers for the degradation reaction of *diuron*, T = 60 °C and μ = 1 M: (●) =

HCO_3^-/CO_3^{2-}, (▲) = $TrisH^+$/Tris, (▼) = 4-methyl-morpholineH^+/4-methyl-morpholine; (■) = $H_2PO_4^-$/ HPO_4^{2-}, (♦) = CH_3COOH/CH_3COO^-.

It is evident that, at least for some buffers, the curves do not pass through the origin of the axes at $\alpha = 0$ and $\alpha = 1$, suggesting that both the acidic and basic forms of the buffers are catalysts of the reaction. It is interesting to note that k^o_{cat} attains higher values with the buffers HCO_3^-/CO_3^{-}, $H_2PO_4^-$/ HPO_4^{2-} and CH_3COOH/CH_3COO^- than for $TrisH^+$/Tris, suggesting that bifunctional acid-base buffers are particularly efficient catalysts. In line with this conclusion are the k_{obs}*-buffer concentration* plots reported in Fig. 4. Also worth of note is the marked difference in the slope at zero buffer concentration between the bifunctional acid-base buffers CH_3COOH/CH_3COO^- and HCO_3^-/CO_3^{2-} and the monofunctional $TrisH^+$/Tris and 4-methyl-morpholineH^+/4-methyl-morpholine, although the pK_a values of the last two buffers are intermediate between those of the two bifunctional acid-base buffers. Table 3 reports the k_{obs} limiting values obtained at high phosphate concentration and, for comparison, the values obtained in 0.1 M HCl and 0.1 M NaOH. For each compound the rate constant is practically independent of the media used.

Table 3. Rate constants ($k \times 10^6 / s^{-1}$) for the hydrolysis of 3-(3,4-dichlorophenyl)-1,1-dimethylurea (I), 3-(4-isopropylphenyl)-1,1-dimethylurea (II), 1-butyl-3-(3,4-dichlorophenyl)-1-methylurea (III), and 3-(3-chloro-p-tolyl)-1,1-dimethylurea (IV) in acidic, basic and high phosphate buffer concentration media: T = 60 °C, μ = 1M (KCl)

Phenylurea	0.1 M HCl	0.1 M NaOH	High phosphate concentration
I	1.3 ± 0.1	1.4 ± 0.1	1.3 ± 0.1
II	1.8 ± 0.1	2.0 ± 0.1	2.3 ± 0.1
III	1.9 ± 0.1	1.7 ± 0.1	2.2 ± 0.1
IV	1.8 ± 0.1	1.7± 0.1	1.6 ± 0.1

Temperature Dependence of Reaction Rate

The Arrhenius plot for the limiting rate-constants reached at high buffer concentration (see the EXPERIMENTAL section) is linear in the temperature range studied (Fig. 7). The pre-exponential factor and the apparent activation parameters for **I** are respectively: ln (A/day^{-1}) = 38 ± 1; E_a = 127 ± 2 kJ mol^{-1} ; $\Delta H^{\ddagger}$ = 124 ± 2 KJ mol^{-1}; $\Delta S^{\ddagger}$ = 14 ± 8 J K^{-1} mol^{-1}.

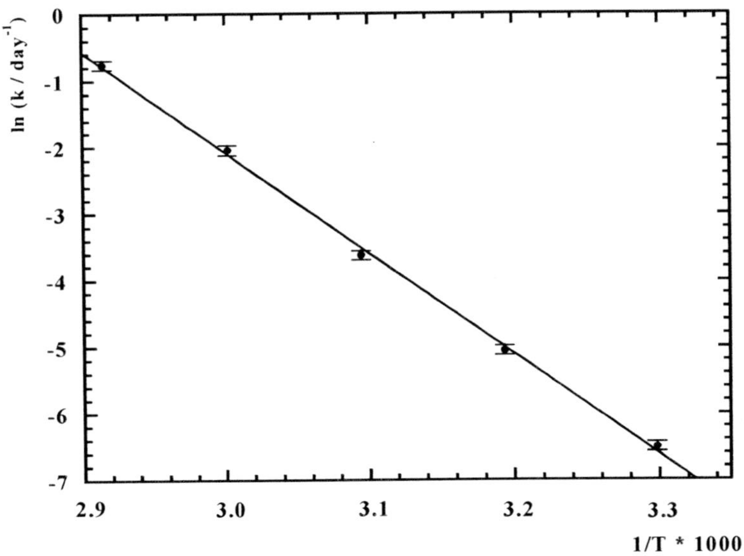

Figure 7. Arrhenius plot for the step that is rate–limiting at high concentration of H^+, OH^- or buffers for the hydrolysis of *diuron*, ionic strength 1 M (KCl)

Sorption Isotherms

Figure 8 shows the sorption isotherms of *diuron* on the soils selected for this study. The common experimental Freudlich equation fits the sorption data well (Table 4)

$$q = K_F (C_e)^N \qquad (1)$$

where q is the amount of adsorbate per unit mass of adsorbent, C_e the solution concentration at the equilibrium, K_F the adsorption coefficient and N the heterogeneity index. All the curves are non-linear, as indicated by N values less than unity.

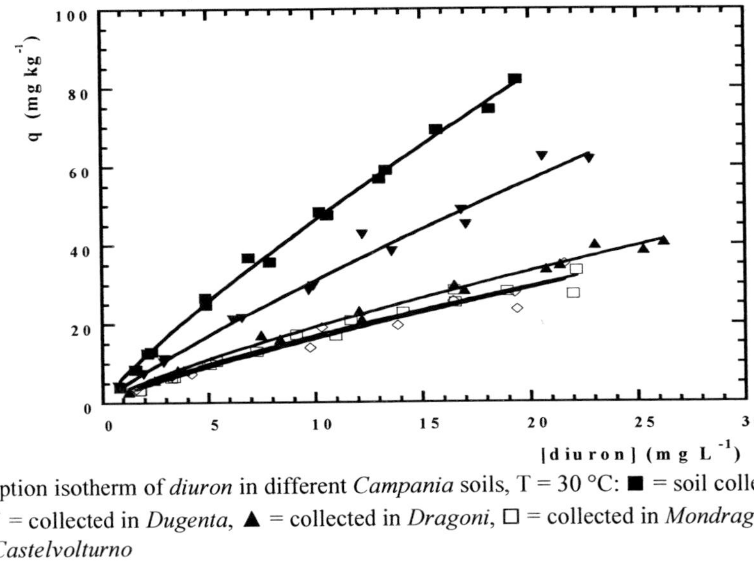

Figure 8. Sorption isotherm of *diuron* in different *Campania* soils, T = 30 °C: ■ = soil collected in *Pignataro Maggiore*, ▼ = collected in *Dugenta*, ▲ = collected in *Dragoni*, □ = collected in *Mondragone*, ◇ = collected in *Castelvolturno*

Table 4. Freundlich isotherm parameters for the adsorption of *diuron* on soil

Soil collected in	Org. Carbon (%)	K_f (μg/g)(μg mL)$^{-N}$	N	R^2
Castelvolturno	0.44	2.2	0.81	0.98
Mondragone	0.54	2.2	0.91	0.99
Dragoni	0.72	2.9	0.88	0.99
Dugenta	1.05	4.6	0.81	0.99
Pignataro Maggiore	2.00	7.9	0.77	0.98

Chemical Degradation in the Soil / Water System

D*iuron* solutions have been stored in the presence of sterile soils characterised by a different content of organic carbon (% OC), or in presence of different amounts of the same soil sample, and monitored for over 3 months. After an initial rapid decrease of *diuron* concentration due to the sorption process, we observed a slower decrease which could tentatively be ascribed to a slow-rate chemical degradation. Figure 9 reports the time-concentration curves obtained for different ratios of *soil weight / water solution.* The full symbols in figure refer to the total amount of the phenylurea in the soil/water systems. They have been obtained by adding, point by point, the phenylurea adsorbed in the soil, computed by the Freundlich equation, to the amount present in solution as derived from experimental data (empty symbols). It is worth noting that the total amount of *diuron* in the reacting system does not depend on the amount of soil in the explored range. In a few experiments the total amount of *diuron* was determined at the end of the kinetic runs by multiple extraction from the soil with ethanol and adding the value obtained to the amount in solution. No relevant difference was find between the experimental and the computed amounts.

The compound 3,4-dichloroaniline, found to be the only aromatic degradation product in homogeneous phase, was detected in much lower amounts with respect to tstoichiometric expectation. This molecule binds covalently to humic acids [16].

As regards the reduction rate of the herbicide concentration in water phase, it must be noted that, as the degradation reaction proceeds in the water phase, more adsorbed *diuron* is released from the soil phase to the aqueous phase according to the repartition equilibrium (scheme 1). This determines an apparent slower reduction of *diuron* concentration in solution than might be expected from the degradation reaction alone.

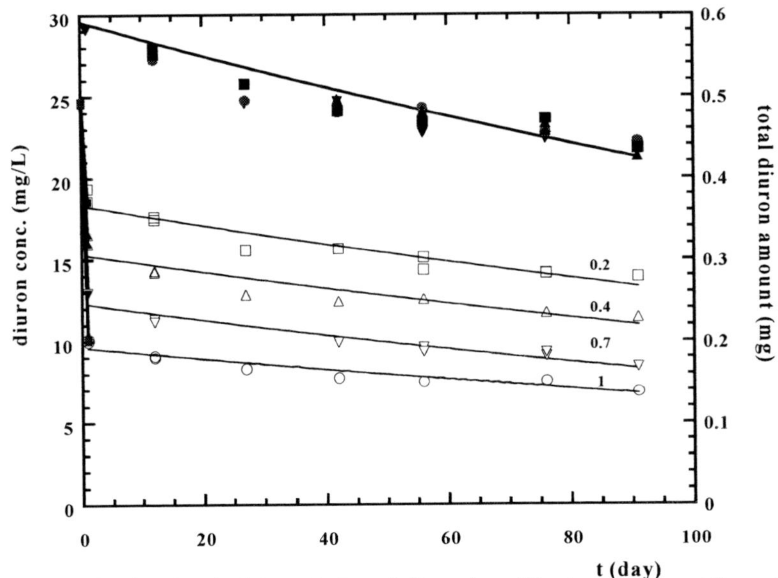

Figure 9. Time profiles for the chemical degradation of *diuron* in soil/water mixtures at the *soil weight/solution volume* ratio indicated (g/mL): *diuron* concentration in the water phase, left scale; total amount of *diuron* in the system, right scale; % OC in the soil = 0.44. The point at t = 0 and [*diuron*] = 25 mg/L belongs to each *diuron* concentration curve

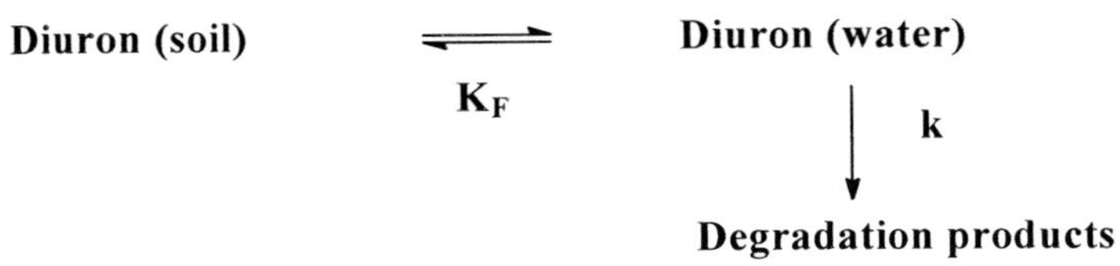

Scheme 1

The microscopic kinetic rate constant (k) in the water/soil mixture was calculated from the time dependence of the substrate concentration in water, according to the equation (2) proposed by us in a former work [3] and derived as reported in the Appendix. The equation is based on the assumption that 1) the chemical reaction occurs only in the water phase and 2) the adsorption-desorption process is much faster and is well described by the Freundlich equation.

$$\ln\frac{C_e}{C_e^0}+\frac{K_F m_s N}{V_w(N-1)}\left(C_e^{N-1}-\left(C_e^0\right)^{N-1}\right)=-k\,t \qquad (2)$$

C_e is the *diuron* concentrations in water phase at the time t; V_w, m_s, N, K_F are the volume of water phase, soil mass, heterogeneity index and partition coefficient, respectively. C_e^0 is the solution concentration at t ≅ 0 after that the adsorption equilibrium has been reached (pre-equilibrium approximation). Its value can be computed either by equation (8) in the Appendix, known K_F and N values of the Freundlich equation, or by extrapolation at t = 0 of time-concentration data in Fig. 9 for t ≥ 1 day. Both procedures gave closely similar values.

The curves in Fig. 9 were drawn using the Freundilich parameters of the soil involved (Table 4) and the kinetic constants (*k*) obtained by least-squares fit of the experimental points to equation (2). In this procedure the datum at t = 0, which indicates the *diuron* concentration before addition of soil, was not used. The dependence of *k* for the hydrolysis of *diuron* on the ratio "*soil weight / solution volume*" is reported in Fig. 10. Within the experimental error the relationship is linear.

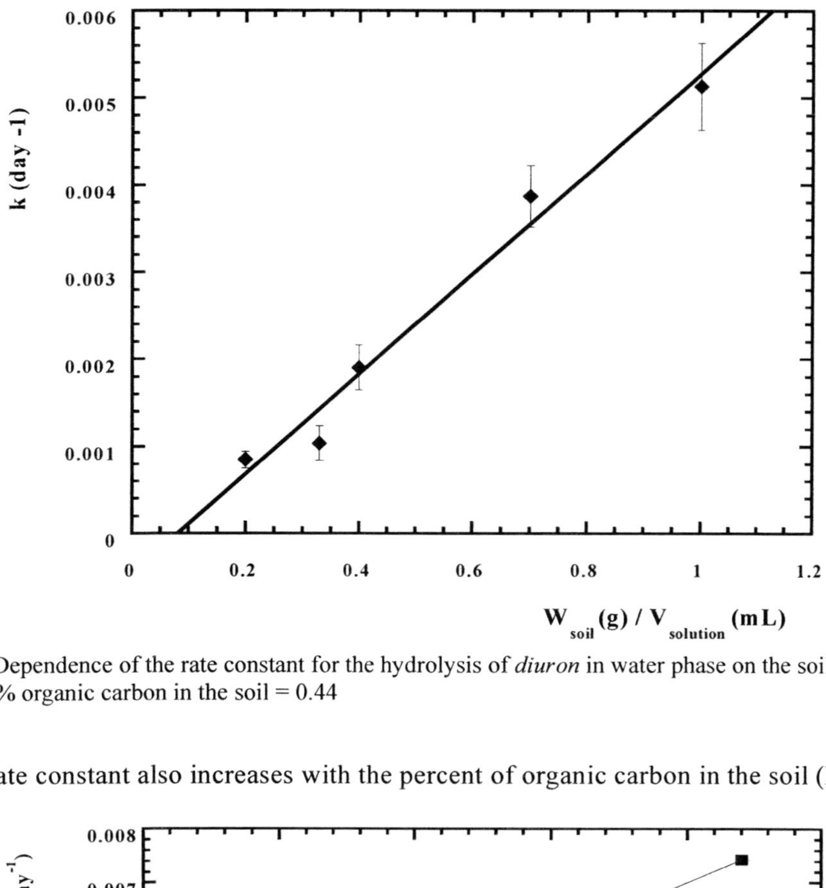

Figure 10. Dependence of the rate constant for the hydrolysis of *diuron* in water phase on the soil/water ratio: T = 40 °C, % organic carbon in the soil = 0.44

The rate constant also increases with the percent of organic carbon in the soil (Fig. 11).

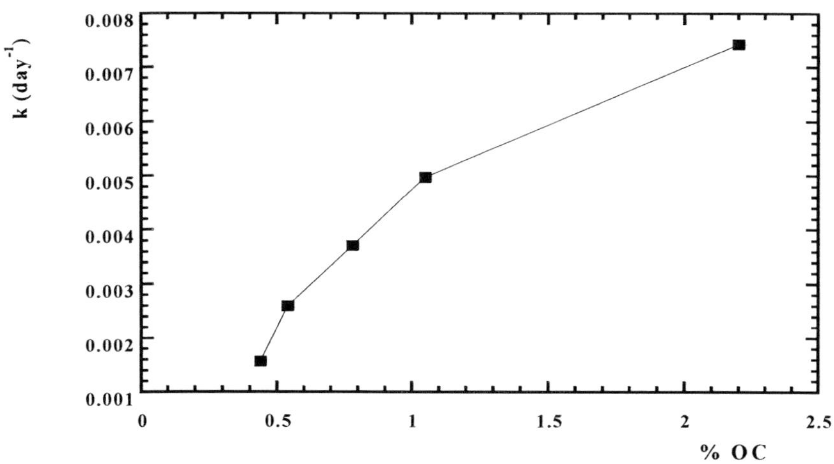

Figure 11. Dependence of the rate constant for the hydrolysis of *diuron* in water phase of a soil/water mixture on the percent of organic carbon in the soil: T = 40 °C, 25 mL water *diuron* solution, 25 g soil

DISCUSSION

Reaction in Homogeneous Water Phase: Kinetics and Mechanism

Because of its environmental implication the most relevant finding of our study of phenylurea herbicides in water solution is that the degradation reaction is catalysed, besides by H^+ and OH^- ions, by bifunctional acid-base buffers. Interestingly, the rate constant increases with buffer concentrations up to a limiting higher value. The result that carboxylic acid buffers are efficient catalysts of the reaction strongly suggests a catalytic role for the carboxylic groups of the soil organic matter in the degradation of phenylureas in soil/water mixture. This heterogeneous material includes the humic acids as a major fraction (up to 80 %). Humic acids are natural macromolecules that bear aliphatic, glucidic, amino acidic, and lipidic surface chains as well as chemically active functional groups of various nature (mainly carboxylic and phenolic groups).

Further features of the kinetic behaviour of phenylureas hydrolysis that are useful for deriving the reaction mechanism can be summarised as follows:

a) both the acidic and basic forms of the buffers are catalysts of the reaction;
b) the non-linear dependence of the rate constant on buffer concentrations suggests the existence of an intermediate, the formation of which is rate determining at some catalyst concentrations, while, at others, the rate-limiting step is the conversion of the intermediate to products [29-31]. Moreover, the enhanced reactivity observed with the polyfunctional acid-base catalysts (HCO_3^-/CO_3^{2-}, $H_2PO_4^-/HPO_4^{2-}$ and CH_3COOH/CH_3COO^-) suggests that they take part in a proton switch step, as it is often observed in cases where such processes are involved [32-33];
c) for each of the phenylureas studied we observed that the rate constant in the pH extreme regions and at high polyfunctional buffer concentration has, within the experimental error, the same value. This is a clear indication that the hydronium and hydroxide ions, and the buffers lead to the same intermediate, whose decomposition at high catalyst concentrations is rate-limiting. Laudien and Mitzner reported [22, 23] the values of the kinetic constants for about thirty phenylureas (including compounds **I** and **II**), recorded in 0.1 M H_2SO_4 and 0.1 M NaOH. Coherently with our results, the rate constant for every compound had an identical value, within the experimental error, both in acidic and basic media;
d) the pH value at the inflection point observed at basic pH is not markedly different from the pK_a value computed for the deprotonation equilibrium of the substrate;
e) finally, the low $\Delta S^{\ddagger}$ value observed for the reaction in high polyfunctional buffer concentrations ($\Delta S^{\ddagger}=14 \pm 8$ J K^{-1} mol $^{-1}$ for **I** in phosphate buffer) suggests that, from the starting substrate to the transition state of the rate-limiting step, there is no major change in the disorder or motional modes. It is worth noting that, for a multistep reaction in which the rate-limiting step is preceded by pre-equilibrium steps, the $\Delta S^{\ddagger}$ observed value is the sum of the activation entropy of the rate-limiting step and the entropy changes of the equilibrium steps [34].

Reaction Mechanism

A pathway for the formation of isocyanate and ammonia derivatives from phenylureas that fits the experimental data is shown in Fig. 12 [35, 36]. Depending on pH and buffer concentrations the reaction starts by three parallel routes, each leading to the zwitterion $\mathbf{Z}^{\ddagger}$. At low pH and buffer concentration the reaction proceeds through pathway shown on the left: protonation of the dialkyl nitrogen atom $_2$N, (fast equilibrium between the substrate and hydrogen ion, for **I** $pK_a = -2.5$), followed by slow deprotonation of the nitrogen atom $_1$N by the general base **B**. This is specific acid-general base catalysis, kinetically equivalent to general acid catalysis. At high pH the reaction proceeds (pathway on the right) prevalently through specific basic–general acid catalysis, kinetically equivalent to general base catalysis: fast reversible deprotonation of $_1$N (for **I** $pK_a = 10.3$), followed by slow protonation of $_2$N by the general acid HA. At high concentration of bifunctional acid-base buffers the reaction proceeds through a proton switch promoted by the buffers. At low catalyst concentrations (pH range 3 – 11 and low buffer concentrations) the formation of $\mathbf{Z}^{\ddagger}$ is rate-limiting, whereas at high catalyst concentrations, its degradation to the final product becomes rate-limiting, consequently the reaction rate no longer depends on pH and buffer concentration. Coherently with this mechanism, the $\Delta S^{\ddagger}$ value recorded at high bifunctional buffer concentrations is very low. Figure 13 reports the proposed role for bifunctional carboxylic acids catalysts. The acidic (O-H) and basic sites (C=O) act a concerted catalysis promoting the proton switch.

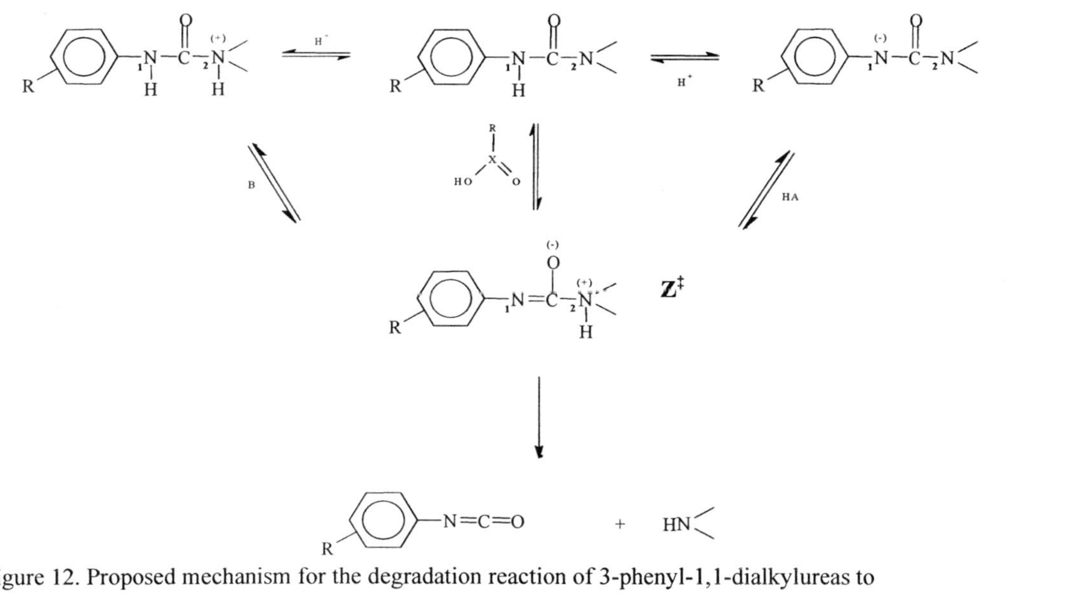

Figure 12. Proposed mechanism for the degradation reaction of 3-phenyl-1,1-dialkylureas to phenylisocyanates and dialkylamine.

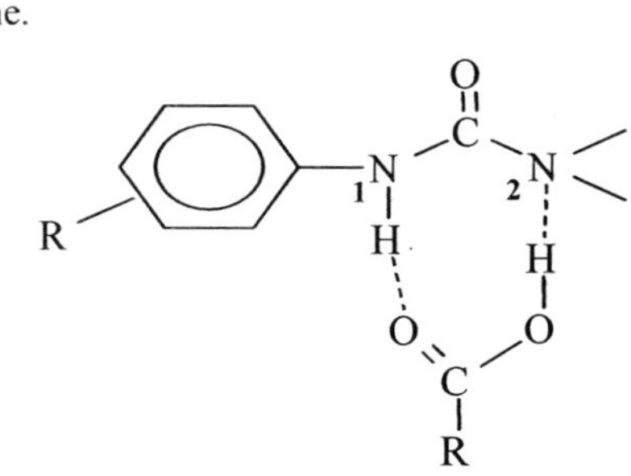

Figure 13. Proton switch assisted by carboxylic acids.

In contrast to the mechanism reported here, Laudien and Mitzner ascribed the rate limiting value at high hydroxide ion concentrations to the formation of an unreactive side product [23]. At low pH, according to these authors, the limiting value is due to the rate determining attack of a water molecule [22]. On the other hand, Sabaliûnas *et al.* postulated the formation of a dianion intermediate as the cause for the limiting value of the rate constant at high pH [19].

Sorption of Phenylureas by Soil

Figure 14 shows the relationship between the partition coefficient of the Freundlich equation (K_F) for the sorption of I and the percent of organic carbon (% OC) in the soils examined.

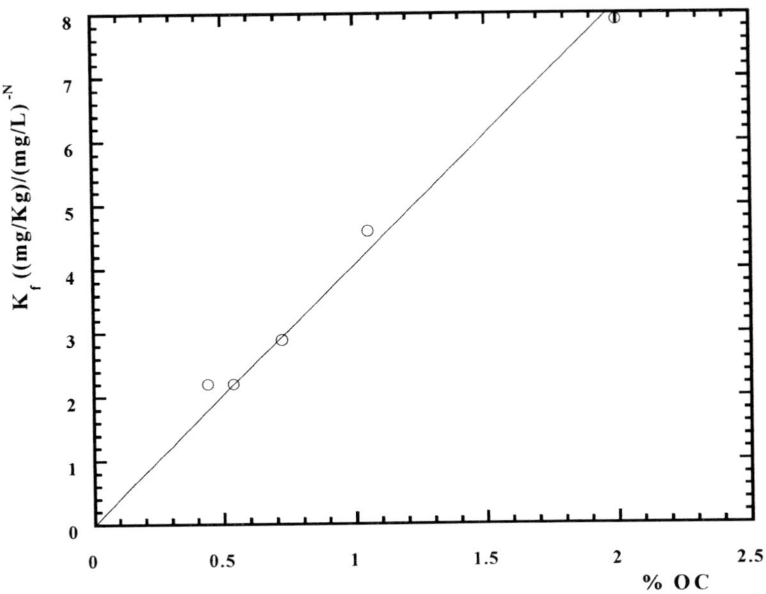

Figure 14. Dependence of the K_F Freudlich parameter for the *diuron* sorption by soil on the percent of soil organic carbon (% OC)

Within the experimental error the relation is linear. A Similar linear relationship has been reported for other phenylurea herbicides [37-39].

It is worth noting that for a N Freundlich parameter equal to 1 ($q = K_F C$), the slope of the straight line in the plot K_F *vs.* % OC gives the organic carbon-water partition coefficient (K_{OC}). For N $\neq$1 the K_F value coincides with the partition ratio (q/C) only for C equal to 1 (in our experiment 1 mg/L). The K_{OC} value determined from the octanol-water partition coefficient (K_{OW}), using the well-known equation "$K_{OC} = 0.41 \times K_{OW}$" [40], is 290 mL/g. This numerical value is markedly lower than the slope of the straight line reported in the previous figure: 410 (μg/g)(μg mL)$^{-N}$. This difference, combined with the non linear pattern observed in the q vs. % OC plot, indicates that the sorption of diuron cannot be considered as a simple repartition between water and the soil organic phase, ruled by the hydrophobicity of the herbicide, but is a more complex phenomenon. According to the Langmuir adsorption model, the absence of linearity may be due to the site saturation, either in the one-type-sites and multiple type sites model [41, 42].

The adsorption isotherms for the multiple-site Langmuir model are described by the equation:

$$q = \sum_i q_{i,MAX} k_i C/(1 + k_i C) \quad (1)$$

where $b_{i,max}$ is the maximum amount of adsorptive that can be adsorbed by the sites of the ith type and k_i is the microscopic equilibrium constant for the reversible binding to the sites i^{th}; C and q have been defined previously.

$$\textbf{adsorbitive (solution) + site } \mathbf{i^{th}} \textbf{ (soil)} \xrightleftharpoons{\mathbf{k_i}} \textbf{adsorbate-site } \mathbf{i^{th}} \textbf{ (soil)}$$

Scheme 2

When data are known only for a small range of concentrations, the determination of the terms of the summatory that make a significant contribution to *q* in (1) is not a straightforward procedure. In these cases, an algebraic manipulation of the data, as used in the study of the interaction of ligand with macromolecules by Scatchard equation [43], can be useful.

For homogeneous sites the equation (1) can be rearranged in the linear form:

$q/C = q_{max}\, k - q\, k$

Linearity in the plot q/C *vs.* q is indicative of a single type of independent site (simple Langmuir isotherm), whilst non-linearity suggests the occurrence of several sites with different affinity, whose k_i and $q_{i\,max}$ values can be obtained by least squares fit of eq. (1) to the experimental data or following one of the procedures reported in literature for decomposing the Scatchard like-plots [44]. The product $q_{i,max}$ k_i corresponds to the macroscopic equilibrium constant (K_i) for the binding of the adsorptive to the sites of type i. Figure 15 shows the dependence of q/C on q. For all the soils tested and throughout the *diuron* concentration range considered, the plots were linear, thereby indicating the presence of a single class of sites. Experiments at higher *diuron* concentration were not possible because of the low water solubility of this molecule.

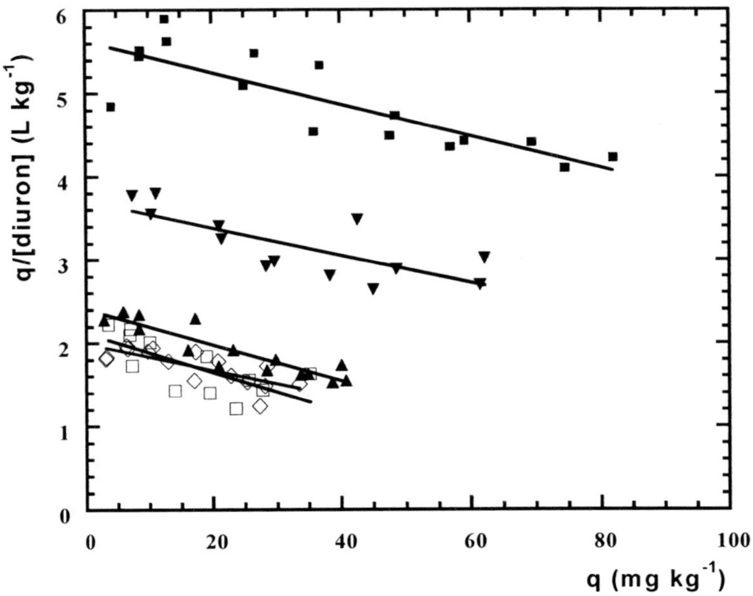

Figure 15. Scatchard-like plot of *diuron* binding to soils. T = 30 °C: ■ = soil collected in *Pignataro Maggiore*, ▼ = collected in *Dugenta*, ▲ = collected in *Dragoni*, □ = collected in *Mondragone*, ◇ = collected in *Castelvolturno*

To the best of our knowledge, this is the first application of the equation above to the study of the interaction of xenobiotic molecules with soils.

Table 6 reports the binding data obtained by the fits.

Table 6. Microscopic binding constant (k) and maximum amount of adsorptive (q_{max}) for the adsorption of diuron by soils

Soil Collected in	OC (%)	k (Kg mg^{-1})	q_{max} 10^{-1} (mg Kg^{-1})	k q_{max}
Castelvolturno	0.44	0.033 (8)	7 (2)	2.23(7)
Mondragone	0.54	0.016(4)	12.5(4)	2.00(8)
Dragoni	0.72	0.021(3)	11.5(2)	2.41(5)
Dugenta	1.05	0.022(3)	17.3(3)	3.81(9)
Pignataro M.	2.00	0.022(3)	26.4(4)	5.81(8)

Estimated standard deviations in the least significant digits are shown in parentheses.

The microscopic equilibrium constant (k), within the experimental error, does not depend on the type of soil, while the maximum amount of *diuron* that can be adsorbed (q_{max}) increases linearly with the % of organic carbon (Fig. 16).

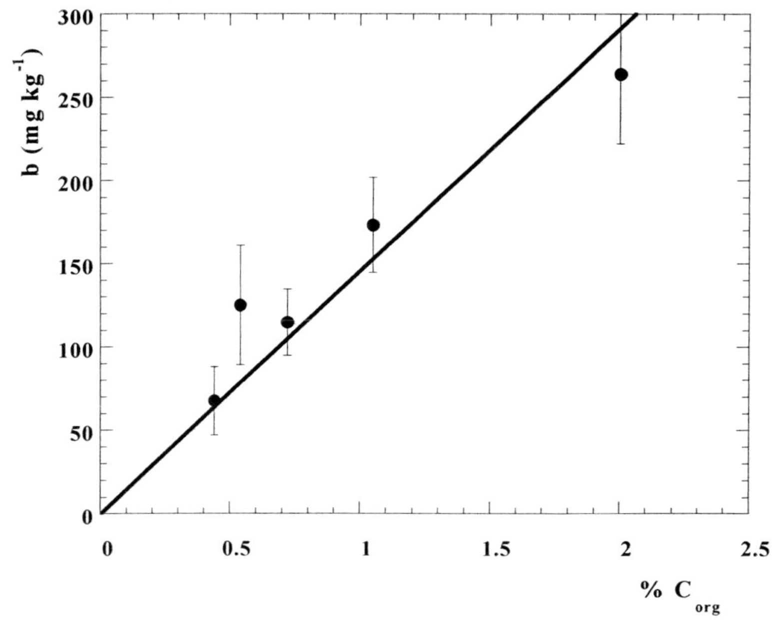

Figure 16. *Diuron* binding to soils: dependence of the maximum amount of adsorptive (q_{max}) on the percentage of organic carbon in the soil (% OC).

The product of k and q_{max} determines the macroscopic equilibrium constants of the binding. Its value coincides with the slope at C = 0 of the Langmuir isotherm and with the partition coefficient of a Freundlich equation with N = 1. It is worth noting that at low concentration a Langmuir isotherm can be described by the Freundlich equation with N = 1. The increase in the organic matter content of the soil causes a proportional increase of the partition equilibrium constant K, with a consequent increase of K_F. Recent studies have shown that K, normalised for the % OC of the soil, has the same value of the corresponding constant determined for the adsorption by dissolved HAs in homogeneous aqueous phase [4]. It has been shown that the interaction between phenylureas and dissolved HAs could be described in terms of a repartition-like equilibrium of the phenylureas between water and the macromolecular compound HAs, considered as microscopic environment [4]. The sorption isotherms for phenylureas on soil reported here show that at low concentration there is a simple repartition between the water and the soil organic matter phase, which constant is determined by the physico-chemical properties of the adsorbent and by the OC content of the soil. At high adsorbent concentration a simple repartition model is no longer suitable probably because of saturation of the adsorbing sites in the soil.

Chemical Degradation in the Soil / Water System

The reaction rate for the hydrolysis of phenylureas in the water phase of a soil/water mixture increases with the relative amount of soil in the system and with the content of organic matter in the soil, thus indicating that organic matter is a catalyst for the reaction. However, as shown by the data reported in Fig. 9, the rate of disappearance of the total phenylurea distributed throughout the soil/water system does not depend on the amount of organic matter. Increasing the organic matter increases the degradation rate in water phase, but also increases the amount of substrate sorbed by soil. The reaction does not take place within the organic matter because of the low dielectric constant.

We think, on the basis of the results obtained in homogeneous water phase, that the catalytic properties of organic matter are due to the carboxylic groups located on the external surface of humic acids (HAs). Dynamic molecular studies have shown that HAs are flexible molecules, whose average structure has hydrophobic groups in the interior and polar groups, like the carboxylic ones, on the surface, in contact with water molecules [45].

CONCLUSION

A phenylurea herbicide added to the water/soil system divides between the two phases according to the partition equilibrium constant. The value of the constant is only partially determined by the hydrophobicity of the herbicide, as shown by the nonlinearity in the sorption isotherm plots and by the weak correlation between the experimental repartition constant and the value computed by means of the repartition constant water/octanol. Moreover, in the water phase the phenylurea molecules are subjected to chemical hydrolysis catalysed mainly by the carboxylic group located on the external surface of the soil organic matter. However, the studies in homogeneous solution have shown that at high catalyst concentrations the rate reaches a limiting value no longer dependent on the catalyst concentration. On the other hand, the phenylurea molecules that are trapped in the "organic matter phase" are protected by the hydrolysis, because of the low value of the local dielectric constant. It has been shown that the rate-limiting step of this reaction proceeds through a zwitterionic intermediate, whose formation in a medium with a low dielectric constant is markedly slowed down.

These two opposite effects of the soil organic matter, catalysing the hydrolysis and sequestering the substrate, make the rate of disappearance of the total amount in the soil/water system practically independent on the amount of the organic mater. According to our studies, at 40 °C, the *diuron* added to the soil/water system has a half-life of about 200 days. These results should be taken into account in predicting the fate of phenylures after their dispersal in the field. Additional work on the catalytic ability of soils on the hydrolysis of other herbicides is in progress.

ACKNOWLEDGEMENTS

We are grateful to Professor L. Mazzarella for helpful comments. This work was financially supported by Italian *Ministero dell'Istruzione, dell'Università e della Ricerca.*

APPENDIX

The reaction rate for the chemical degradation of phenylureas, assuming that it occurs only in water solution and is first-order with respect to the phenylurea, can be expressed by the equation:

$$v = C_W V_W k \quad (1)$$

where v is the reaction rate defined as – dn/dt, n = number of moles of the phenylurea in solution, C_W is the molar concentration of the reactant in solution, V_w is the volume of the aqueous solution, and k is the first-order rate constant.

Moreover:

$$n_{tot} = C_W V_W + C_S V_S \quad (2)$$

where n_{tot} is the total amount of moles of the phenylurea in the system (water + soil), C_S and V_S are the concentration in the soil and the soil volume, respectively.

Assuming that the soil/water partition rate of the phenylurea is markedly higher than its hydrolysis (pre-equilibrium hypothesis), and considering the Freundlich equation $q_e=K_F(C_e)^N$, where K_F and N are constants, q_e is the concentration of the adsorbate molecule in the soil phase (w/w) and C_e is the concentration at the equilibrium in the water phase (w/V), we have:

$$n_{tot} = C_e V_W + m_s K_F (C_e)^N \quad (3)$$

where m_s is the soil mass and $C_e = C_W$, according to the pre-equilibrium hypothesis. In this hypothesis the Diuron concentration in solution is considered to be determined solely by the partition equilibrium (Scheme 1), being unperturbed by the degradation reaction (Connors 1990).

Differentiating with respect to time:

$$dn_{tot}/dt = V_w dC_e/dt + K_F m_S dC_e^N/dt \quad (4)$$

Moreover, supposing that Diuron undergoes hydrolysis only in water phase, the variation of total moles dn_{tot} corresponds to the variation of moles present in the aqueous solution dn defined in eqn. (2):

$$dn_{tot} = - (C_e V_w k)dt \quad (5)$$

Combining eqn. (4) and (5) we get:

$$-(C_e V_w k) = V_w dC_e/dt + K_F m_S dC_e^N/dt \quad (6)$$

which, rearranged and integrated, gives:

$$\ln \frac{C_e}{C_e^0} + \frac{K_F m_s N}{V_w (N-1)} \left(C_e^{N-1} - \left(C_e^0\right)^{N-1} \right) = -k\,t \quad (7)$$

where C_e^0 is the Diuron concentrations in water phase after the partition equilibrium has been reached at t ≅ 0. Its values can be derived by the Freundlich equation, knowing the constants and the initial Diuron concentration (C_0) in the in water before its contact with the soil.

$$\frac{q_e^0}{\left(C_e^0\right)^N} = \frac{\dfrac{(C_0 - C_e^0)}{m_s} \cdot V_w}{\left(C_e^0\right)^N} = K_F \quad (8)$$

It is worth nothing that for N ≠ 1 the implicit equation (2) cannot be inverted respect to C_e^0.

REFERENCES

[1] Farran, A.; Ruiz, S. *J. Chromatogr.* A 2004, 1024, 267.

[2] Bonnemoy, F.; Lavédeine, B.; Boulkamh, A. *Chemosphere* 2004, 54, 1183

[3] Salvestrini, S.; Coppola, E.; Capasso, S. *Chemosphere* 2004, 55, 333.

[4] Salvestrini, S.; Coppola, E.; Capasso, S. *FEB* 2004, 13, 233.

[5] Gooddy, D.C.; Chilton, P.J.; Harrison I. *Sci. Total Environ.* 2002, 297, 67.

[6] Sesia, E. Gruppo di Lavoro ANPA-ARPA-APPA. *Fitofarmaci-Elaborazione dati di vendita prodotti fitosanitari.* 2000.

[7] Kallies, B.; Kleinpeter, E.; Koch, A.; Mitzner, R. *J. Mol. Struct.*1997, 435, 123.

[8] Cadiergue, H.; Pèpe, G.; Astier, J.P.; Boistelle, R. *Acta Cryst.*1993, C49, 1078.

[9] Baughman, R:G.; Sams, D.D.; Helland, B.J:; Jacobson, R.A: *Cryst. Struct. Com.* 1980, 9, 885.

[10] Wauchope, R.D.; Buttler, T.M., Hornsby, A.G.; Augustijn-Beckers, P.W.; Burt, J.P. *Rev. Environ. Contam. Toxicol.* 1992, 123, 1.

[11] Berger, B.M. *J. Agric. Food Chem.*1999, 47, 3389.

[12] Cullington, J.E.; Walker, A. *Soil Biology & Biochemistry* 1999, 31, 677.

[13] Ellis, P.A.; Camper, N.D. *J. Environ. Sci. Health* 1982, B17, 277.

[14] Attaway, H.H.; Camper, N.D.; Paynter, M.J.B. *Pestic. Biochem. Physiol.* 1982, 17,96.

[15] Tixier, C.; Bogaerts, P.; Sancelme, M.; Bonnamoy, F.; Landoald, T.; Cuer, A.; Bohatiet, J.; Veschambre, H. *Pest Manag. Sci.* 2000, 56, 455.

[16] Bollag, J.M.; Minard, R:D.; Liu, S.Y. *Environ. Sci. Technol.* 1983, 17, 72.

[17] Simmons, K.E.: Minard, R.D.; Bollag, .M. *Environ. Sci Technol.* 1989, 23, 115.

[18] Adrian, P.; Lahaniatis, E.S.; Andreux, F.; Mansour, M.; Scheunert I.; Korte, F. *Chemosphere* 1989, 18, 1599.

[19] Sabaliûnas, D.; Ellington, J.; Lekevicius, R. *Intern. J. Environ. Anal. Chem.*, 1996, 64, 123.

[20] Williams, A.; Jencks, W.P. *J. Chem. Soc., Perkin Trans. 2* 1974, 1753.

[21] Hegarty, A.F.; Frost, L.N. *J. Chem. Soc., Perkin 2* 1973, 1719.

[22] Laudien, R.; Mitzner, R. *J. Chem. Soc., Perkin Trans. 2*, 2001, 2230.

[23] Laudien, R.; Mitzner, R. *J. Chem. Soc., Perkin Trans. 2*, 2001, 2226

[24] M.I.P.A.F. (Ministero per le Politiche Agricole e Forestali – Osservatorio Nazionale Pedologico e per la Qualità del Suolo) *Metodi di Analisi Chimica del Suolo*, No. 1124.2, Franco Angeli Ed., Milano, Italy 2000.

[25] Connors, K.A. *Chemical Kinetics: The Study of Reaction Rates in Solution,* VCH, New York, USA, 1990, pp 187 – 243.

[26] Carreira, L.A.; Rizk, M.; El-Shabrawy, Y.; Zakhari, N.A.; Hilal, S. *Talanta,* 1996, 43, 607.

[27] Hilal, S.; Karickhoff, S.W.; Carreira, L.A. *Quant. Struc. Act. Rel.*, 1995, 14, 348.

[28] Castro, E.A., Moodie, R.B., Sansom, P.J.,. *J. Chem. Soc. Perkin II* 1985, 737.

[29] S.A. Hand, W.P. Jencks, *J. Am. Chem. Soc.*, 1975, 97, 6221.

[30] R.W. Nagorski, T. Mizerski, J.P. Richard, *J. Am. Chem. Soc.*, 1995, 120, 4718.

[31] Capasso; S; Vergara, A.; Mazzarella, L. *J. Am. Chem. Soc.*, 1998, 117, 1990.
[32] Hibbert, F. *Adv. Phys. Org. Chem.* 1986, 22, 113.
[33] Kirby, A.J. *Adv. Phys. Org. Chem.* 1990, 28, 820.
[34] Isaac, N.S: *Physical Organic Chemistry*; Longman Scientific & Tecnical: New York, 1992, pp 77 – 113.
[35] Salvestrini, S.; Di Cerbo, P.; Capasso, S. *J. Chem. Soc., Perkin Trans 2*, 2002, 1889.
[36] Salvestrini, S., Di Cerbo, P., Capasso, S. *Chemosphere* 2002, 48, 69. 220.
[37] Fouqué-Brouard, C.M.; Fournier J.M. Talanta 1996, 43, 1793.
[38] Fernández- Pérez, M.; González-Pradas, E.; Villafranca-Sánchez, M.; Flores-Céspedes, F. Chemosphere 2000, 41, 1495.
[39] Cox, L.; Walker, A. Chemosphere 1999, 38, 2707.
[40] Makay, D. *Multimedia Environmental Models- The fugacity approach*; Lewis: New York, 1991, pp 67 - 114.
[41] Valsaraj K.T.; Thibodeaux L.J. *Fate of pesticides and chemicals in the environment.* Schnoor J.L. Ed.; John Willey & Sons, New York, 1992 pp 155 – 174.
[42] Veith J.A; Esposito, G. *Soil Sci. Soc. Am. J.* 1977, 41, 497.
[43] Cantor C.R.; Schimmel, P.R. *Biophysical Chemistry*; W.H. Freeman and Company: United States of America, 1980; pp 849-886.
[44] Giardina B.; Amiconi G. *Methods in Enzymology*, vol. 76., Antonini, E.,; Rossi-Berbnard, L.; Chiacone E. Eds. Academic Press: New York, 1981.
[45] Schulter H:R. *Fresenius J. Anal. Chem.*1995, 351, 62.

In: Trends in Agriculture and Soil Pollution Research
Editor: James P. Livingston, pp. 121-144
ISBN: 1-59454-325-9

Chapter 5

ADSORPTION OF TRACE METALS ONTO NATURAL SURFACE COATINGS

Deming Dong**, Fan Yang, Yu Li, Xiuyi Hua, Rong Zhang, Xiaojun Lü, Jingjing Zhang, Botao Zhang, Na Zheng and Liang Liu

College of Environment and Resources, Jilin University, Changchun, 130012, People's Republic of China

ABSTRACT

Surface coatings (biofilms and mineral associated) were developed in natural water to investigate the effect of the physicochemical nature of trace metals, the competition of cations and organic ligands coexistent, along with the solution chemistry (i.e. pH and ionic strength) on trace metals adsorption to the surface coatings. Adsorption of trace metal(s) was measured under controlled laboratory conditions and analyzed using nonlinear least squares fitting and linear regression analysis to estimate the adsorption parameters. It was observed that metals adsorption to the surface coatings followed a decreasing order of Pb, Cd, Cu, Zn and Ba, relating with the characters of the adsorbate, including standard electrode potential, covalent radius and atomic radius, etc. There was significant interference between metals when several cations present in solution, while the adsorption was dramatically restrained for certain metal. Among the metals studied, the competition adsorption between Co and Ni was the most remarkable. The presence of other metals had little effect on the selectivity sequence of metal adsorption. Likewise, certain organiochlorine pesticide in natural aquatic systems act on the adsorption of heavy metals, for Pb, the main influence was competition and restriction; while for Cd the main influence was acceleration with the organic ligands present. The maximum adsorption of Pb and Cd on the surface coatings increased linearly with increase of pH in the solutions significantly at pH range from 4.5 to 7.0, and decreased dramatically with the increase of concentration of $NaNO_3$ as supporting electrolyte, but when the concentration of $NaNO_3$ was higher than 0.5 M, the amount of adsorption kept relatively

* Corresponding author phone: 86-431-5166401; fax: 86-431-5166401; e-mail: dmdong@mail.jlu.edu.cn.

steady, implying the simultaneous existence of electrostatic adsorption and specific adsorption during trace metal adsorption.

Keywords: trace metal, adsorption, natural water, surface coatings.

INTRODUCTION

Trace heavy metals are considered as hazardous pollutants due to their bioaccumulation potential and toxicity even at low concentration. Many researches have been devoted to develop models to describe trace meals behavior and transport in aquatic environments, particularly their adsorption to heterogeneous materials [Dzombak and Morel, 1990; Schindler, 1991; Westall et al., 1995]. The natural surface coatings, including biofilms and mineral deposit associated, consist of various microorganisms, forming on all surfaces of solid phase in rivers, lakes and wetland, and account for a wide range of microbial organisms on earth [Costerton et al., 1987]. Metal oxides and organic materials with high specific surface areas are main components of the coatings. Some researchers have reported that certain metals were accumulated more significantly in biofilms compared to the surface sediments [Schorer and Eisele, 1997; Fara et al., 1998], and Pb, Cd have been found associated strongly with natural surface coatings in some aquatic systems [Dong et al., 2002a, 2003a, b, 2004]. An adsorption additive model as well as a novel selective extraction technique has been employed to identify the adsorption sites of trace metals on different surface coatings by Nelson et al. [1999, 2002] and Dong et al. [2000, 2002a, 2004] respectively, and the predominant contributions of hydrous ferromanganese oxides were certified in the preceding studies.

Metals adsorption to surface coatings is expected to be governed by the composition of the solid phase, and the associative characters of the solution, such as pH [Anderson and Rubin, 1981; Templeton et al., 2003], ionic strength [Achterberg, et al., 2003] and the adsorbed metals [Luoma et al., 1981; Lead et al., 1998]. It is suggested that metals adsorption property to soil is relevant with many parameters, such as pH, valence, ion exchange ability, distribution coefficient, and so on [Paulo et al., 2001].

In this chapter, surface coatings developed in natural water were used to study the effect of some parameters of trace metals, the presence of competitive cations and some organic ligands, as well as the condition of the experiment media (i.e. pH and ionic strength) on trace metals adsorption onto the surface coatings. The purpose of the present work is to determine the contributive factors of metals adsorption to solid phase in aquatic environments, and devote to mechanistic understanding of trace metals association with the heterogeneous surface coatings.

Experimental Section

Development and Characterization of Natural Surface Coatings

One lake and a wetland in Jilin Province, P. R. China were chosen as the aquatic environments to develop natural surface coatings in this research. Nanhu Lake is a eutrophic lake with significant Mn deposition, where most of our field samples were obtained. A reed pond in Xianghai wetland was used to cultivate surface coatings to investigate the effect of pH on Pb and Cd adsorption.

At each aquatic field site, precleaned glass microscope slides (5.0×7.5×0.1 cm) were fixed on polypropylene racks and submerged in the water at a depth of about 30 cm for a period of 10 - 14 days, while in the wetland the sample depth was about 40 cm to avoid considerable evaporation of the water. This method has been used extensively to acquire representative heterogeneous assemblages in surface water [Dong et al., 2000, 2002a, 2003a, b, 2004; Nelson et al., 1999], and previous study showed significant correlations between the contents of some important components (e.g., Fe, Mn and Al etc.) in surface coatings collected using this method and those in the water [Dong et al., 2002b]. After exposure in the field, the glass slides with visible surface coatings presented were transported to the laboratory for microscopic examination, chemical characterization and measurement of Pb, Cd and other metals adsorption.

Organic material in the surface coatings was quantified by measuring the chemical oxygen demand (COD) according to a modification of Standard Method # 5220 B [Dong et al., 2000; APHA, 1995]. Total acid extractable metal concentrations (Fe, Mn, Al) in the surface coatings were determined by extraction with 25 ml of 15 % HNO_3 for 24 h (it was assumed here that Fe, Mn and Al speciation existed as the forms of oxides in surface coatings). Fe and Mn in the extracts were analyzed by FAAS, while Al was measured by ICP-AES. Since insignificant role in binding Pb and Cd etc., Al oxides was provided only for certain reference here [Dong et al., 2000, 2002a, 2004].

Measurement of Metals Adsorption to Surface Coatings

Cd, Pb, Cu, Zn and Ba adsorption to intact surface coatings were performed respectively in solutions with chemically defined metal speciation, with Cd and other metals initial concentrations ranging from 0.2 to 2.5 μM. This concentration of Pb and Cd would have no significant effect on most aquatic microorganism with short-term contact [Carr et al., 1998; Prasad et al., 1998]. The solutions were prepared by dilution of 1000 mg/L metals nitrate using a minimal mineral salts (MMS) solution which contained 200 μM $CaCl_2$, 140 μM $MgSO_4$, 910 μM $(NH_4)_2SO_4$, 150 μM KNO_3 and 10 μM $NaHCO_3$. The ionic strength of the solution was adjusted to 0.05 M with $NaNO_3$ and the pH was adjusted to 6.0 ± 0.1. Such adsorption media was used to minimize possible metals complexation and maintain the activity of organisms in the coatings. The pH of 6.0 was chosen to avoid the possible formation of Cd and Pb precipitates, and favor the comparison with some similar studies [Nelson et al., 1999, 2002; Dong et al., 2000, 2004]. The calculations using MINEQL+ (William Schecher Environmental Research Software, 1994) of Cd and Pb speciation in the solution at 1 μM showed that the free cations comprise 89 % of the total dissolved metals because of low ligand concentrations. For surface coatings collected in each body of water, sets of two slides were placed in polypropylene racks, then submerged into each of five 800-mL solutions contained in 1-L beakers with different metal concentrations. The solutions were stirred continuously with magnetic stirrers for 24 h under a condition of 25 ± 1 □, maintaining the pH at 6.0 ± 0.1 manually using 0.01 M HNO_3 and NaOH. Metals adsorption to blank glass slides was negligible as shown in previous studies [Dong et al., 2000].

After equilibrium, each slide was extracted into 25 mL of 15 % HNO_3 (trace metal grade) for 24 h. Cd, Pb, Cu, Zn and Ba etc as well in the acid extracts were analyzed by FAAS to determine metal(s) adsorbed onto the surface coatings. Cd and other metals concentrations in equilibrium solutions were also measured by GFAAS or FAAS.

With regard to Pb, Cd, Cu, Co and Ni competitive adsorption experiments onto natural surface coatings, the initial concentration level of adsorbed metals ranged from 0.5 to 4.0 μM, and the adsorption was conducted separately and simultaneously with other metal (s) present respectively. Pb, Cd adsorption to surface coatings treated with 0 - 5.0 μg/L organiochlorine pesticide (mixture of α-BHC, β-BHC, γ-BHC, δ-BHC, O, P-DDT, O, P-DDE, P, P-DDD, P, PP-DDT and PCNB) under preceding adsorption condition for 24 h, likewise Pb, Cd and organiochlorine pesticide adsorption were carried out together, so were Pb and Cd adsorption in different pH levels.

When discussing the effect of pH and ionic strength on trace metals adsorption, there were some exceptions that the defined pH level ranged from 4.5 to 7.0, besides of one initialized at 6.0 and not controlled, and the ionic strength ranged from 0 to 1.0 mol/L expressed by the concentrations of $NaNO_3$, respectively, with the operation and other experiment conditions regulated at the identical state with those described above.

Statistical Analyses

The classical Langmuir and Freundlich adsorption isotherms were applied to estimate metal(s) adsorption parameters to natural surface coatings. One model used for the regression analyses was expressed as a Langmuir adsorption isotherm:

$$\Gamma = \frac{\Gamma_{max} \cdot K \cdot [M]}{1 + K \cdot [M]},$$

where Γ is metal adsorption by surface coatings per unit surface area (μmol/m^2) for each metal, Γ_{max} is the maximum adsorption of metal by surface coatings, K is the Langmuir equilibrium coefficient (L/μmol), and [M] is the concertration of metal in solution. Here metal adsorption to the coatings is expressed as per unit nominal surface area of the glass slides containing the surface coatings, not the total surface area of the adsorbing phase. Taking reciprocal both sides of the equation gives a linear expression of Langmuir equation:

$$\frac{1}{\Gamma} = \frac{1}{\Gamma_{max} \cdot K \cdot [M]} + \frac{1}{\Gamma_{max}},$$

correspondingly, $\frac{1}{\Gamma_{max} \cdot K}$ is equal to the slope and $\frac{1}{\Gamma_{max}}$ is equal to the intercept of the line of $\frac{1}{\Gamma}$ vs. $\frac{1}{[M]}$.

Another model used for the regression analyses was Freundlich adsorption isotherm:

$$\Gamma = K_d \cdot [M]^N,$$

where K_d is the Freundlich constant (L/m^2) and N is a measure of the nonlinearity. Taking natural logarithm both sides of the equation gives:

$$\ln\Gamma = N \cdot \ln[M] + \ln K_d.$$

Hence Cd, Pb and other metals adsorption data were analyzed using nonlinear least squares fitting (NLSF) by Microcal Origin (Version 6.0) or linear regression of the equations introduced selectively to estimate the adsorption parameters of metal (s) to surface coatings in different experiments [Dong et al., 2003a, 2004].

RESULTS AND DISCUSSION

Relationship between Metal Properties and Their Adsorption to Natural Surface Coatings

Character and Components Analysis of Surface Coatings

The sage green surface coatings with dominative organisms of zoogloea and some algae, which were developed in Nanhu Lake, were adopted to investigate the variability of adsorption property of Pb, Cd, Cu, Zn and Ba to the freshwater surface coatings [Dong et al., 2003a]. Surface concentrations of important components in the coatings were 146 ± 6 μmol Fe/m^2 for Fe oxides, 44 ± 2 μmol Mn/m^2 for Mn oxides, 240 ± 15 μmol Al/m^2 for Al oxides and 429.27 ± 23.10 mg COD/m^2 for organic material, respectively. Surface coatings developed were consistent from slide to slide (component concentrations varied less than 6 %), allowing the use of different slides for characterization and measurement of Pb and other metals binding.

Pb, Cd, Cu, Zn and Ba Adsorption to Surface Coatings

The experiment data of Pb, Cd, Cu, Zn and Ba adsorption to surface coatings are shown in Figure 1. Estimates Langmuir parameters (Γ_{max} and K) for each metal adsorption were provided using linear regression analysis of equilibrium concentration and adsorption data for surface coatings (Table 1), as well as the curves fitted were depicted in Figure 1.

Table 1. Estimated Langmuir parameters for Pb and other metal adsorption to natural surface coatings based on a linear regression analysis of adsorption*

Parameter	Pb	Cd	Co	Zn	Ba
Γ_{max} (μmol Me/m^2)	40.0	10.8	166.7	29.4	1.8
K (L/μmol)	62.5	1.4	1.5	3.0	0.4
r*	0.9651	0.9875	0.9818	0.9951	0.9983

*The coefficient statistically significant at p=0.01.

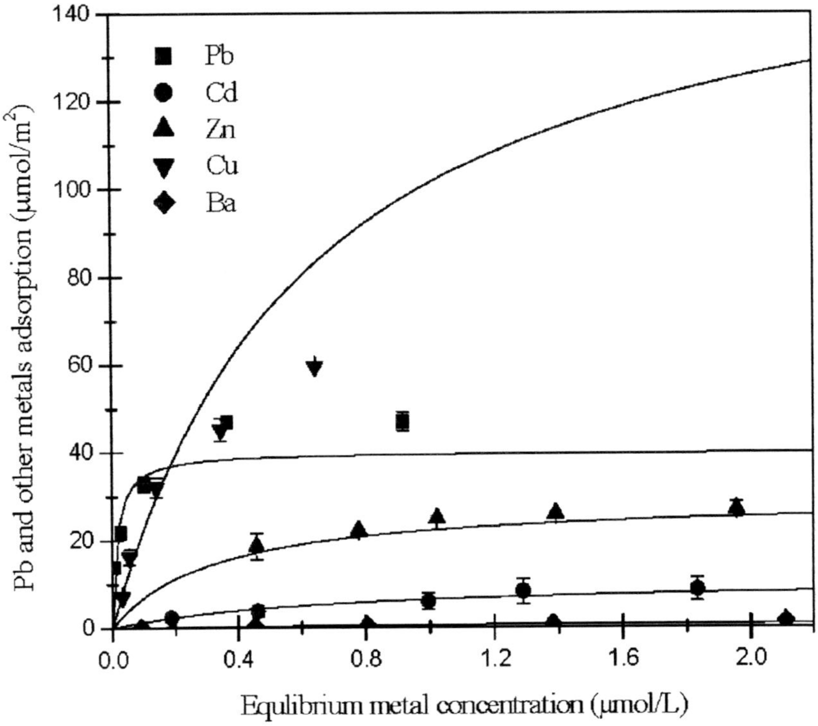

Figure 1. Lead and other metals adsorption to surface coatings developed in Nanhu Lake. Error bars indicate ± one deviation (n=2).

The values of the correlation coefficient (r) showed all metals adsorption to surface coatings at pH 6.0 in MMS solution followed Langmuir adsorption isotherms at the confidence level of p=0.01. Generally, the adsorption of heavy metals to the surface coatings decreased in the order: Cu, Pb, Zn, Cd and Ba (Figure 1). Comparison of the isotherms for Pb and other metals adsorption to the surface coatings manifested that Cu was adsorbed more strongly at an concentration above 0.2 μmol/L metal in solution, while Ba was adsorbed by the coatings least strongly. The curves for Cu and Pb adsorption crossed at lower ion concentrations (< 0.2 μmol/L), therefore, the adsorption of Pb to the surface coatings would exceed that of Cu at lower ion concentrations, and to be the most efficiently adsorbed element among metals studied. And it was suggested that freshwater surface coatings was likely to act as an important sink for Pb, particularly at environmentally relevant low ion concentration.

As presented in Table 1, the estimated Langmuir parameters of Γ_{max} of Pb and other metals adsorption to surface coatings, which were developed in the identical aquatic environments at the same period, varied remarkably. Factors that caused this variability may essentially originate from of nature of metals themselves, and the characters of aquatic environment might have some subordinate influence.

Correlation Analyses between $1/\Gamma_{max}$ and Some Metal Properties

Correlation analyses between $1/\Gamma_{max}$ and some metal properties, as shown in Table 2, reveal that some physicochemical parameters of metals might have notable relation with their adsorptive behavior. There is a tendency for the values of $1/\Gamma_{max}$ to increase linearly with decrease in the values of standard electrode potential of the elements significantly ($n=5$, $p=0.01$). Therefore, it seems reasonable that the standard electrode potential is a primary parameter for the variability of Pb and other metals adsorption to the surface coatings. While the covalent radius and atomic radius are expected to be another two less important factors, as there is also a general trend for the values of $1/\Gamma_{max}$ to increase linearly with increase in the values of both covalent radius and atomic radius at the confidence level of $p=0.05$ and $p=0.10$, respectively (Table 3).

Metal oxides such as iron and manganese oxides in surface coatings play a prominent role in controlling Cd and Pb adsorption to surface coatings as reported in previous studies. In the present discussion, it will be assumed that functional groups such as O and OH may exist on the surfaces of iron and manganese oxides in the layers of surface coatings in aquatic solutions by hydroxylation reaction. One plausible explanation for the difference in adsorption is that Pb and other metal ions might reach the surface of iron and manganese oxides, and adsorb to surface coatings by forming covalent bond with O and OH function. Abilities of Pb and other metal ions, as electron acceptors, decrease with the decreasing in values of the standard electrode potential. On the other hand, with an increasing in values of both covalent radius and atomic radius, it is resulted in decreasing in ions binding with O and OH groups and the amount of lead and other metal adsorption to surface coating.

Therefore the significant decreasing trend of Pb, Cd, Cu, Zn and Ba adsorption to the identical heterogeneous surface coatings in Nanhu lake can be explained, at least partly, by the different characters of the adsorbate, such as their standard electrode potential, covalent radius and atomic radius, etc.

Table 2. Some physicochemical parameters of Pb and other metals studied

Parameter	Pb	Cd	Co	Zn	Ba
Atomic radius (pm)	175	149	128	133	217
Covalent radius (pm)	154	141	117	125	198
Divalent ion radius (pm)	120	97	72	74	135
Hydration heat (kJ/mol)	697	373	83	135	873
Ion charge density *	1.7059	5.3718	6.6546	6.2517	0.3520
Standard electrode potential (V)	-0.1263	-0.4029	0.3370	-0.7620	-2.9060

* Ratio of ion charge and divalent ion radius.

Table 3. Correlation coefficients between $1/\Gamma_{max}$ and some parameters of heavy metals studied

Parameters	Coefficients (r)
Atomic radius - $1/\Gamma_{max}$	0.8296***
Covalent radius - $1/\Gamma_{max}$	0.9087**
Divalent ion radius - $1/\Gamma_{max}$	0.8320
Hydration heat - $1/\Gamma_{max}$	0.7253
Ion charge density - $1/\Gamma_{max}$	0.1897
Standard electrode potential - $1/\Gamma_{max}$	0.9614*

*The coefficient statistically significant at p=0.01.
** The coefficient statistically significant at p=0.05.
*** The coefficient statistically significant at p=0.10.

Competitive Adsorption of Five Metals onto Natural Surface Coatings

Components Analysis of Natural Surface Coatings

Surface coatings developed in Nanhu Lake consisted of assemblages of microorganisms in a biofilm matrix associated with mineral deposits. The concentrations of metal oxides in surface coatings were 254.29 ± 46.23 μmol Fe/m^2 ; 231.28 ± 16.34 μmol Mn/m^2, 346.83 ± 39.42 μmol Al/m^2 and organic material quantified by chemical oxygen demand in the coatings was 160.67 ± 29.37 mg COD /m^2.

Metal Adsorption to Surface Coatings with Other Element(S) Present

The Langmuir adsorption isotherms of five kinds of heavy metal (i.e. Pb, Cd, Cu, Co and Ni) adsorption to natural surface coatings, which were conducted in solutions separately or with random another metal present, as well as five kinds of heavy metal adsorption together, resulted from NLSF are described in Figure 2-6.

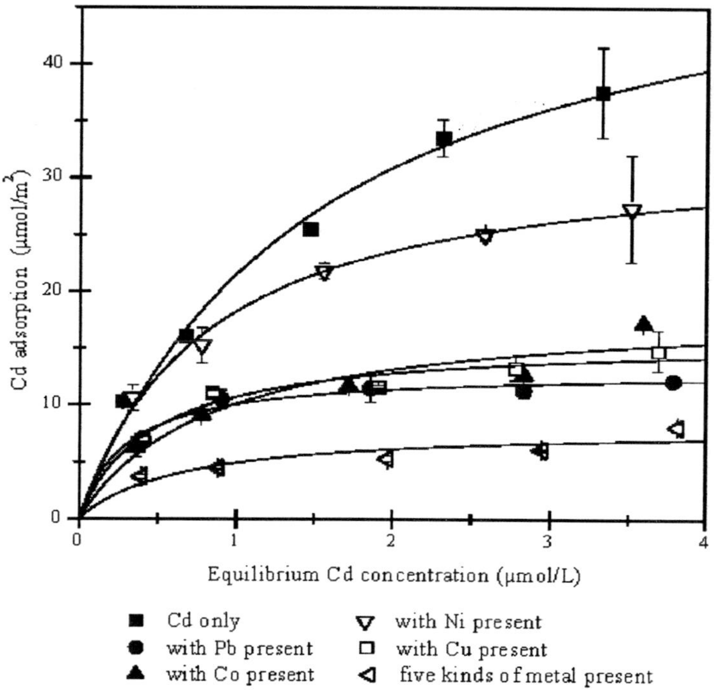

Figure 2. Cd adsorption to natural surface coatings in different solution.

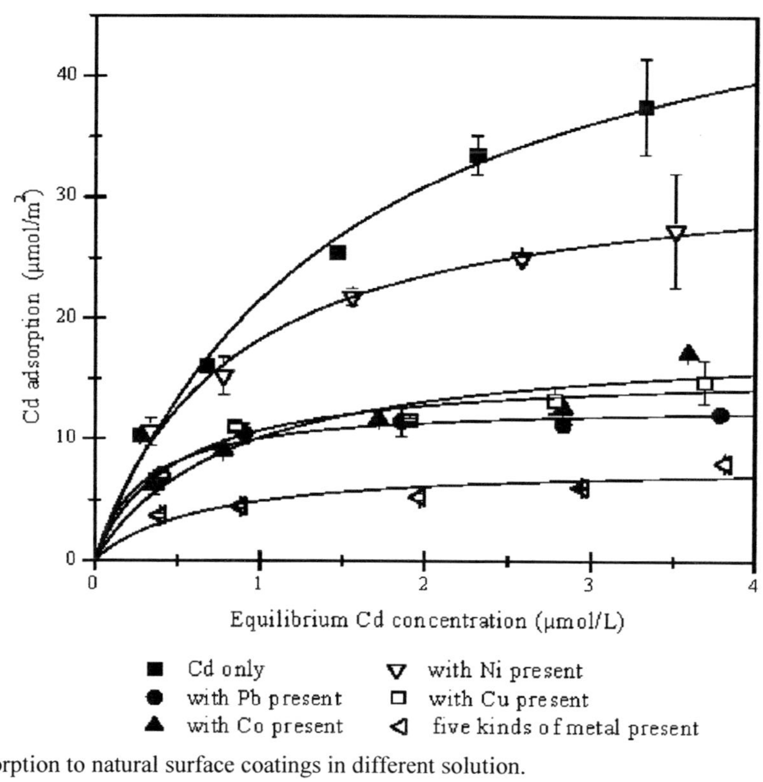

Figure 3. Co adsorption to natural surface coatings in different solution.

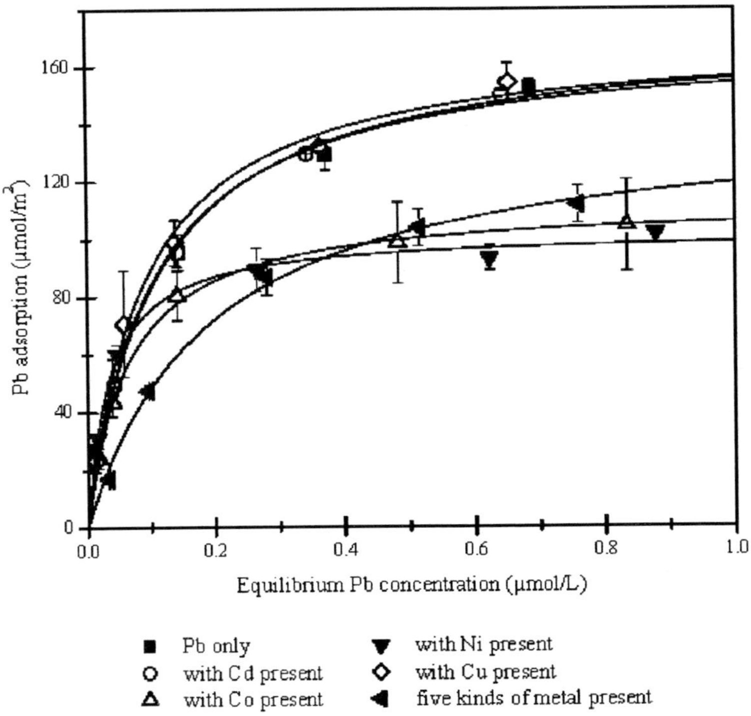

Figure 4. Pb adsorption to natural surface coatings in different solution.

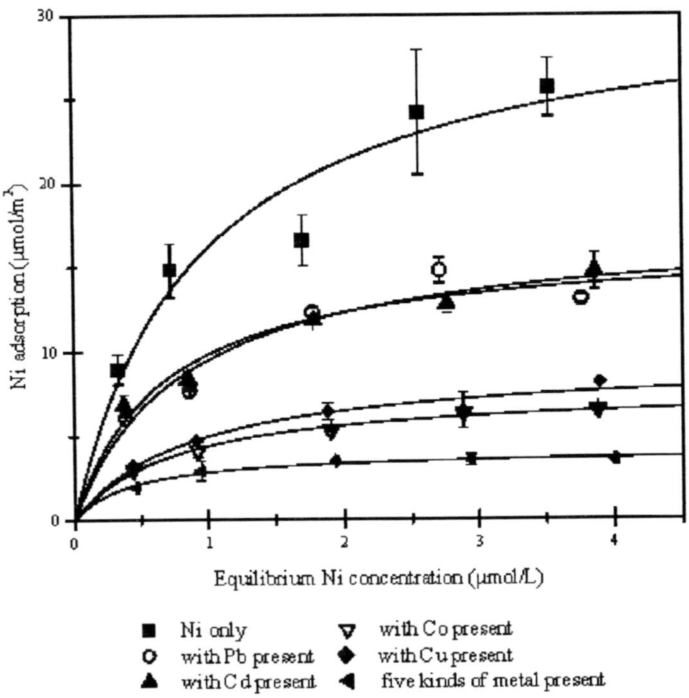

Figure 5. Ni adsorption to natural surface coatings in different solution.

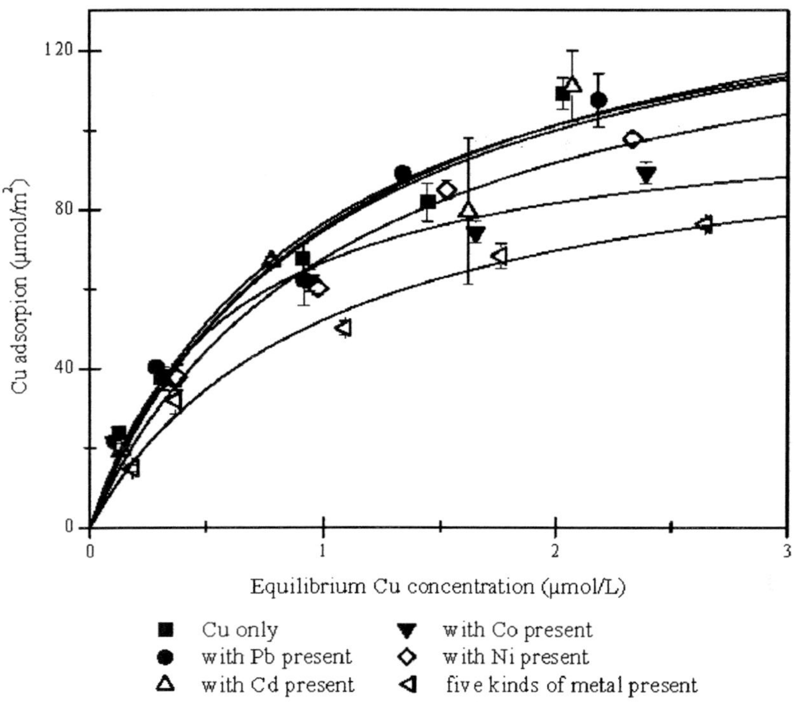

Figure 6. Cu adsorption to natural surface coatings in different solution.

Mutual interfering phenomenon between metals coexisted was observed in the current study. As shown in the figures, for particular heavy metal studied, the competitive reaction of elements existed in certain solution system reduced its amounts of binding to the surface coatings, due to the possible competition to occupy surface adsorption sites on the coatings among all these metals. Generally, the interference of other metals to the individual element adsorption was minor when their concentrations were low, while the intensity of the influence varied increasingly with the increase of the concentrations to different extent. And within a specific concentration range, certain one of the metals could contribute most to the competition, which was related with the nature of the metal, such as its diameter, electrode potential, atomic valence state etc [Sánchez-Martín and Sánchez-Camazano, 1993].

Heavy metal adsorption was interfered differently by various elements. For Cd adsorption to the surface coatings (see Figure 2), the intensity of competition followed the order of Ni < Co < Cu and Pb, with the interference of Pb, Cu and Co were almost the same, implicating the similar adsorptive mechanisms. Regarding Co adsorption (Figure 3), little effect was observed from Cd because of the different dominant adsorbents in the surface coatings of Cd, Co binding [Dong et al., 2004; Nelson et al., 1995] and the slight competition of the adsorption sites. Cu and Pb showed similar moderate interference on Co adsorption, while Ni had the greatest significant influence on Co adsorption, especially in low concentration, which might derive from the proximity of Ni, Co physicochemical characteristic. As for Pb adsorption onto the mixed mineral-organic assemblages (Figure 4), Cd and Cu had little influence on it, as well as the perceptible but insignificant interference of Ni and Co present was obtained. During the process of Ni adsorption to surface coatings (Figure 5), coexistent trace metals had strong influence on Ni adsorption, especially when the five metals existed simultaneously in solution the amounts of Ni adsorption were reduced pronouncedly. The extent of competition adsorption effect of metals on Ni adsorption was in the order Pb, Cd < Cu < Co. Due to the dissimilarity of elements nature, Pb, Cd exhibited approximate slight influence on Ni adsorption. The interaction between Co and Ni were the most severe among all the competition adsorption between random two kinds of metals, since they had similar physicochemical parameters though they could not replaced the other completely. It was plausible that the adsorption mechanisms of Co and Ni might be parallel to certain extent. Concerning Cu binding to the coatings (see Figure 6), the impact of Cd and Pb was insignificant relatively, similar to the case of Cd and Cu impact on Pb adsorption. Hence the adsorption mechanism of Pb and Cu might be homoplastic, both of which had peculiar strong affinity with the components in the surface coatings, therefore other metals would have little influence on their adsorption. Conversely, the complexation of Cd to components of surface coatings was labile, subject to be substituted by other metals. Similar phenomenon was observed by Echeverria et al. [1998] in research of Pb, Cu, Ni, Cd and Zn adsorption to soil.

When the five kinds of metal were present together, metal adsorption was limited and decreased apparently compared to adsorption by itself, and the total interference was higher than that with any other metal coexistent. Nevertheless, it was not the summation of intensity of competition of the other four metals between two of them, yet. Since trace metal adsorption trended to saturate after some time, and the adsorption sites on surface coatings would be occupied by diverse metals simultaneously to restrain the increase of metal adsorption with the rise of cations concentration in solution.

Selectivity of Metal Adsorption onto Surface Coatings

Fitting curves of metal adsorption individually and with five metals coexisted to the surface coatings were summarized and depicted in Figure 7 and Figure 8, respectively. Pb, Cd, Cu, Co and Ni adsorption to surface coatings fitted Langmiur adsorption equation very well, which were analyzed via the NLSF method (see Table 4).

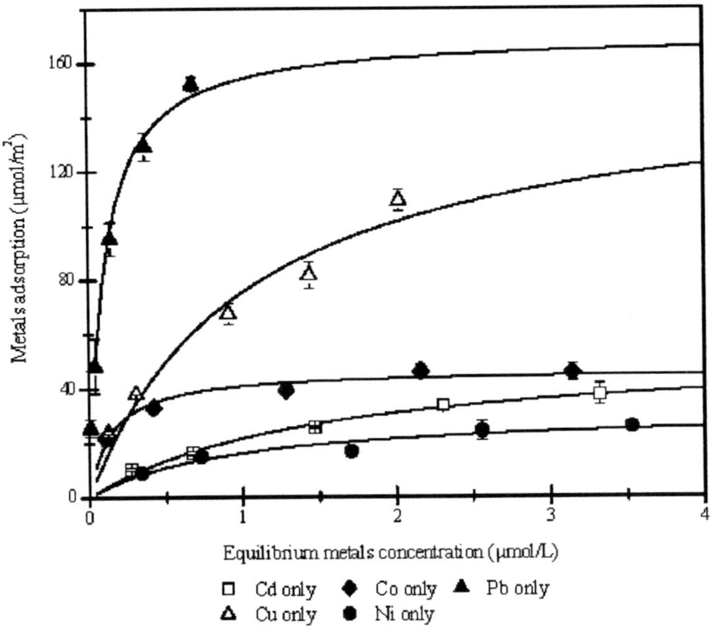

Figure 7. The isotherms of metals adsorption to surface coatings without other metal (s) present.

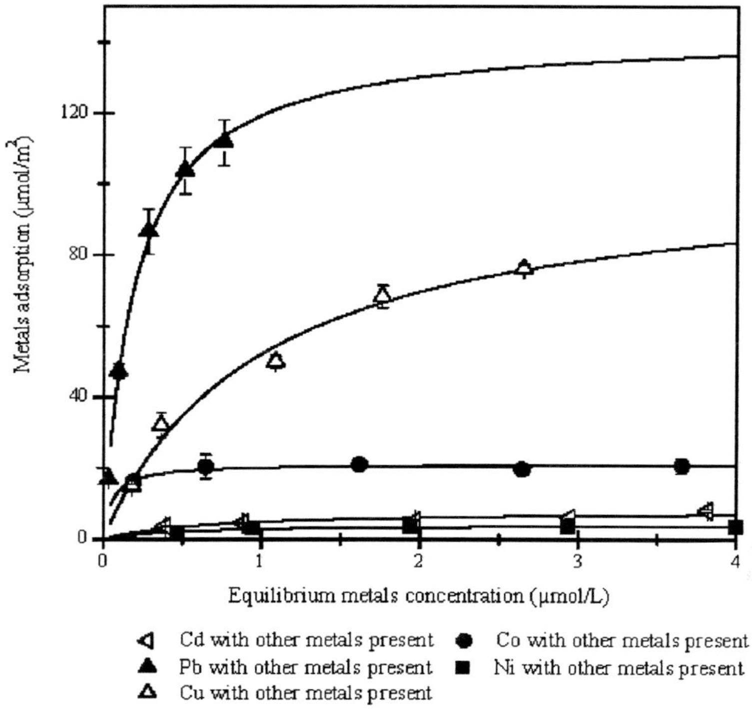

Figure 8. The isotherms of metals simultaneous adsorption to surface coatings

In this study, the affinity between metals and solid phase adsorbents could be represented by the distribution coefficient of K_d, with higher K_d means more remaining on the adsorbents. As shown in Table 5, while Pb existed by itself, its K_d was highest, indicating that Pb had strong affinity with surface coatings; K_d of Cu and Co were a little lower than that of Pb, suggesting their secondarily higher binding abilities; the K_d of Cd and Ni were the lowest relatively, implying that they would be easily replaced by other metals during the process of competitive adsorption. Echeverria et al. [1998] have confirmed that other cations had little influence on the easily adsorbed Pb and Cu, which was quite different from that of Ni, Cd and Zn in soil.

According to Table 5, when five cations coexisted, the K_d of each heavy metal reduced considerably compared with metals existed separately, and the K_d reduced with the increase of original concentration of the metals, on account of several metals coexistence increased the total metals concentration, so as to reduce the K_d of cations because of metals competition of adsorption sites.

Figure 7 and Figure 8 manifested that the five metals adsorption decreased in the order: Pb > Cu > Co > Cd > Ni when they coexisted, resembling that when they were adsorbed alone, which meant that the presence of other metals didn't change the adsorption mechanisms of metals adsorption to surface coatings, but only reduced their adsorption capabilities.

In summary, there was interference took place in solution with two kinds of heavy metal adsorption together and five kinds of heavy metal adsorption together, while the intensity of competition varied with metals including Pb, Cd, Cu, Co and Ni. Among these metals, the competition adsorption between Co and Ni was the most remarkable. When the five kinds of metals coexisted, the total impact from others was reinforced for particular trace metal, due to the mutual interference between metals, but not the summation of competition between two kinds of metals. The selectivity sequence of metal adsorption separately was similar to that of with five kinds of metal present, and the K_d between solid phase and liquid phase reduced with the increase of original concentration.

Table 4. Estimated Langmuir parameters for metal adsorption to the surface coatings separately and with other four metals present

The state of metal		Parameters				
		Γ_{max} ($\mu mol/m^2$)	Fit error	K (L/mol)	Fit error	r
One metal present	Cd	54.81	4.74	0.65	0.13	0.9940*
	Co	55.88	7.13	0.62	0.16	0.9924*
	Pb	170.14	10.42	9.03	2.05	0.9928*
	Ni	31.50	4.60	1.04	0.44	0.9544**
	Cu	154.78	30.48	0.95	0.41	0.9788*
Five metal coexisted	Cd	8.13	1.33	1.52	0.89	0.8684**
	Co	21.11	0.56	8.25	5.35	0.9248**
	Pb	143.05	5.77	5.06	0.60	0.9979*
	Ni	4.07	0.19	2.27	0.47	0.9921*
	Cu	104.77	9.88	0.99	0.23	0.9768*

* The coefficient statistically significant at p=0.01, n=5.

** The coefficient statistically significant at p=0.1, n=5.

Table 5. The distribution coefficients of different metals adsorption to surface coatings

Initial concentration ($\mu mol/L$)		Distribution coefficient of K_d (L/m^2)				
		Pb	Cu	Co	Cd	Ni
One metal present	0.5	2550.00	190.40	204.16	37.37	26.53
	1	1207.50	123.47	78.15	23.64	20.26
	2	657.93	74.09	30.58	17.35	9.72
	3	346.38	56.70	21.32	14.50	9.44
	4	222.16	53.85	14.53	11.30	7.27
Five metals coexisted	0.5	497.06	80.38	82.99	9.34	4.17
	1	488.66	86.95	31.38	5.03	3.06
	2	307.45	45.87	12.93	2.67	1.80
	3	201.55	38.82	7.44	2.06	1.20
	4	147.11	28.82	5.66	2.11	0.88

Table 6. Estimated results of Pb and Cd adsorption to surface coatings treated with organiochlorine pesticide using linear analysis of Langmuir equation

Organiochlorine pesticide concentration (μg/L)	Pb			Cd		
	Γ_{max}	K	r	Γ_{max}	K	r
0	99.01	50.50	0.9961	15.67	4.17	0.9920
0.1	103.09	17.76	0.9277	21.79	2.94	0.9680
0.5	97.09	103.09	0.9833	22.32	3.73	0.9733
2.0	74.07	45.05	0.9822	24.04	2.00	0.9708
5.0	58.48	243.90	0.9638	25.58	1.90	0.9667

Table 7. Estimated Langmuir parameters of Pb, Cd and organiochlorine pesticide simultaneous adsorption to surface coatings

Metal	0 μg/L organiochlorine pesticide present			1.0 μg/L organiochlorine pesticide present		
	Γ_{max}	K	r	Γ_{max}	K	r
Pb	30.3951	14.31	0.9709	28.5714	6.14	0.9718
Cd	20.5761	2.60	0.9827	24.0385	2.32	0.9787

Table 8. Estimated parameters of Pb and Cd adsorption to the surface coatings at different pH

pH	Freundlich parameter			Langmuir parameter		
	K_d	N	r	K	Γ_{max}	r
Pb adsorption						
4.5	147.30	0.6577	0.9784	2.00	192.31	0.9922
5.0	230.07	0.5535	0.9954	5.00	222.22	0.9964
5.5	415.34	0.4808	0.9896	8.00	344.83	0.9957
6.0	731.79	0.4584	0.9862	13.33	625.00	0.9825
6.5	1078.80	0.4249	0.9848	14.50	833.33	0.9814
7.0	1516.71	0.4543	0.9951	16.67	1000.00	0.9921
Cd adsorption						
4.5	8.68	0.2665	0.9943	6.88	22.03	0.9720
5.0	23.02	0.2922	0.9883	6.38	27.47	0.9789
5.5	30.62	0.2919	0.9997	6.02	36.90	0.9775
6.0	31.72	0.5920	0.9952	5.41	54.34	0.9906
6.5	48.88	0.3162	0.9851	5.57	59.88	0.9827
7.0	73.80	0.4607	0.9952	3.78	94.34	0.9769

Effect of Organiochlorine Pesticide on the Adsorption of Pb and Cd to Natural Surface Coatings

Components Analysis of Natural Surface Coatings

Nanhu Lake was selected as the representative water to develop surface coatings. Concentrations of acid extractable Fe oxides and Mn oxides in the surface coatings used to evaluate Pb, Cd adsorption to the surface coatings treated with the organiochlorine pesticide were 447 ± 40 μmol Fe/m^2 (n=15) and 424 ± 32 μmol Mn/m^2 (n=15), and the concentrations of those used to exam the competitive adsorption between organiochlorine pesticide and Pb, Cd were 517 ± 46 μmol Fe/m^2 (n=15) and 330 ± 28 μmol Mn/m^2 (n=15). Because of the variation of the water temperature, there was some difference between the contents of metal oxides in the surface coatings while the concentration of Fe oxides was higher than that of Mn oxides in the natural coatings developed at the same period.

Pb and Cd Adsorption Characteristics of Surface Coatings Treated with Organiochlorine Pesticide

The fitting results of Langmuir adsorption isotherms of Pb and Cd adsorption to surface coatings treated with the organiochlorine pesticide were illustrated in Table 6. As shown, the rules of Pb and Cd adsorption to surface coatings treated with the organiochlorine were as follows: as to Pb, while the concentration of the organiochlorine pesticide was low, the maximum adsorption of Pb to treated surface coatings was similar to Pb adsorption to those not treated with the organiochlorine pesticide, and with the increase of the concentration of organiochlorine pesticide, the maximum adsorption of Pb to surface coatings decreased gradually, which revealed that the organiochlorine pesticide reduced Pb adsorption ability of the coatings. Since surface coatings could adsorb the organiochlorine pesticide easily, while little Pb would complex with organic materials, there was severe adsorption competition between Pb and organiochlorine pesticide to natural coatings. So the organiochlorine pesticide took up many adsorption sites to restrain Pb adsorption to the coatings. In contrast, as to Cd, when the concentration of organiochlorine pesticide was low, the maximum adsorption of Cd to surface coatings treated with the organiochlorine pesticide was proximate to those not treated. Nevertheless with the increase of organiochlorine pesticide concentration, the maximum adsorption of Cd to natural coatings increased gradually. That was because that Cd could chelate with organic polymer easily or form organic Cd with the organic materials, and the presence of organiochlorine macromolecule, which had lower solubility and higher adsorption ability, improved Cd adsorption to the coatings. Wherefore Cd would be adsorbed to surface coatings obviously, particularly when the adsorbents adsorbed nonionic macromolecule organic polymer by means of molecule adsorption. However, the variation of Cd adsorption ability to natural coatings with the increase of organiochlorine pesticide concentration was relatively less than that of Pb adsorption.

Competitive Adsorption Characteristics between Pb, Cd and Organiochlorine Pesticide to Surface Coatings

Table 7 illustrated the fitting results by Langmuir adsorption isotherms of Pb, Cd and organiochlorine pesticide simultaneous adsorption to natural coatings. As referred, when they coexisted in the solution, the organiochlorine pesticide restrained Pb adsorption capability to natural coatings, while for Cd, the presence of organiochlorine pesticide improved Cd adsorption capability to surface coatings, which proved the founding about the interference adsorption between Pb, Cd and organiochlorine pesticide. Considering its low concentration in natural aquatic environment (about 0.2-0.5 μg/L) [Lang, 1998], compared with Pb, organiochlorine pesticide remained in the environment would contribute a lot to the transference and transformation of Cd.

In all, certain organiochlorine pesticide in natural aquatic systems could affect the adsorption of heavy metals onto natural surface coatings, for some cations, e.g. Pb, the main influence was competition and restriction; while for other cations, e.g. Cd, apparently intense acceleration was found with the organic ligands present.

Effect of pH on Representative Metals Adsorption to Natural Surface Coatings

Contents of Main Components on Natural Surface Coatings

surface coatings used to study the effect of pH on Pb and Cd adsorption were cultivated in a reed pond in Xianghai wetland, and the contents of main components were 73.02 ± 4.66 μmol Fe/m^2 for Fe oxides, 13.06 ± 0.63 μmol Mn/m^2 for Mn oxides and 475.21 ± 38.82 mg COD/m^2 for organic materials.

Relationship between Solution pH and Time of Pb and Cd Adsorption to Surface Coatings

The pH of adsorption solution initialized at 6.0 varied during the proceeding of Pb and Cd adsorption to surface coatings, which were drawn in Figure 9. There were three stages for both Pb and Cd adsorption: first, in the adsorption stage the pH of solution increased rapidly; during the time of 50 - 200 min, the solution pH increased slowly, indicating the occurrence of adsorption-desorption stage; after 200 min, the solution pH decreased slightly, tending towards the stable status with little desorption observed, and Pb and Cd adsorption got to equilibrium condition gradually.

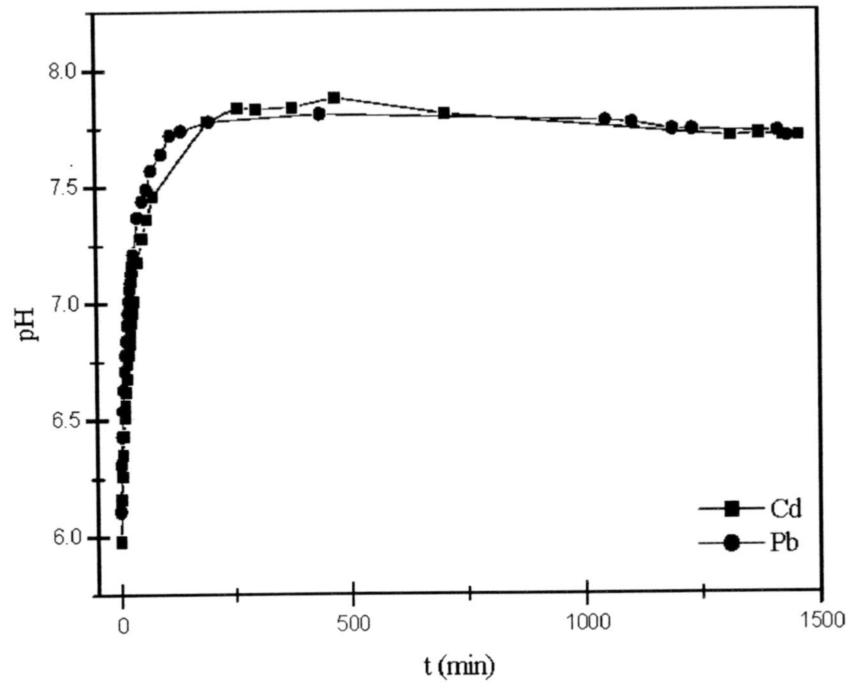

Figure 9. Relationship between values of pH in solutions and adsorption time.

According to the surface binding pattern of oxides proposed by Stumm et al. [1981], the adsorption of H^+, OH^-, metal cations and anions on oxides surface were considered as surface binding reactions. The ≡MeOH groups on the surface of metal oxide, which were formed by the binding behavior of water to the unsaturated surface of metal oxides via dissociation and adsorption of water, could react with metal cations and anions to form surface bindings, hence exhibiting amphoteric characters and corresponding electric charges variation. These reactions could be described as the following:

$$\equiv MeOH + Me^{z+} \rightleftharpoons \equiv MeOM^{(z-1)+} + H^+$$

$$2 \equiv MeOH + Me^{z+} \rightleftharpoons (\equiv MeO)_2M^{(z-2)+} + 2\,H^+$$

$$\equiv MeOH + A^{z-} \rightleftharpoons \equiv MeA^{(z-1)-} + OH^-$$

$$2 \equiv MeOH + A^{z-} \rightleftharpoons (\equiv Me)_2A^{(z-1)-} + 2\,OH^-$$

It can be deduced that the exchange reaction of surface coatings with anions in the solution happened simultaneously with the heavy metals adsorption to the coatings: in the first stage of adsorption, the anions adsorption rate of natural surface coatings was much faster than cations, which made the solution pH increase rapidly; in the second stage of adsorption-desorption, the binding rate of anions was a little faster than that of cations, so the solution pH increased relatively slowly; finally, in the adsorption equilibrium stage, with the slight desorption occurring, the desorption rate of anions on the natural coatings was a little faster than that of cations, resulting in the decline of the solution pH. When the adsorption of cations and anions in the solution came to equilibrium, the pH became stable.

Effects of pH on Pb and Cd Adsorption to Natural Surface Coatings

The results of Pb and Cd adsorption at different pH to natural coatings developed in the wetland calculated using Langmuir and Freundlich equation were illustrated in Table 8. As shown in the table, the data of Pb and Cd adsorption to the surface coatings fitted Freundlich and Langmuir isotherms very well (n=5, p=0.01). In both of the isotherms, the estimated Pb adsorption capability to the surface coatings was one order of magnitude greater than that of Cd (K_d and Γ_{max}).

In this study, there was a systematic increase in Pb and Cd adsorption parameter (K_d and Γ_{max}) with increasing of pH ranged from 4.5 to 7.0 in the adsorption media. The statistical analyses between solution pH and the maximum Pb and Cd adsorption revealed significant linear correlation: the correlation equation of Pb was Y=351.54x-1458.10, and the equation of Cd was Y=27.21x-107.32 (Y represented the maximum adsorption of Pb or Cd, x represented solution pH and the correlation coefficients were 0.9785 and 0.9567, respectively).

The phenomena of the maximum adsorption of Pb and Cd on the surface coatings increased with increase of solution pH are explained here extensively. On one hand, because the metal oxides and organic materials possess some positive charges, when the solution pH is low, considerable positive charges make it difficult to adsorb metal ions. The amount of charges equals to zero when pH come to a critical value, exceed which the negative charges increase with the increasing pH, leading to the higher trace metals adsorption capacities. On the other hand, the higher pH is favorable for the hydrolysis reaction of metal cations or the formation of hydroxyl complex, alluring more ions adsorbed to the surface of the adsorbents by Coulomb force and short-term gravitation. When pH exceeds some certain critical value, the main reaction in the solution is precipitation. Moreover, the media pH has greater influence on Pb adsorption than that of Cd to natural coatings, which still need further study.

On the whole, both Freundlich and Langmuir isotherms could be used to described Pb and Cd adsorption to the surface coatings developed in aquatic environments of a wetland. Pb adsorption capability to the surface coatings was greater than that of Cd, which was similar to previous research [Dong et al., 2000], and the maximum adsorption of Pb and Cd on the surface coatings increased linearly with increase of pH in the solutions significantly.

Effects of Ionic Strength on Pb and Cd Adsorption to Natural Surface Coatings

Contents Analysis of Main Components in Surface Coatings

In the research of Cd and Pb adsorption thermodynamics in solution with different ionic strengths, the average contents of Fe oxides, Mn oxides and organic matters in surface coatings were 111.98 ± 9.49 μmol Fe/m^2 (n=10), 47.98 ± 4.93 μmol Mn/m^2 (n=10) and 583.00 ± 53.65 mg COD/m^2 (n=3), respectively.

Pb and Cd Adsorptive Thermodynamics of Surface Coatings under Different Ionic Strength

In these experiments, the ionic strengths of the solution were adjusted with $NaNO_3$ as supporting electrolyte. Figure 10 and Figure 11 illustrated the experiment data and the fitting curves by nonlinear analysis of Langmuir adsorption isotherms of Pb and Cd adsorption under different ionic strengths to natural surface coatings. The adsorption parameters and corresponding correlation coefficients of the fitting were shown in Table 9. From the figures and the table, Langmuir isotherm was suitable to describe Pb and Cd adsorption to the surface coatings under a series of ionic strengths involved here (n=5, p=0.01).

From the parameters and plot provided in Table 9, Figure 10 and Figure 11, while the concentration of supporting electrolyte was less than 0.5 M, the maximum adsorption amounts of Pb and Cd onto the surface coatings decreased rapidly with the increase of ionic strength; while the concentration of supporting electrolyte was more than 0.5 M, Pb and Cd adsorption onto the surface coatings varied mildly, trending to be stable.

Table 9. Estimated Langmuir parameters for Pb and Cd adsorption to the surface coatings based on a non-linear regression analysis at different ionic strengths

Concentration of $NaNO_3$ (mol/L)	Pb adsorption			Cd adsorption		
	Γ_{max} (μmol/m^2)	K (L/μmol)	r	Γ_{max} (μmol/m^2)	K (L/μmol)	r
0.0	110.94	1.2498	0.9956	15.04	0.7871	0.9778
0.1	96.54	1.0811	0.9972	11.27	0.8251	0.9639
0.3	64.68	0.7556	0.9432	8.59	0.5400	0.9503
0.5	49.36	0.8828	0.9808	6.56	0.3179	0.9894
1.0	46.39	0.9427	0.9550	6.25	0.5636	0.9645

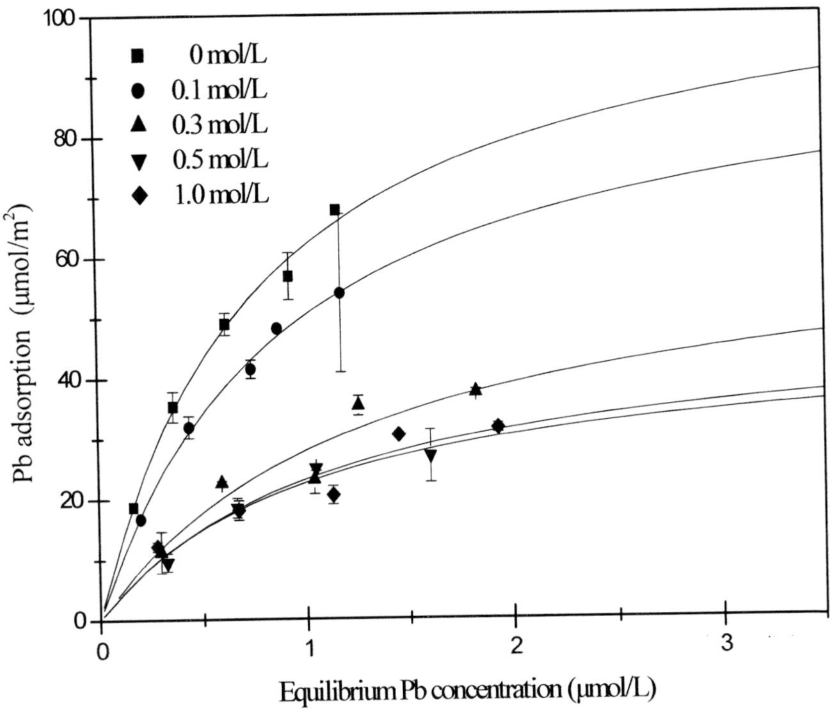

Figure 10. Pb adsorption to surface coatings at different $NaNO_3$ concentrations.

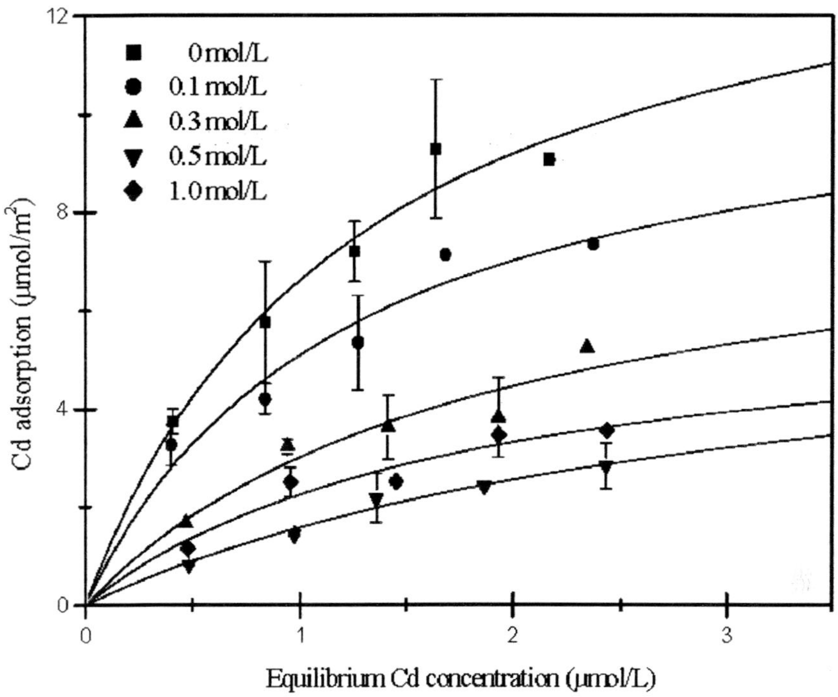

Figure 11. Cd adsorption to surface coatings at different $NaNO_3$ concentrations.

Effects of Ionic Strength on Pb and Cd Adsorption to Natural Surface Coatings

Stumm suggested that the effect of ionic strength on ions adsorption was a simple and convenient way to distinguish electrostatic adsorption and specific adsorption [1992]. Ionic strength has strong effects on electrostatic adsorption. As to the cations, which have slight exchange affinity, if substantive in solution they could displace the cations with lower concentration but higher exchange affinity. If the ionic strength is high enough, cations electrostatic adsorption could be ignored completely [Chen et al., 1987]. However, the specific adsorption will still exist when the adsorbents have no charge, even when they have some positive charges in some case.

Therefore, it was indicated that the electrostatic adsorption and specific adsorption coexisted during Pb and Cd adsorption to natural surface coatings. Correspondingly, with the increase in concentration of supporting electrolyte (e.g. < 0.5 M), lots of the negative charge sites on the surface of natural coatings were taken up by Na^+ to reduce the electrostatic adsorption distinctively; while the insignificant change of Pb and Cd adsorption under ionic strength above 0.5 mol $NaNO_3$/L suggested that the adsorption of cations were saturated by electrostatic adsorption, and the adsorption of Pb and Cd came to the stage of specific adsorption. The adsorption amounts of Pb and Cd to natural coatings reduced slightly in the gentle part, which was probably because that the increase of ionic strength resulted in the reinforcement of the reciprocity force between the ions, the reduction of ionic activity and the decrease of the efficient concentration, leading to the decrease of Cd and Pb adsorption to the surface coatings.

It was also found that there was evident competition between Pb and Cd adsorption to natural coatings when Pb and Cd adsorption was performed simultaneously, which might due to the similarity of their adsorption sites or the adsorption mechanism. Therefore during the course of adsorption, Pb with higher binding ability won the ascendant in the competition to take up most of the adsorption sites. For this reason, Cd present had little effects on Pb adsorption, whereas Pb present had great influence on Cd adsorption to natural coatings. Moreover, when the ionic strength was lower, because of the coexistence of electrostatic adsorption and specific adsorption, Pb had greater effects on Cd adsorption to natural surface coatings, and the effect increased with the increase of ionic strength.

In one word, the results showed that the amount of Pb and Cd adsorption to surface coatings dropped dramatically with the increase of concentration of $NaNO_3$ as supporting electrolyte (0 - 0.5 M), but when the concentration of $NaNO_3$ was higher than 0.5 M, the amount of adsorption kept relatively steady, which implied the possible simultaneous existence of the electrostatic adsorption and specific adsorption during trace metal adsorption.

CONCLUSION

Taking into account all these discussed above, the conclusion may be reached that adsorption of trace metals to natural surface coatings developed in aquatic systems is influenced by a series of factors besides of the dominant components of the coatings, such as the internal characteristics of the adsorbate, the composition and conditions of the adsorption medium etc.

It was founded in this study that the maximum adsorption of trace metals to natural surface coatings was in connection with the standard electrode potential, covalent radius and atomic radius of trace metals, while the particular physicochemical nature of Pb led to its more active transport behavior in water. Both of organic and inorganic factors involved in the solution interfered trace metals adsorption significantly. Coexistent cations, which competed with each other, restrained trace metal adsorption onto surface coatings, while some organic ligands such as organiochlorine pesticide present in the water should decrease or increase trace metals adsorption by means of competition or coordinate reaction with the trace metal, respectively. Electrostatic adsorption and specific adsorption took place simultaneously during trace metal adsorption, consequently Pb and Cd adsorption to the surface coatings increased with increase of pH in the solutions, and decreased with the increase of ionic strength in the experimental solution system.

This chapter explores certain conditions implemental for trace metals adsorption onto natural surface coatings, and the resulted information is conducive to determine favorable parameters of surface coatings behavior in determination of transportation and bioavailability of trace metals in natural waters in further research.

ACKNOWLEDGMENTS

This research was supported by the National Natural Science Foundation of China (NSFC, No. 20077011) and the Specialized Research Fund for the Doctoral Program of Higher Education (SRFDP, No. 20020183056) provided by the Ministry of Eduction of P.R. China.

REFERENCES

[1] Achterberg, E. P.; Herzl, V. M. C.; Braungardt, C. B.; Millward, G. E. *Environ. Pollut.* 2003, *121*, 283-292.

[2] Anderson, M. A.; Rubin, A. J. *Ann. Arbor., Mich.: Ann. Arbor. Science.* 1981, 1-357.

[3] Carr, H. P.; Carino, F. A.; Yang, M. S.; Wong, M. H. *Bull. Environ. Contam. Toxicol.* 1998, *60*, 433-440.

[4] Chen, J. *Aquatic chemistry.* Beijing, China: Higher Education Press, 1987, 135-143.

[5] Costerton, J. W.; Chen, K. J.; Geesey, G. G.; Ladd, T. I.; Nickel, J. C.; Dasguta, M.; Marrie, T. J. *Ann. Rev. Microbil.* 1987, *41*, 435-464.

[6] Dong, D.; Derry, L. A.; Lion, L. W. *Water Res.* 2003, *37*, 1662-1666.

[7] Dong, D.; Hua, X.; Li, Y.; Li, Z. *Environ. Pollut.* 2002a, *119*, 317-321.

[8] Dong, D.; Li, Y.; Hua, X.; Zhang, J. *Chem. J. Chinese Univ.* 2002b, *23*, 1507-1509.

[9] Dong, D.; Li, Y.; Zhang, J.; Hua, X. *Chemosphere.* 2003a, *51*, 369-373.

[10] Dong, D.; Derry, L. A.; Lion, W. L. *Water Res.* 2003b, *37*, 1662-1666.

[11] Dong, D.; Nelson, Y. M.; Lion, L. W.; Shuler, M. L.; Ghiorse, W. C. *Water Res.* 2000, *34*, 427-436.

[12] Dzombak, D. A.; Morel, F. M. M. *Surface complexation modeling. Hydrous ferric oxides.* New York: John Wiley, 1990.

[13] Echeverria, J. C.; Morera, M. T.; Mazkiaran, C.; Garrido, J. J. *Environ. Pollut.* 1998, *101*, 275–284.

[14] Farag; A. M.; Woodward, D. F.; Goldstein, J. M.; Brumbaugh, W.; Meyer, J. S. *Arch. Environ. Contam. Toxicol.* 1998, *34*, 119–127.

[15] Lang, P. *Variability and toxicity of organic material in Songhua Rive.* Changchun, China: Jilin Science and Technology Press, 1998, 1-9.

[16] Lead, J. R.; Hamilton–taylor, J.; Davison, W. *Sci. Total Environ.* 1998, *209*, 193-199.

[17] Luoma, S. N.; Bryan, G. W. *Sci. Total Environ.* 1981, *17*, 165-196.

[18] Nelson, Y. M.; Lion, L. W.; Shuler, M. L.; Ghiorse, W. C. *Environ. Sci. Technol.* 2002, *36*, 421-425.

[19] Nelson, Y. M.; Lion, L. W.; Shuler, M. L.; Ghiorse, W. C. *Limnol. Oceanog.* 1999, *44*, 1715-1729.

[20] Nelson, Y. M.; Lo, W.; Lion, L. W.; Shuler, M. L.; Ghiorse, W. C. *Water Res.* 1995, *29*, 1934-1944.

[21] Paulo, C. G.; Mauricio, P. F. F.; Aderbal G. S.; Eduardo, S. M.; André, R. N. *Soil Sci. Soc. Am. J.* 2001, *65*, 1115-1121.

[22] Prasad, M. N.; Drej, K.; Skawinska, A.; Stratka, K. *Bull. Environ. Contam. Toxicol.* 1998, *60*, 306-311.

[23] Sánchez-martín, M. J.; Sa´nchez-Camazano, M. *J. Environ. Qual.* 1993, *22*, 737–742.

[24] Schindler, P. W. In: Vernet, J. P. *Heavy Metals in the Environment.* Amsterdam: Elsevier, 1991, 95-123.

[25] Schorer, M.; Eisele, M. *Water, Air, Soil Pollu.* 1997, *99*, 651–659.

[26] *Standard Methods for the Examination of Water and Wastewater*; APHA. AWWA, WEF: Washington, DC, 1995.

[27] Stumm, W.; Morgan, J. J. *Aquatic Chemistry.* New York: John Wiley & Son. Inc., 1981, 249-502.

[28] Stumm, W. *Chemistry of the Solid-water Interface*. New York: John wiley, 1992, 23.
[29] Templeton, A. S.; Spormann, A. M.; Brown, G. E., Jr. *Environ. Sci. Technol.* 2003, *37*, 300–307.
[30] Westall, J. C.; Jones, J. D.; Turner, G. D.; Zachara, J. M. *Environ. Sci. Technol.* 1995, *29*, 951-959.

In: Trends in Agriculture and Soil Pollution Research
Editor: James P. Livingston, pp. 145-163
ISBN: 1-59454-325-9

Chapter 6

DESORPTION OF LEAD AND CESIUM FROM KAOLINITE AND ILLITE USING CATIONIC SURFACTANTS

Zhaohui Li[1,*], Christine D. Reardon[1] and Christine V. Evans[1]
[1]Department of Geosciences, University of Wisconsin - Parkside,
900 Wood Road, Kenosha, WI 53141

ABSTRACT

Clay minerals have large surfaces and high affinity for transition metals, making the removal of sorbed metals from clays more difficult. This study focused on exploring the possibility of using cationic surfactant as desorbing agents to remove Cs^+ and Pb(II) from kaolinite and illite. The Cs^+ and Pb(II) sorption capacities were determined to be 14 and 5 mmol/kg on kaolinite and 24 and 87 mmol/kg on illite, respectively. About 90 to 100% of the adsorbed Pb(II) on kaolinite at the loading levels corresponding to 6, 32, and 76% of its Pb(II) sorption capacity can be readily removed using alkylammonium surfactant solution at concentrations of 6 to 10 mM. In contrast, essentially no Pb(II) removal from illite at the loading levels corresponding to 3, 17, and 77% of its Pb(II) sorption capacity was achieved at initial surfactant concentrations of 2 to 30 mM. At a fixed surfactant concentration, the percentage of Pb(II) removal from kaolinite increased as the chain length of the surfactant increased from C12 to C16, reflecting a stronger affinity for clay surfaces with a longer chain length and thus, a better Pb(II) desorption. FTIR spectra showed asymmetric and symmetric C-C stretching vibration of alkylammonium chain on kaolinite, confirming the affinity of alkylammonium for kaolinite. A higher percentage of Pb(II) desorption was achieved when initial Pb(II) loading was high and solution pH was low. As the solution pH increased to 11, Pb(II) removal efficiency reached zero. Added NaCl had less effects on Pb(II) desorption from kaolinite, but a higher Pb(II) desorption from illite was observed. A liquid to solid ratio of 7 to 10 is necessary to remove 80% of sorbed Cs^+ and Pb(II) from kaolinite. Column leaching tests confirmed the batch results

---

* Corresponding author, Phone: (262) 595-2487, Fax: (262) 595-2056, E-mail: li@uwp.edu

that alkylammonium ion is a better desorbing agent to remove Cs^+ and Pb(II) from kaolinite than from illite. The results from this study indicated that different binding mechanisms attributed to Cs^+ and Pb(II) sorption on and desorption from kaolinite and illite. Thus, cationic surfactants may be a good candidate for Cs^+ and Pb(II) removal from kaolinite but not from illite.

Keywords: Alkylammonium, Cs^+, Pb(II), desorption, illite, kaolinite, sorption.

1. INTRODUCTION

Lead intake by children of six or under contributes significantly to their health and intelligence development. Due to lead mining, smelting and use of lead-based paints, soils contaminated with lead impose a great threat to the environment. Lead is relatively immobile due to its low aqueous solubility and great affinity for soil surfaces. Thus, remediating lead-contaminated soil presents a great challenge. ^{137}Cs and ^{135}Cs are fission products typically present in nuclear wastes. Because of their long half-lives and high solubility, these radionuclides are considered among the most hazardous in the spent nuclear fuel (de Llano et al., 1998).

Since kaolinite, illite, and montmorillonite are the major components of different soils, Detailed studies have been conducted to investigate the sorption and desorption of lead by these clay minerals. Lead sorption on these minerals followed the order: Mn-oxides > montmorillonite > kaolinite > Fe-oxides > illite (Kabata-Pendias, 1980). Tests on lead sorption and desorption from three soils of different types showed that the soil containing significant amount of montmorillonite had the highest amount of lead sorption Pardo (2000). Addition of nitrilotriacetic acid (NTA) to the soil suspension decreased metal sorption on the three soils by roughly 50% (Pardo, 2000). In studying the lead retention by kaolinite, Coles and Yong (2002) found that the pH of the kaolinite suspension had little influence on Pb(II) uptake. Sorption of Pb(II) was essentially the same at pH 4 and 6 with the sorption capacity around 12 mmol/kg for a kaolinite whose cation exchange capacity (CEC) is 20 meq/kg. Businelli et al. (2003) showed that the adsorption capability of different types of sorbents followed: illite > montmorillonite > kaolinite > goethite, and the desorption isotherms demonstrate the irreversibility of the bonds formed.

The Cs^+ sorption on kaolinite is a reversible process as revealed from the similarity between Cs^+ sorption and desorption (Erten et al. 1988). Sorption of Cs^+ on illite was attributed to three types of sorption sites (Type I, II, and III) based on Cs^+ selectivity coefficient (K_c), and these sites corresponded to 0.5 %, 3 %, and 96.5 % of the illite's CEC with the former exhibiting an exceedingly high selectivity for Cs^+ (Brouwer et al. 1983). These three sites were denoted as frayed edge, type II, and planar sites and were used to describe the concentration dependent uptake of Cs^+ on natural argillaceous rock systems (Bradbury and Baeyens, 2000). Equal amount of Cs^+ sorption on kaolinite accompanied with cation desorption suggested that ion exchange was the dominant mechanism (Shahwan et al. 1999). Soil polymers, often macro organic molecules, could cause a decrease in Cs^+ sorption and contribute to the unexpectedly high bioavailability of Cs^+ in organic soils (Dumat and Staunton, 1999). However, the interference of macro organic molecules on decreased Cs^+

sorption may present a way of using such molecules to selectively remove Cs^+ from contaminated soil.

The influence of chelating agents or other present organic acids on heavy metal sorption and desorption on soils and clays have been studied extensively. Studies of the influence of three organic acids, p-hydroxybenzoic acid, o-toluic acid and 2,4 dinitrophenol on aqueous and surface complexation reactions between kaolinite and Pb(II) showed that aqueous lead-organic complexation reduced sorption of Pb(II) by kaolinite and thus the subsurface transport of Pb(II) may be enhanced via complex interactions with organic wastes as well as their degradation products (Puls et al. 1991). Pb(II) extraction efficiencies from fractions of carbonate, Fe/Mn oxides, humic acid, and clays containing 50% montmorillonite and 50% kaolinite were 50, 90, 85, and 100% using 100 mM ethylenediaminetetraacetic acid (EDTA) solution at a liquid to solid ratio of 20 (Lo and Yang, 1997). The presence of fulvic acid in the landfill leachate resulted in enhanced Pb(II) sorption by kaolinite below pH 5 (Schroth and Sposito, 1998). Batch and dynamic leaching experiments to evaluate the Pb(II) sorption and subsequent extraction using EDTA and carboxymethyl β-cyclodextrin (CMCD) indicated that EDTA and CMCD were effective in removing Pb(II) from surface soil, but the Pb(II) extraction efficiency was drastically reduced for subsurface soil, in which significant amount of Al^{3+} and Fe^{3+} were present since both EDTA and CMCD complex more strongly with Al^{3+} and Fe^{3+} compared with Pb(II) (Vulava and Seaman, 2000). Removal of Pb(II) from bentonite using L-cysteine and L-penicillamine demonstrated that at a molar Pb(II) to penicillamine ratio of 1:20, 30% of Pb(II) could be removed in 2 hours. However, the percentage of Pb(II) removal became zero at 192 hours, which was attributed to a fast oxidation of the amino acids to their corresponding disulfide dimers, and thus forming insoluble precipitates of Pb(II) disulfide complexes (Fischer, 2002), which may limit the use of these chelating agents for soil washing. NTA was also effective in removing Cu, Pb, and Zn from soils, but it may not be suitable for soil remediation due to its Class II carcinogen (Peters, 1999).

Parallel to the study of using chelating agents to remove heavy metals, effects of surfactants on heavy metals and radionuclides sorption on soils or clay minerals were also studied in great detail. Beveridge and Pickering (1983) found that several cationic surfactants were able to reduce metal ion uptake by montmorillonite, but the influence was to a much small scale for kaolinite and illite. Hayes et al. (1995) studied the effects of quaternary ammonium ions on trace metal sorption and showed that significant Sr^{2+} desorption from montmorillonite was achieved using alkyltrimethylammonium surfactants after the clays were pre-equilibrated with Sr^{2+}. Brusseau et al. (1997) demonstrated that a great enhancement of removal of Cd(II), Ni(II), and Sr^{2+} from soils could be obtained in the presence of cyclodextrin, a compound that can solubilize small low-polarity organic compounds and complex cationic heavy metals. Bors et al. (1997) revealed that the metal distribution coefficient, K_d, for Cs^+ and Sr^{2+} sorption on vermiculite was greatly reduced after treating the clay with hexadecylpyridinium chloride. Malakul et al. (1998) showed that Cd(II) sorption capacity was totally lost after montmorillonite was modified by an anchoring long chain cationic surfactant alone, suggesting competition between Cd(II) and the cationic surfactant was the major cause for the decreased Cd(II) sorption. Doong et al. (1998) tested removal of Cd(II) from soils using different types of surfactants and found that addition of anionic and nonionic surfactants enhanced the desorption rates of Cd(II), Pb(II), and Zn(II), whereas addition of cationic surfactant decreased the desorption efficiency of heavy metals. Furthermore, the desorption efficiency was found to increase linearly with the increasing surfactant concentration below the critical micelle concentration (CMC) and remained relatively constant above the CMC. Under acidic conditions, cationic surfactant was more effective to extract heavy metals than anionic and nonionic surfactants (Doong et al. 1998). Column leaching tests using two cationic surfactants showed that 20 pore volumes (PV) of 25 mM isostearamidopropyl morpholine lactate or 230 PVs of 25 mM lapyrium chloride were needed to remove 50% of the sorbed Pb(II) (Kornecki et al. 1999). In the presence of a cationic surfactant cetyltrimethylammonium bromide at a concentration of 5 mM, the Freundlich coefficient (K_f) values for Cd (II) sorption decreased to 28.7, 61 and 62.9% of the values for surfactant-free clay, calcareous, and sandy soils respectively (Elkhatib et al. 2001). However, the influence of an anionic surfactant sodium dodecyl sulfate (SDS) on Cd(II) sorption from soils was not significant. At an SDS concentration equivalent to 6.1 CMC, the K_f values for Cd (II) sorption decreased from 2.58, 2.05, and 1.86 to 2.19, 1.84, and 1.66 for clay, calcareous, and sandy soils, respectively (Elkhatib et al. 2001). Hexadecyltrimethylammonium (HDTMA), a cationic surfactant with 16 carbons on its tail group, was able to desorb significant amounts of Sr^{2+} and Zn(II) from clinoptilolite (Li et al. 2002). In the presence of HDTMA, Pb(II) sorption on bentonite was also greatly affected. Without HDTMA, Pb(II) sorption capacity on a bentonite was 100 mmol/kg. With the addition of HDTMA equivalent to 100% of the bentonite's CEC, the Pb sorption decreased to 35 mmol/kg (Lee et al. 2002). Desorption of Cs^+ from kaolinite increased as the initial Cs^+-loading on kaolinite decreased and 90% of initial sorbed Cs^+ could be removed by HDTMA with a minimal HDTMA to Cs^+ ratio of 30 (Willms et al. 2004). These results showed that the competition between surfactant cations and metal cations for the sorption sites could reduce sorption of metal cations and create a way of metal extraction.

This research was intended to study the batch desorption of Pb(II) and column leaching efficiency of Cs^+ and Pb(II) from a kaolinite and an illite using alkylammonium surfactant. The hypothesis was that the strong affinity between the clay surface and long chain alkylammonium ions can significantly increase Cs^+ and Pb(II) desorption kinetics and desorption efficiency. If ion exchange is the dominant reaction of metal sorption in soils, due to a stronger affinity of cationic surfactants for clay surfaces than do most metal cations, it is expected that the desorption of Cs^+ and Pb(II) by alkylammonium ions could present a way to remove heavy metals and radionuclides from contaminated soils.

2. MATERIALS AND METHODS

2.1 Clay Minerals

The kaolinite used was KGa-1b (a low-defect kaolinite and a substitute for KGa-1) from the Source Clay Minerals Repository (Purdue University, West Lafayette, IN) and was used as received without further purification. The illite used was Morris illite (Morris, IL) obtained from Ward's Natural Science (Rochester, NY), and was ground to < 200 mesh (<0.065 mm) using a Spex 8000 Mixer/Mill (Spex Industries Inc., Edison, NJ). The CEC determined by the ammonia electrode method (Borden and Giese, 2001) was 30 and 70 meq/kg for the kaolinite and illite, respectively.

2.2 Pb(II) and Cs^+ Sorption

Two grams of clay minerals and 20.0 mL of aqueous solution of $PbCl_2$ or CsCl (from Fisher) were combined in a 50-mL centrifuge tube, as a liquid to solid ratio of 10 was able to produce satisfied results (Li et al. 2002). The initial concentrations of $PbCl_2$ or CsCl ranged from 0 to 4.0 mM and from 0 to 6.0 mM at 0.5 mM increment. The mixture was equilibrated on a shaker table for 24 h, centrifuged, and the supernatant analyzed.

2.3 Loading Pb(II) and Cs^+ on Clay Minerals

After the Pb(II) sorption maxima on kaolinite and illite were determined, the clay minerals were loaded with Pb(II) at targets of 6, 30, and 100 percent of their sorption plateaus. To each 250-mL bottle, 50 g of kaolinite and 150 mL of 0.1, 0.6, and 2 mM $PbCl_2$ solution, or 50 g of illite and 150 mL of 0.8 and 5.0 mM $PbCl_2$ solution, or 20 g of illite and 200 mL of 6.0 mM $PbCl_2$ solution were mixed for 24 h, centrifuged, the supernatant removed, and then the mixture washed with two portions of deionized (DI) water. The Pb(II) concentrations in the equilibrium and each washing solutions were analyzed. The final amounts of Pb(II) loading were 0.3, 1.6, and 3.8 mmol/kg on kaolinite, corresponding to 6, 32, and 76% of its Pb(II) sorption capacity, and were 2.4, 15, and 67 mmol/kg on illite, corresponding to 3, 17, and 77% of its Pb(II) sorption capacity. Selected metal-loaded clays were independently tested on their Pb(II) content using acid digestion followed by lead analysis of the digested solution using ICP. The results were 0.37±0.01, 1.57±0.06, 2.3±0.1, and 16.7±0.3 mmol/kg for kaolinite with Pb(II) loading level of 0.3 and 1.6 mmol/kg and for illite with Pb(II) loading level of 2.4 and 15 mmol/kg, respectively.

The above procedure was also used to load Cs^+ on kaolinite and illite. The final amounts of Cs^+ loading were 0.8, 3.6, and 14 mmol/kg on kaolinite, corresponding to 6, 25, and 100 % of its Cs^+ sorption capacity, and were 0.9, 4.5, and 24 mmol/kg on illite, corresponding to 4, 20, and 100 % of its Cs^+ sorption capacity.

2.4 Desorption of Pb(II) by the Alkylammonium Surfactant

To each 50-mL centrifuge tube, 2.00 g of Pb(II)-loaded clay minerals and 20.0 mL of HDTMA bromide (from Aldrich) or dodecyltrimethylammonium (DDTMA) bromide (from Acros) aqueous solution were combined. The initial concentrations of alkylammonium ions varied from 3.6 to 12 mM for kaolinite and 1.8 to 30 mM for illite. The mixture was shaken for 24 h, centrifuged, and the supernatants analyzed.

For Pb(II) desorption kinetic study, 2.00 g of Pb-loaded clay minerals (at different initial Pb(II) loading levels) and 20.0 mL of HDTMA solutions with initial concentrations of 2.0 and 6.0 mM for Pb (II) desorption from the kaolinite and 6.0 and 30.0 mM for Pb(II) desorption from the illite were combined in a 50-mL centrifuge tube. The influence of solution pH on Pb(II) desorption was tested using waters with pH 3, 5, 7, 9, and 11 made from 0.75 mM HCl, 0.01 mM HCl, 0.006 mM NaOH, 0.027 mM NaOH, and 2 mM NaOH solutions, respectively, before HDTMA was dissolved. To test the influence of added salts on

Pb(II) desorption, 3, 10, and 30 mM NaCl solutions were used to make the HDTMA solution. The mixture was shaken for 0.25, 0.5, 1.0, 2.0, 4.0, and 8.0 h, centrifuged, and the supernatants analyzed.

All samples were prepared in duplicate for each individual experimental condition. The equilibration was performed on a shaker table at 22°C and 150 rpm for 24 hours for sorption study and specified time for kinetic study. Centrifugation was performed at 3200 *g* for 20 min. The supernatant was filtered through a 0.22 μm syringe filter before analyses.

2.5 Column Leaching Experiment

Column tests were performed in four 30-mL plastic syringe columns, each with 2 cm in diameter and 12 cm in length. Metal-loaded clays were homogenized with fine clean sand at a ratio of 1:1 before being packed into columns in order to enhance the permeability. The dry bulk densities were 2.0±0.1 g/cm^3 for the illite columns and 1.4±0.1 g/cm^3 for the kaolinite columns. The packed dry weights were 60±2 g for the illite and 43±2 g for the kaolinite. The pore volumes were 13.4±1.5 cm^3 for the illite columns and 18.0±0.7 cm^3 for the kaolinite columns. The porosities were 0.45±0.05 for the illite columns and 0.60±0.02 for the kaolinite columns. Two columns were fed with regular DI water while the other two with 6 mM HDTMA solution at a flow rate of 0.01 PV per minute. Samples were taken every 5 to 6 mL, corresponding to 1/3 PV. The metal concentration was analyzed by either AA or ICP.

2.6 Chemical Analysis

Lead was analyzed using a Perkin Elmer (Norwalk, CT) Plasma 400 ICP emission spectrometer. Cs^+ was analyzed by atomic emission using either a Perkin-Elmer 360 or 560 Flame Atomic Absorption Spectrophotometer (Shelton, CT). To overcome the influence of background electrolyte, 1 mL of 5000 mg/L K^+ from KCl was added to each 4 mL of samples. All calibrations were prepared using three to five standards covering the range of sample concentrations. Proper dilutions were made to samples over the linear response range. The coefficients of determination (r^2) for the standard curves were no less than 0.99.

2.7 Determination of Point of Zero Charge (pzc) of Kaolinite

The pzc was determined using a titration method. To each 50 mL beaker, 2 g kaolinite and 20 mL of NaCl solution with concentrations of 0.001, 0.01, and 0.1 M were stirred until the pH stabilized. Then, aliquots of 0.2 mL of 0.01 M NaOH or HCl were added into the mixture, and the pH, when stabilized, was recorded. The pH at the intercept of titration curves at different ionic strength indicates the pzc (Coles and Yong, 2002).

2.8 FTIR Spectroscopy

Fourier transform infrared (FTIR) spectra were acquired on a Perkin Elmer Spectra One Spectrometer equipped with Attenuated Total Reflection (ATR) accessory. The spectra were obtained by accumulating 256 scans at a resolution of 4 cm^{-1}.

3. RESULTS AND DISCUSSION

Sorption of Pb(II) on kaolinite and illite is plotted in Figure 1. The data were fitted to the Langmuir sorption isotherm:

$$C_S = \frac{K_L C_m C_L}{1 + K_L C_L} \qquad (1)$$

where C_L is the equilibrium solution concentration, C_S, the amount of Pb(II) sorbed, C_m the Pb(II) sorption maximum, and K_L the Langmuir sorption coefficient reflecting the affinity of Pb(II) to the surface (sorption intensity). Langmuir type of sorption isotherms was previously used to describe Pb(II) sorption on three types of calcareous soils (Elkhatib et al. 1991), on smectite under uncontrolled pH (Shen et al. 1999), on bentonite (Lee et al. 2002), on three tropical soils (Mollisol, Oxisol, and Ultisol) from Puerto Rico (Appel and Ma, 2002), and on Thai kaolin and ballclay, whose major mineral component is illite (Chantawong et al. 2003).

The fitted parameters are listed in Table 1. The Pb(II) sorption capacities on kaolinite and illite are 5.0 and 87 mmol/kg, respectively. The Pb(II) sorption capacity on kaolinite is comparable to 10 mmol/kg on a kaolinite obtained from Dry Branch Kaolin Company, whose CEC is 166 meq/kg with a surface area of 16.41 m^2/g (Miranda-Trevino and Coles, 2003) and to 3.2 mmol/kg on a Thai kaolin whose CEC is 110 meq/kg (Chantawong et al. 2003). The Pb(II) sorption capacity on the illite is greater than its CEC. A similar phenomenon was noticed in studying Pb(II) sorption by an oxisol soil and the excess Pb(II) sorption was attributed to inner-sphere surface reactions (Appel and Ma, 2002). The Pb(II) sorption maximum on the illite is greater than that of 23 mmol/kg for the Silver Hill illite whose CEC was 330 meq/kg (Rybicka et al. 1995) and that of 17 mmol/kg for a Thai ballclay whose major clay component is illite with a CEC of 150 meq/kg (Chantawong et al. 2003). The Pb(II) sorption intensities on kaolinite and illite based on Langmuir isotherm are 70 and 21 L/mmol, respectively (Table 1), comparable to 23 L/mmol by a Thai kaolin and 51 L/mol by a Thai ballclay (Chantawong et al. 2003).

Table 1. Langmuir sorption parameters for Pb(II) and Cs^+ sorption on kaolinite and illite

	Pb(II) Sorption			Cs^+ Sorption		
	K_L L/mmol	S_m (mmol/kg)	r^2	K_L L/mmol	S_m (mmol/kg)	r^2
Kaolinite	70	5	0.93	2.1	14	0.99
Illite	21	87	0.98	2.3	30	0.98

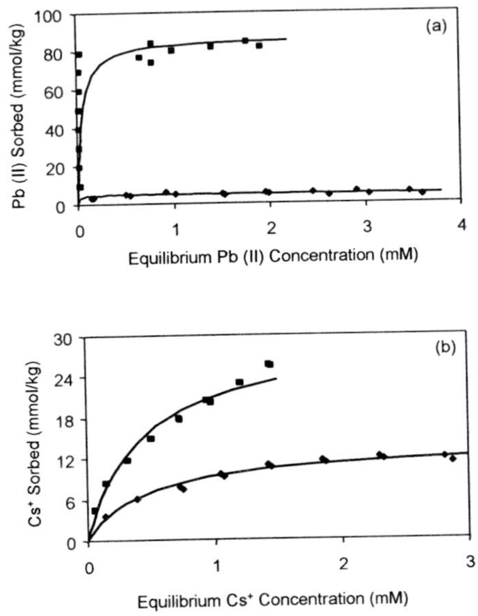

Figure 1. Sorption of Pb(II) (a) and Cs^+ (b) on kaolinite (◆) and illite (■). The lines are Langmuir fit to the observed data

After the Pb(II) sorption maximum on kaolinite and illite was determined, the kaolinite was loaded with Pb(II) to 0.3, 1.6, and 3.8 mmol/kg, corresponding to 6, 32, and 76% of its Pb sorption capacity, and the illite to 2.4, 15, and 67 mmol/kg, corresponding to 3, 17, and 77% of its Pb(II) sorption capacity. A comparison of surfactant chain length on the percentage of Pb(II) removed as a function of initial alkylammonium concentration is illustrated in Figure 2. For kaolinite, the percentage of Pb(II) removal was 93 and 81, respectively, when DDTMA and HDTMA were used. In addition, the percentage of Pb(II) removal from kaolinite is almost independent from the initial DDTMA concentration, while the percentage of Pb(II) removal decreased as the HDTMA concentration exceeded 6 mM (Figure 2). From the similarity between Pb(II) sorption capacity and the CEC, Coles and Yong (2002) concluded that Pb(II) retention by kaolinite is mainly due to cation exchange and specific sorption may not occur or may be negligible. The input amount of alkylammonium surfactant even at the lowest initial concentration is large enough to fully compensate the CECs. Thus, when DDTMA was used, due to primarily cation exchange, the lead desorption remained invariable to input concentration. The difference in Pb(II) removal by DDTMA and HDTMA could be attributed to the states of surfactant in aqueous solution and sorbed on kaolinite surfaces. DDTMA, whose CMC is 25 mM (Rosen, 1989), exists as monomers in solution, while HDTMA, whose CMC is 0.9 mM (Rosen, 1989), exists as micelles in the given surfactant concentration range. When sorbing on kaolinite surfaces, DDTMA forms mainly as monolayer primarily by cation exchange while HDTMA forms admicelles due to combination of cation exchange and hydrophobic interaction. Previous results showed that Pb(II) sorption on a zeolite modified by HDTMA to a bilayer surfactant configuration still retained 60% of its original Pb(II) sorption capacity, indicating the presence of specific interaction between HDTMA admicelles and sorbed Pb(II) (Li et al. 2002). Similarly, a bentonite modified with HDTMA to 100% of its CEC still has 40% of its original Pb(II) sorption capacity (Lee et al. 2002). Doong et al. (1998) also found that Pb(II) and Zn(II) desorption remained constant at surfactant concentration greater than the CMC.

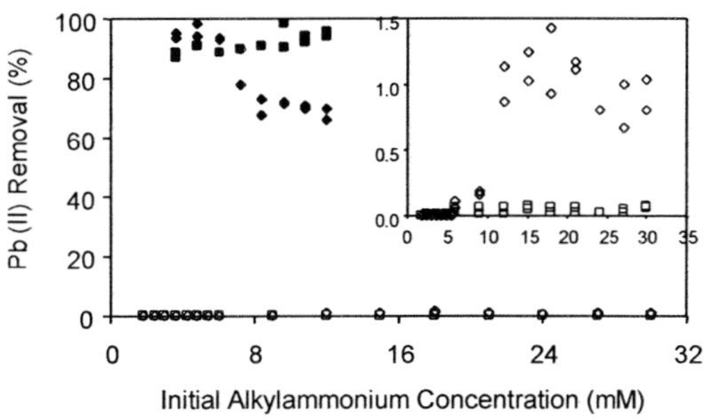

Figure 2. Percentage of Pb(II) removal from kaolinite at an initial Pb(II) loading of 3.8 mmol/kg (solid symbols) and from illite at an initial Pb(II) loading of 67 mmol/kg (open symbols) as a function of initial HDTMA (◆, ◇), DDTMA (■, □) concentration after equilibrated on a shaker table at 150 rpm for 24 hours. The insert is the enlargement of percentage of Pb(II) removal from illite surfaces.

The removal of Pb(II) from illite is totally different from that from kaolinite. HDTMA only removed about 1% of the initially sorbed Pb(II) while DDTMA essentially removed none of the initially sorbed Pb(II). Previous studies of influence of Pb(II) on alkylammonium ions sorption by clinoptilolite zeolite showed that the removal of Pb(II) from zeolite was not significant, indicating that specific sorption played a significant role in Pb(II) retention by the zeolite (Li et al. 2002). The removal of Pb(II) from illite in this study followed a similar trend as with the removal of Pb(II) from the zeolite.

At the initial surfactant concentrations of 2 and 6 mM, the desorption of Pb(II) by HDTMA and DDTMA from kaolinite was less affected by desorption time (Figure 3a and 3b). Thus, the Pb(II) desorption at each individual time was determined and the average percentage of Pb(II) desorbed at a particular initial Pb(II) loading and with a particular alkylammonium salt with a particular initial concentration was calculated. The results are tabulated in Table 2. It can be seen from Table 2 that a slightly higher percentage of Pb(II) desorption was achieved with HDTMA compared to DDTMA. In addition, a higher percentage of Pb(II) removal was achieved when the initial alkylammonium concentration was 6 mM compared to 2 mM.

From the perspective of HDTMA used vs. Pb(II) desorbed, 2 mM solution is more economic than 6 mM solution. Figure 3 also shows that a higher percentage of Pb(II) desorption was achieved when the initial Pb(II) loading level on kaolinite was high. Contrast to Pb(II) desorption from kaolinite, the desorption of Pb(II) from illite was completely different. First, no desorption of Pb(II) was observed when the initial alkylammonium concentration was 6 mM regardless of the initial Pb (II) loading on illite. Second, even at an initial DDTMA concentration of 30 mM, only negligible amount of Pb(II) was removed from illite with initial Pb(II) loadings of 15 and 67 mmol/kg (Table 3). Third, at 67 mmol/kg of Pb (II) loading level, the percentage of Pb(II) removal by 30 mM HDTMA is time dependent (Fig. 3c). Fourth, the highest percentage of Pb(II) removal from illite surface was less than 1.5%, similar to the Pb(II) desorption from zeolite (Li et al. 2002).

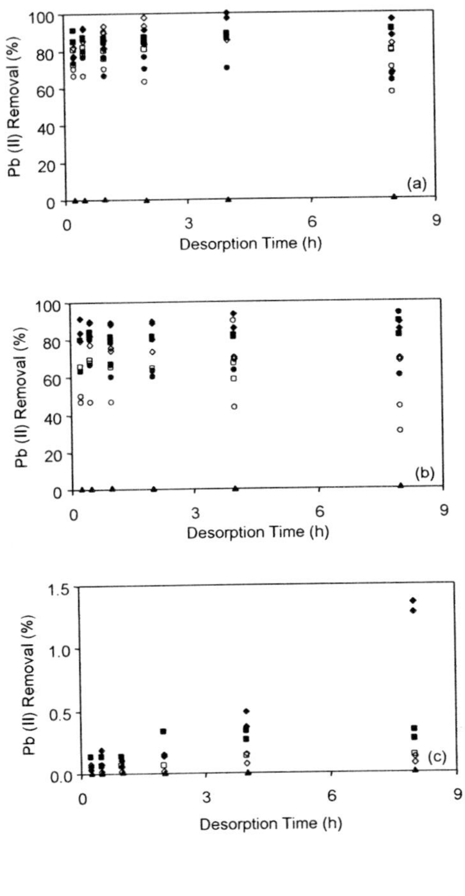

Figure 3. Percentage of Pb(II) removal from kaolinite at initial Pb(II) loadings of 0.3 (●), 1.6 (■), and 3.8 mmol/kg (◆) as a function of desorption time by HDTMA (a) and DDTMA (b) at initial alkylammonium concentrations of 2 mM (open symbols) and 6 mM (solid symbols). Percentage of Pb(II) removal from illite at initial Pb(II) loadings of 15 (■) and 67 mmol/kg (◆) as a function of desorption time by HDTMA (solid symbols) and DDTMA (open symbols) at an initial alkylammonium concentration of 30 mM (c). Desorption of Pb(II) by water (▲) is negligible.

Table 2. Percentage of Pb(II) removal from kaolinite

Initial Pb(II) Loading (mmol/kg)	Initial HDTMA Conc.		Initial DDTMA Conc.	
	2 mM	6 mM	2 mM	6 mM
0.3	70±8	72±5	51±15	67±10
1.6	83±8	88±2	66±6	82±3
3.8	86±8	93±7	75±5	89±3

FTIR spectra of kaolinite treated with HDTMA showed absorptions at 2850 and 2920 cm^{-1} (Figure 4). The former was attributed to symmetric and the latter to asymmetric stretching vibrations of C-C in the alkyl chain (Kung and Hayes, 1993). At a liquid to solid ratio of 10 and initial concentration of 2 mM, the amount of HDTMA sorbed on the kaolinite is about 20 mmol/kg less than 30 meq/kg, the CEC of the kaolinite. On the other hand, at an initial concentration of 6 mM and the same liquid to solid ratio, the amount of HDTMA sorbed will be 50 mmol/kg, almost twice of the CEC of kaolinite (Li, 1999). The stronger absorption at 2850 and 2920 cm^{-1} when the initial HDTMA concentration was 6 mM further verified the high surfactant sorption on kaolinite due to bilayer formation.

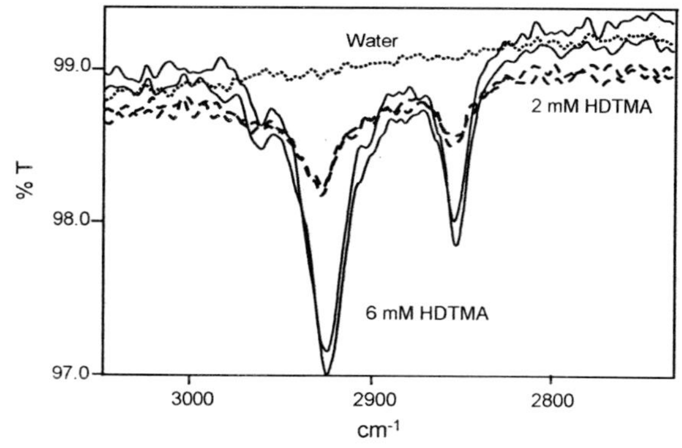

Figure 4

Figure 4. FTIR spectrum of kaolinite after equilibrated with water (dotted line), 2 mM HDTMA (dashed lines) and 6 mM HDTMA (solid lines). Sorption bands at 2850 and 2920 cm^{-1} are due to symmetric and asymmetric stretching vibrations of C-C in the alkyl chain (Kung and Hayes, 1993).

Table 3. Percentage of Pb (II) removal from illite

Initial Pb (II) Loading (mmol/kg)	Initial HDTMA Conc.		Initial DDTMA Conc.	
	6 mM	30 mM	6 mM	30 mM
2.4	–	–	–	–
25	–	0.10±0.03	–	0.22±0.10
67	–	0.05±0.05	–	0.36±0.47

Note: "–" represents zero.

The influence of solution pH on desorption of Pb(II) can be seen in Figure 5. At pH 3, a higher percentage of Pb(II) desorption was achieved. In contrast, at pH 11, the percentage of

Pb(II) desorption was almost zero. The desorption of Pb(II) is less affected at pH 5, 7, and 9. The pzc determined in this study are about 4 (Figure 6) contrast to 5 as determined by Sutheimer et al. (1999), and to 3.2 of a different kaolinite determined by the same titration method (Coles and Yong, 2002). Below the pzc, the surface would be positively charged and resulted in higher Pb(II) desorption. As pH increases, the surface would be more negatively charged, which should result in a lower Pb(II) desorption. However, the surface charge increased at a much slower rate in the pH 4-9 range (Figure 6), which is possible the reason that the Pb(II) desorption is less dependent in pH 5-9 range (Figure 5a). At pH 11, the percentage of Pb(II) removal reached zero. Similar observation is found when DDTMA was used as the desorbing agent (Figure 5b). The low Pb (II) desorption at pH 11 indicated that a precipitation may have occurred. Pb(II) sorption on montmorillonite showed that at a pH lower than the hydrolysis point the sorption is primarily due to ion exchange, while at a pH higher than the hydrolysis point, it is a combination of both ion exchange and precipitation (Businelli et al. 2003).

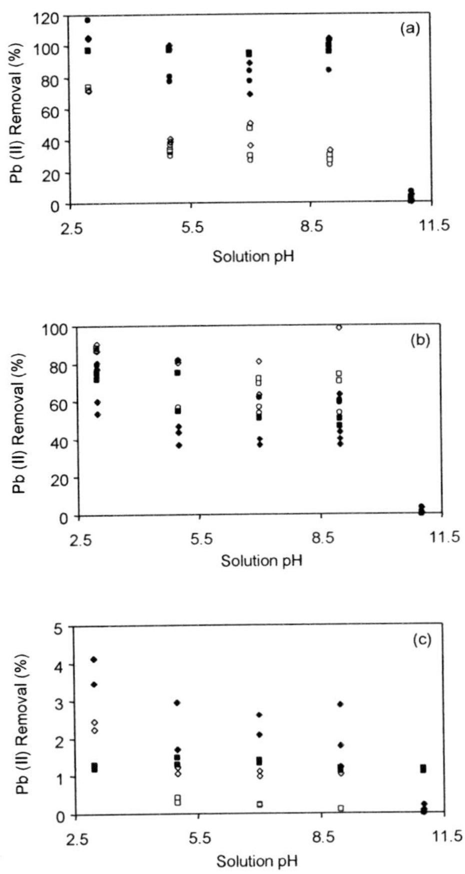

Figure 5. Percentage of Pb(II) removal from kaolinite by HDTMA (a) and DDTMA (b) as a function of pH and initial Pb (II) loadings of 0.3 (●), 1.6 (■), and 3.8 (◆) mmol/kg. The initial surfactant concentrations were 2 mM (open symbols) and 6 mM (solid symbols). Removal of Pb (II) from illite (c) at a Pb(II) loading level of 67 mmol/kg as a function of pH by HDTMA (solid symbols) and DDTMA (open symbols) at alkylammonium concentrations of 6 mM (■) and 30 mM (◆).

Also noticed from Figure 5a is that a higher Pb(II) desorption was achieved when a higher HDTMA concentration was used in tested pH ranges. Similar trend for Pb(II) desorption from illite is seen in Figure 5c. The only difference is that the percentage of Pb(II) removal is less than 4% from illite, much lower than the percentage of Pb(II) removal from kaolinite. Added NaCl in the desorbing solution had negligible effect on Pb(II) desorption using HDTMA (Figure 7). Although a higher Na^+ concentration could achieve a higher Pb (II) desorption due to cation exchange, the strong affinity of alkylammonium ions for kaolinite and illite minimized the concentration effect of Na^+. Pierangeli et al. (2001) also found that changing in ionic strength had a little effect on Pb desorption from most Oxisols, which was attributed to an inner-sphere sorption. Shen et al. (1999) found that the pH-edge of Pb(II) sorption on smectite shifted to a higher pH as the Pb concentration increased, while the ionic strength only affected Pb(II) sorption in the low pH range.

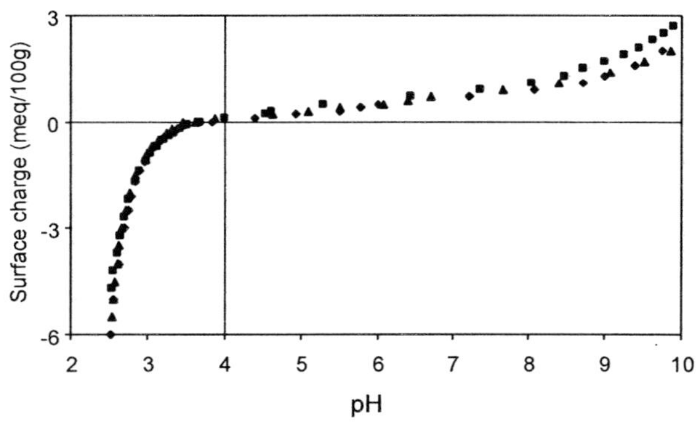

Figure 6. Titration curves used to determine the pzc of the kaolinite. Titration solutions used were NaCl at concentrations of 0.001 M (◆), 0.01 M (▲), and 0.1 M (■).

Removal of Pb(II) using an initial HDTMA solution of 6 mM as a function of liquid to solid ratio can be seen in Figure 8. The percentage of Pb(II) removal from kaolinite increases as the liquid to solid ratio increases. In contrast, Pb(II) removal from illite was negligible for all tested liquid to solid ratios, indicating that the HDTMA solution at 6 mM is not a good candidate to desorb lead from illite surface (Figure 8a). Similar to Pb(II) desorption, removal of Cs^+ from kaolinite is more efficient than that from illite (Figure 8b). Up to 20% of initially sorbed Cs^+ could be removed from illite at an initial HDTMA concentration of 6 mM and a liquid to solid ratio of 7.

Column leaching tests using 6 mM HDTMA solution to remove Pb(II) from kaolinite and illite loaded to 100% Pb(II) sorption capacity can be seen in Figure 9. After three PVs, corresponding to a liquid to solid (L/S) ratio of 2/3 for illite and 5/4 for kaolinite, the percentage of Pb(II) removal from kaolinite reached about 9% (Figure 9a), similar to that determined from a batch test (Figure 8a). In contrast, water was able to remove 3% of the initially sorbed Pb(II). This amount of Pb(II) removal could be originated from the residual Pb(II) after the lead modification. The difference between Pb(II) removal by water and by 6 mM HDTMA solution shows that 6 mM HDTMA is at least 3 times more efficient than water. Contrast of Pb(II) removal from kaolinite, Pb(II) removal from illite is totally different. After 5 PVs of flushing, only 0.1% and 0.03% of initially sorbed Pb(II) was removed from illite by 6 mM HDTMA solution and water, respectively (Figure 9b). However, the percentage of Pb(II) removal by 6 mM HDTMA solution was three time more than that by water, again reflecting the enhanced Pb(II) removal by cationic surfactant. In a column leaching test, nearly100% and 56% of the residual Pb(II) were extracted from a surface soil with EDTA and CMCD, while 80% and 2% of the residual Pb(II) were extracted from a subsurface soil with EDTA and CMCD (Vulava and Seaman, 2000). However, the Pb(II) loading in their study is different from that in this study. In their study the soils column were fed with Pb (II) solution until full breakthrough, the extraction agents were applied thereafter. The Pb(II) sorption on the soil never experienced a wet-dry cycle. The extraction of Pb(II) in this study was from kaolinite and illite that were previously loaded with Pb(II) and dried afterward before they were packed into the column.

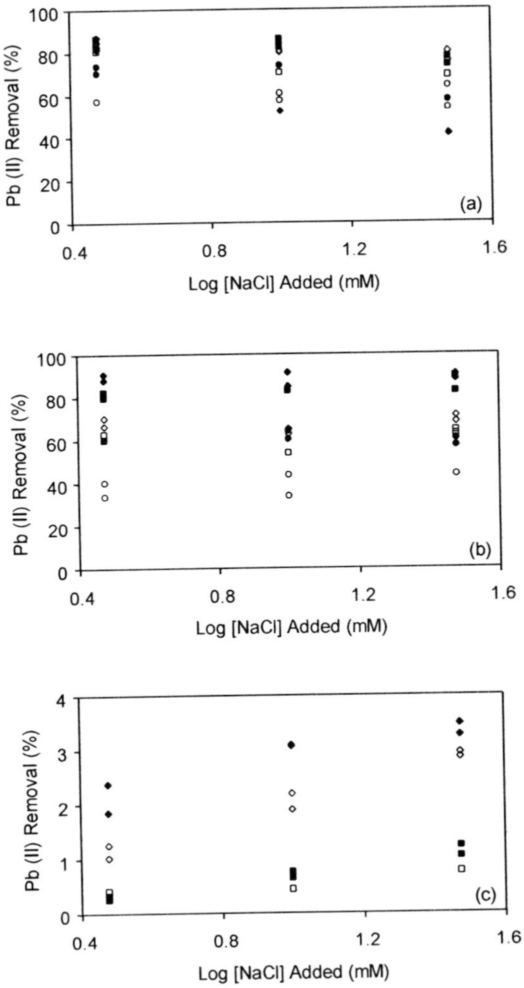

Figure 7. Removal of Pb(II) from kaolinite by HDTMA (a) and DDTMA (b) as a function of added NaCl concentrations at initial Pb (II) loading of 0.3 (●), 1.6 (■), and 3.8 (◆) mmol/kg. The initial surfactant concentrations were 2 mM (open symbols) and 6 mM (solid symbols). Removal of Pb(II) from illite (c) at a Pb(II) loading level of 67 mmol/kg as a function of added NaCl concentrations by HDTMA (solid symbols) and DDTMA (open symbols) at alkylammonium concentrations of 6 mM (■) and 30 mM (◆).

Column leaching tests using 6 mM HDTMA solution vs. water to remove Cs^+ from kaolinite and illite can be seen in Figure 10. In general, more Cs^+ is removed from kaolinite than from illite. In addition, after flushing with 4 PVs, corresponding to a L/S ratio of 5/3, twice amount of Cs^+ was removed from kaolinite using HDTMA solution compared to water at the 6% and 100% Cs^+ loading levels. At this L/S ratio the percentage of Cs^+ removal from 100% loaded kaolinite is about 40% (Figure 10b), which is similar to that determined from the batch test (Figure 8b). For illite, at 100% Cs^+ loading level, after flushing with 8 PVs, corresponding to a L/S ratio of 7/4, about 12% of initial Cs^+ was removed (Figure 10d), which is again, similar to that determined from a batch test (Figure 8b). Compared to Pb(II) removal, removal of Cs^+ using cationic surfactant is much more effective.

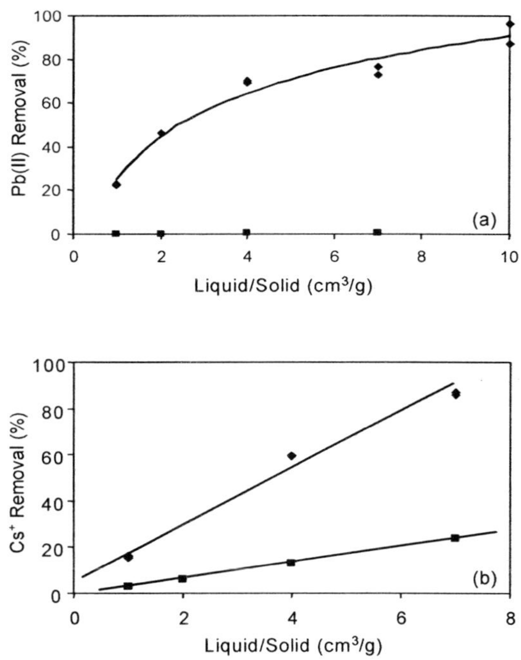

Figure 8. Removal of Pb(II) from 100% Pb(II)-loaded kaolinite (◆) and illite (■) using 6 mM HDTMA solution (a) and removal of Cs^+ from 100% Cs^+-loaded kaolinite (◆) and illite (■) using 6 mM HDTMA solution (b) as a function of liquid to solid ratio to assess the feasibility of column leaching tests.

It has been long noticed that Cs^+ is preferred to sorb on illite over kaolinite (Komarneni, 1978; Maguire et al. 1992; Campbell and Davies, 1995). Previous results showed that little desorption of Cs^+ from a clinoptilolite-rich zeolite surface was achieved using HDTMA (Li et al. 2002) and a higher Cs^+ desorption was obtained from kaolinite than from illite in batch studies (Willms et al. 2004). The column Cs^+ leaching test in this study further confirmed that Cs^+ is more easily removed from kaolinite than from illite. The less efficiency of Cs^+ desorption from illite by HDTMA suggest that Cs^+ sorption on illite is very selective and is primarily locked in the "frayed edge" sites characterized as high affinity/low capacity (Poinssot et al. 1999, Willms et al. 2004).

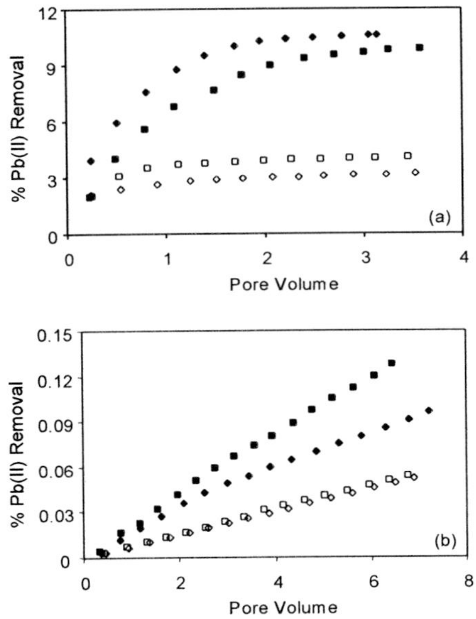

Figure 9. Lead leaching from columns packed with kaolinite loaded with 100% Pb(II) (a) and from columns packed with illite loaded with 100% Pb(II) (b). The open symbols represent Pb(II) desorption by water and solid symbols by 6.0 mM HDTMA solution.

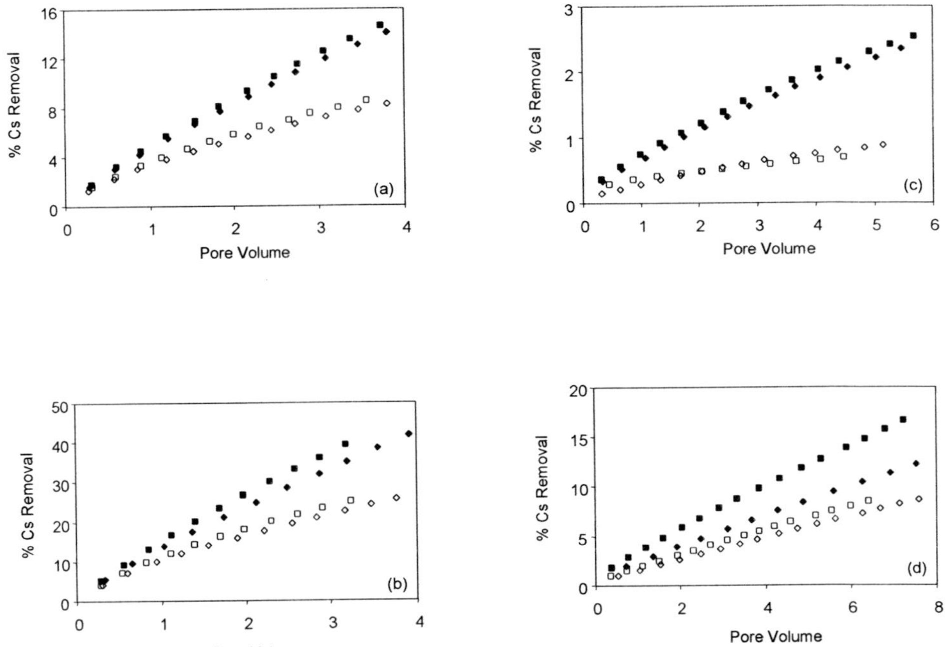

Figure 10. Cesium leaching from columns packed with kaolinite loaded with 6% (a) and 100% (b) Cs^+ and from columns packed with illite loaded with 22% (c) and 100% (d) Cs^+. The open symbols represent Cs^+ desorption by water and solid symbols by 6.0 mM HDTMA solution.

4. Conclusion

The desorption of Pb(II) from a kaolinite (KGa-1b) and an illite (Morris, IL) by alkylammonium ions followed different trends. Pb(II) could be more easily removed from kaolinite than from illite. The percentage of Pb(II) removal from kaolinite reached up to 70-90% while percentage of Pb(II) removal from illite was only 1-3%. Pb(II) desorption is less affected by pH when the pH is less than 9. At pH 11 the removal of Pb(II) reached to about zero, indicating that precipitation at the high pH was the main cause for the higher Pb(II) retention. Column leaching tests further verified batch results. The results from this study show that the types of clay minerals are the most important factor to consider before soil flushing with cationic surfactant can be initiated.

References

Appel, C., and Ma, L., 2002. Concentration, ph, and surface charge effects on cadmium and lead sorption in three tropical soils. *J. Environ. Qual.* 31, 581-589.

Beveridge, A., and Pickering W. F., 1983. The influence of surfactants on the adsorption of heavy metal ions by clays. *Water Res.* 17, 215-225.

Borden, D., Giese, R.F., 2001. Baseline studies of the clay minerals society source clays: cation exchange capacity measurements by the ammonia-electrode method. *Clay Clay Miner.* 49, 444-445.

Bors, J., Gorny, A., Dultz, S., 1997. Iodide, caesium and strontium adsorption by organophilic vermiculite. *Clay Miner.* 32, 21-28.

Bradbury, M. H., Baeyens, B., 2000. A generalised sorption model for the concentration dependent uptake of caesium by argillaceous rocks. *J. Contam. Hydrol.* 42, 141-163.

Brouwer, E., Baeyens, B., Maes, A., Cremers A., 1983. Cesium and rubidium ion equilibria in illite clay. *J. Phys Chem.* 87, 1213-1219.

Brusseau, M.L., Wang, X., Wang, W., 1997. Simultaneous elution of heavy metals and organic compounds from soil by cyclodextrin. *Environ. Sci. Technol.* 31, 1087-1092.

Businelli, M., Casciari, F., Businelli, D., Gigliotti, G., 2003. Mechanisms of Pb (II) sorption and desorption at some clays and goethite-water interfaces. *Agronomie* 23, 219-225.

de Llano, A. Y., Benitez, A. H., Gutierrez, M. G., 1998. Cesium sorption studies on Spanish clay materials. *Radiochimica Acta* 82, 275-278.

Campbell, L. S., and Davies, B. E., 1995. Soil sorption of cesium modeled by the Langmuir and Freundlich isotherm equations. *Appl. Geochem.* 10, 715-723.

Chantawong, V., Harvery, N. W., and Bashkin, V. N., 2003. Comparison of heavy metal adsorptions by Thai kaolin and ballclay. *Water, Air, and Soil Pollution* 148, 111–125.

Coles, C. A., and Yong, R. N., 2002. Aspects of kaolinite characterization and retention of Pb and Cd. *Appl. Clay Sci.* 22, 39-46.

Doong, R.-A., Wu, Y.-W., and Lei, W.-G., 1998. Surfactant enhanced remediation of cadmium contaminated soils. *Water Sci. Technol.* 37, 65-71.

Dumat, C., Staunton, S., 1999. Reduced absorption of caesium on clay minerals caused by various humic substances. *J. Environ. Radioactiv.* 46, 187-200.

Elkhatib, E.A., Elshebiny, G.M., Balba, A.M., 1991. Lead sorption in calcareous soils. *Environ. Pollution* 69, 269-276.

Elkhatib, E.A., Thabet, A.G., Moharem, K.L., 2001. Sorption of cadmium in surfactant-amended soils, *Arid Land Research and Management* 15, 385-394.

Erten, H. N, Aksoyoglu, S., Hatipoglu, S., Gonktork, H., 1988. Sorption of caesium and strontium on montmorillonite and kaolinite. *Radiochemica Acta* 44/45, 147-151.

Fischer, K., 2002. Removal of heavy metals from soil components and soil by natural chelating agents. Part I: Displacement from Clay Minerals and Peat by L-Cysteine and L-Penicillamine. *Water, Air, and Soil Pollution* 137, 267-286.

Hayes, K.F., Chen, C., McAvoy, D.C., 1995. Quaternary ammonium surfactant effects on sorption of trace metals onto quartz and aluminosilicates. *Soil Sci. Soc. Am. J.* 59, 380-387.

Kabata-Pendias, A. 1980. Heavy metals sorption by clay minerals and oxides of iron and manganese. *Mineralogia Polonica* 11, 3-13.

Komarneni, S., 1978. Caesium sorption and desorption behaviour of kaolinites. *Soil Sci. Soc. Am. J.* 42, 531-532.

Kornecki, T. S., Brown, G. O., and Allred, B., 1999. Saturated column feasibility study using cationic surfactants for *In Situ* removal of lead from soil. *Environ. Geosci.* 6, 42-52.

Kung, K. S. and Hayes, K. F., 1993. Fourier transform infrared spectroscopic study of the adsorption of cetyltrimethylammonium bromide and cetylpyridinium chloride on silica. *Langmuir* 9, 263-267.

Lee, J.-J., Choi, J., and Park, J.-W., 2002. Simultaneous sorption of lead and chlorobenzene by organobentonite. *Chemosphere* 49, 1309-1313.

Lee, S. Y. and Kim, S. J., 2002. Adsorption of naphthalene by HDTMA modified kaolinite and halloysite. *Appl. Clay Sci.* 22, 55-63.

Li, Z., 1999. Sorption of oxyanions and surface anion exchange by surfactant-modified clay minerals. *J. Environ. Qual.* 28, 1457-1463.

Li, Z., Alessi, D., Allen, L., 2002. Influence of cationic surfactant on the sorption of metal cations onto zeolite. *J. Environ. Qual.* 31, 1106-1114.

Li, Z., 2004. Cationic surfactant in mineral surface modification and its environmental application *In Recent Research Developments in Surface & Colloids,* Research Signpost, Kerala, India, 79-95.

Lo, I. M. C. and Yang, X. Y., 1999. EDTA extraction of heavy metals from different soil fractions and synthetic soils, *Water, Air, & Soil Pollution* 109, 219-236.

Maguire, S., Pulford, I. D., Cook, G. T., and Mackenzie, A. B., 1992. Caesium sorption-desorption in clay-humic acid systems. *J. Soil Sci.* 43, 689-696.

Malakul, P., Srinivasan, K.R., Wang, H.Y., 1998. Metal adsorption and desorption characteristics of surfactant-modified clay complexes. *Ind. Eng. Chem. Res.* 37, 4296-4301.

Miranda-Trevino, J. and Coles, C. A., 2003. Kaolinite properties, structure and influence of metal retention on pH. *Appl. Clay Sci.* 23, 133-139.

Pardo, M.T., 2000. Sorption of lead, copper, zinc, and cadmium by soils: Effect of nitriloacetic acid on metal retention. *Communications in Soil Science and Plant Analysis* 31, 31-40.

Pierangeli, M. A. P., Guilherme, L. R., G., and Oliveira, L. R., 2001. Effect of ionic strength of the equilibrium solution upon lead adsorption/desorption in Brazilian Oxisols. *Pesq. Agropec. Bras.* 36, 1077-1084.

Peters, R. W., 1999. Chelant extraction of heavy metals from contaminated soils. *J. Hazard. Mater.* 66, 151-210.

Poinssot, C., Baeyens, B., and Bradbury, M. H., 1999. Experimental and modelling studies of caesium sorption on illite. *Geochim. Cosmochim. Acta* 63, 3217-3227.

Puls, R. W., Powell, R. M., Clark, D., Eldred, C. J., 1991. Effects of pH, solid/solution ratio, ionic strength, and organic acids on Pb and Cd sorption on kaolinite. *Water, Air, & Soil Pollution* 57-58, 423-430.

Rosen, M. J., 1989 *Surfactants and Interfacial Phenomena*, 2nd ed., John Wiley & Sons, New York, 431 pp.

Rybicka, E. H., Calmano, W., and Breeger, A., 1995. Heavy metals sorption/desorption on competing clays: an experimental study. *Appl. Clay Sci.* 9, 369-381.

Schroth, B. K., and Sposito, G., 1998. Effect of landfill leachate organic acids on trace metal adtiontion by kaolinite. *Environ. Sci. Technol.* 32, 1404-1408.

Shahwan, T., Erten, H. N., Black, L., Allen, G. C., 1999. TOF-SIMS study of Cs^+ sorption on natural kaolinite. *Sci. Total Environ.* 226, 255-260.

Shen, S., Taylor, R. W., Bart, H., Tu, S.-I., 1999. Equilibrium and spectroscopic studies of lead retention in smectite. *Communications in Soil Sci. and Plant Analysis* 30, 2711-2730.

Sutheimer, S. H., Maurice, P. A., Zhou, Q., 1999. Dissolution of well and poorly crystallized kaolinites: Al speciation and effects of surface characteristics. *Am. Mineral.* 84, 620-628.

Vulava, V. M, and Seaman, J. C., 2000. Mobilization of lead from highly weathered porous material by extracting agents. *Environ. Sci. Technol.* 34, 4828-4834.

Willms, C., Li, Z., Allen, L., Evans, C.V., 2004. Desorption of cesium from kaolinite and illite using alkylammonium salts, *Appl. Clay Sci.25*, 125-133.

ACKNOWLEDGMENTS

This research was partially sponsored by the State of Wisconsin Groundwater Coordinating Council. Funding from Creative Research Activity, Professional Opportunity Fund, and Collaborative Undergraduate Research Project of University of Wisconsin – Parkside is greatly appreciated.

In: Trends in Agriculture and Soil Pollution Research
Editor: James P. Livingston, pp. 165-190
ISBN: 1-59454-325-9

Chapter 7

DERIVING CRITICAL LOADS FOR INDIA, WHERE DO WE STAND TODAY?

Ajay Taneja and Gur Sumiran Satsangi
School of Chemical Sciences, Department of Chemistry,
St. John's College, Agra- 282 002, India
email: ataneja5@hotmail.com, gursumiran@hotmail.com

ABSTRACT

New fundamental research, faster and more effective transmission of new and existing knowledge to policy and decision makers and better communication of this knowledge to the public will all be required to meet the challenge for society to move towards a more sustainable biosphere. One of the most pressing issues for over a decade has been the need to predict the effects of acid deposition on terrestrial and aquatic ecosystems. Emphasis on cost-effective strategy based on scientific criteria has led to the development of the *Critical Loads Concept*. They provide an effect-based approach whereby an acid deposition flux greater than the critical load implies that long-term harmful effects on a selected target species will occur. They are intended to thresholds below which pollutant deposition does not damage ecosystems. The different methods and criteria to derive critical loads are discussed here highlighting their strengths and weakness.

In India, tropical climatic conditions, alkaline rich soil, desert and sea derived aerosols etc. are the for most favourable factors in preventing acid rain problem however the pace of current development activities has overtaken these natural favourable factors and many cities are now showing decreasing pH or acidic trends in precipitation. Projections made on the basis of RAINS-Asia model to identify the areas of prospective impact on the ecosystem will be discussed. Moreover, the results of the study done on the assessment of the potential of soil acidification in North India using the critical load approach and locally derived data for acidic and basic inputs with respect to Ashoka (*Saraca indica*), Eucalyptus (*Eucalyptus hybrid*), Anjan grass (*Cenchrus ciliaris*), Lemon (*Citrus argentifolium*), Hibiscus and Black siris (*Albizza odoratissima*) has been discussed, emphasizing the formidable difficulties and uncertainties in critical load calculations.

It has been concluded that nevertheless, as a starting point, the RAINS-Asia model can be applied in combination with critical load calculations at local levels to make policies, which are aimed at reducing acidic emissions such that the excess of critical loads in future can be controlled. This technique will be to anticipate pollution risks to a particular receptor. However, the calculation of critical loads for India with its highly varying environmental conditions and ecosystems needs a considerable effort at local levels.

During the last few decades, humans have emerged as a new faces of nature, modifying physical, chemical and biological systems in new ways at faster rates and over larger spatial scales than ever recorded on earth. This will have profound implications for all of life on earth. Among the most critical challenges facing humanity are the conservation, restoration and wise management of the earths resources [Lubchenco 1998]. Pollution control strategies need for gradual shifting of the focus from the reactive strategy to preventive one in order to reduce burden on the ecosystem and environment. Sustainable development cannot be achieved overnight. The philosophy is well established but the road to this achievement is quite lengthy as it passes through various developmental processes like awareness building, strategy formulation, financial resource mobilization for technology renovation, human resource development for attitudinal change, fixing of goals in a phased manner for each of the segmental activities and above all, the commitment at all levels to achieve the respective goals.

The future is quite likely to involve increasing rates of change, greater variance in system parameters, greater uncertainty about responses of complex ecological system and more surprises. The challenges for society are formidable and will require substantial information, knowledge wisdom and energy from the scientific community to reduce the rate at which we alter the Earths systems and how they interact with the numerous components of human caused global change and knowledge to manage the planet. In this regard studies like Intergovernmental Panel on climate change [IPCC, 1996] the ozone assessment and the Global Biodiversity Assessment [Heywood, 1995] have provided excellent guidance to policymakers, especially when they summarize certainties and uncertainties and specify the likely outcomes of different options.

Concern about the undesirable effects of air pollution in many areas of Northern hemisphere has led to a considerable amount of national and international research into the impact of acidic deposition on ecosystem functioning and vitality. Deposition of S and N compounds, which are mainly the result of urbanization cause the acidification of surface water, ground water and soil. Due to atmospheric deposition the major change in water chemistry are increase in SO_4, NO_3 and Al concentrations and decrease in pH and alkalinity values. Population of fish and other biota in water are decreasing due to high concentration of labile Al and low pH of water [Henriksen et al., 1992]. Apart from acidification, excess inputs of nitrogen through deposition also cause high level of NO_3 in water and change of vegetation [Boumans, 1991]. The research on the effects of air pollution on terrestrial ecosystem often yields results, which are not quantitative or regionally relevant. As a result, this information may be of little value to policy makers and resource managers. In order to facilitate the transfer of scientific information for policy decisions concerning emissions of air pollutants the concept of Critical Loads and Target Loads have been developed. Critical loads are the maximum amount of pollutant that ecosystem can tolerate without being damaged. The definition has been redrafted in order to fit specialist areas of interest, most particularly the acidification of fresh water, ground water and soil.

CRITICAL LOAD

The most general definition of critical load is "***quantitative estimate of an exposure to one or more pollutants below which significant harmful effects on specified sensitive elements of the environment do not occur according to our present knowledge***". It can also be defined as the maximum deposition of acidifying compounds that will not cause chemical changes leading to harmful effects on ecosystem structure and function [Nilsson 1986]. Some aspects of these definitions are worth special attention. Firstly, this concept focuses on chemical changes in the environment, and establish these as the link between emissions and the biological ecosystem. Secondly, the phrase long-term expresses the fact that there is a time lag between a change in emissions and a change in chemical effects in the environment. Establishing critical loads involve a new angle to ecological work. While most ecological sub-disciplines previously focused on describing the structure and function of ecosystems and its components as effected by environmental conditions, the concept for critical loads calls for another strategy to define the environmental conditions that ensures a certain structure and function. While the former way of working is a natural scientists angle to ecology, the latter is an engineering approach. The engineering approach to the task of adjusting levels of pollution to the critical loads is as a design exercise. If the ecosystem should provide certain "ecological services" the corresponding external conditions should be specified. Therefore one should work the way back to an industrial economic and legislative system that will provide us with these services. The difference between this angle and the assessment of different alternatives based on causality is describes in Fig.1. To make it possible to apply the concept of critical loads additional terms are necessary to understand, which are:

Receptor: It is the type of ecosystem, which is considered for study like surface water (lakes and streams), ground water and forest soils.

Biological indicator: is the type of organism that has been selected to represent to receptor. Typical biological indicators would be fish for surface water, trees and vegetation for forest soil and human health for ground water.

Chemical criteria: is a chemical measure affected by the acid deposition that is used to predict the risk of damage to biological indicators.

Critical limit: are the most unfavorable values that the chemical criteria may attain without large term harmful effects on ecosystem structure and function (these effects are identified by damage to the biological indicators).

For each receptor ecosystem, a biological indicator is chosen. A suitable chemical criterion is related for that biological indicator and a critical chemical limit is assigned. By using appropriate methods, a maximum value of atmospheric deposition/concentration (critical load/level) is calculated at which the critical limit will not be exceeded. In this way, ecosystem response is linked to atmospheric inputs. In deriving critical loads, a long-term (decades) perspective has to be considered, as most of the effects are results of accumulated deposition. The term critical load should not be confused with target load (Figure 2), which is related to political decision. The critical load is an inherent property of an ecosystem, where as the target load is less restrictive and implies other factors such as economical (cost benefit) or emotional consideration.

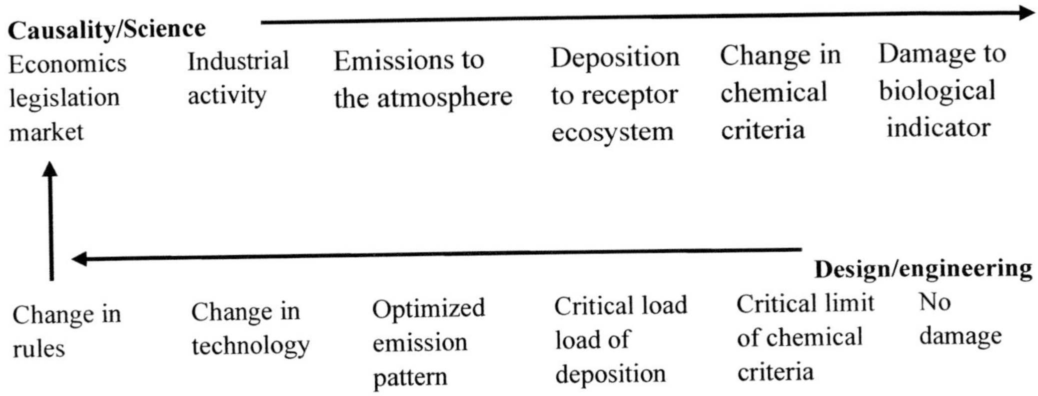

Figure 1. A comparison of the scientific approach and the engineering approach
Source: Warfvinge and Sverdrup, 1995

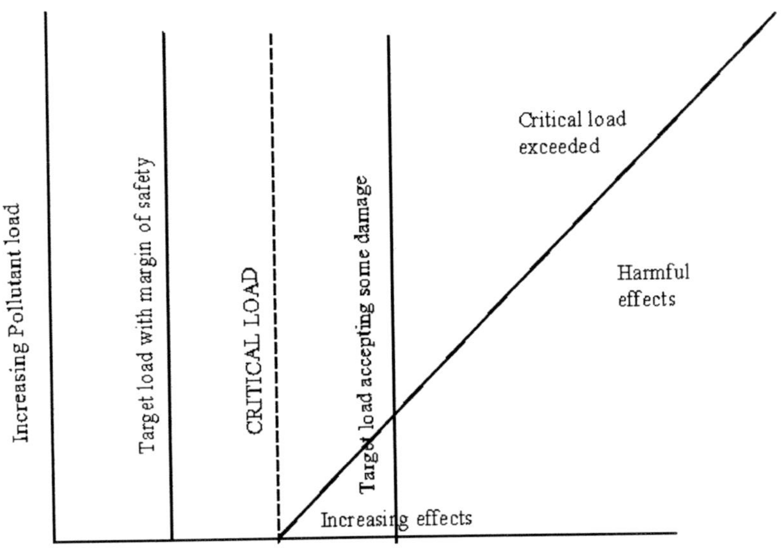

Fig 2: Relationship between critical load and target load [source: Acid News, 1995]

The necessity of determining critical loads is to set goals for future acid deposition such that the environment is protected. Computation of critical load is an important tool for assessing acceptable levels of acid deposition throughout the region. It is also intended to represent the long-term capacity of ecosystems to absorb pollutants. It is important instrument in balancing the acidification and nutrient stress imposed on ecosystem by land use, such as forestry against acid deposition. The calculated value of critical load serves as an indicator that how much acidification can be loaded on to the system through deposition of S and N compounds that cannot lead to harmful effects.

METHODS TO DERIVE CRITICAL LOADS

The critical loads for total acidity of sulphur and nitrogen need to be determined so that a coherent agreement can be reached with regard to abatement policies. Critical loads for ecosystem can be derived in a direct and an indirect way as indicated in Figure 3. Thus the two methods by which the Critical load can be calculated are

I. Relative Sensitivity approach

Critical loads can be derived directly from the relationship between atmospheric deposition and effects on 'specified sensitive element' with in a ecosystem (ecosystem status) by correlative or experimental research. This method is based upon the climate factors, factors relating to geology and soil characteristics, other geophysical features of an area, plant species and their tolerance ranges.

II. Steady State Mass Balance Method

Critical load can be derived indirectly from critical values for ion concentrations or ion ratios in the ecosystem, based on dose response relationships between these chemical criteria and the ecosystem status. This method computes the critical load on the basis of the ability of a system to buffer the acidity due to supply of acidic compounds. Many countries in Europe have derived critical loads in an indirect way because a direct assessment generally leads to an overestimation, since effects on ecosystems may occur before harmful effects are visible [Sverdrup et al., 1990, de Vries, 1991].

Both these methods for assessing the risk of ecosystem acidification and association impacts from acidifying deposition can be shown in a block diagram (Fig. 4). These both methods are discussed in detail under below:

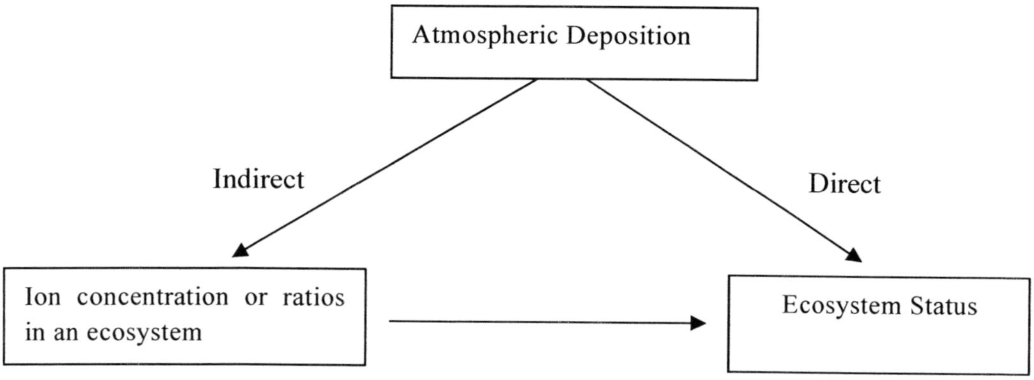

Figure 3. Research methods to derive critical loads for N and S (arrows indicate a relationship that can be assessed by correlative approach, process research and/or model research)

Source: de Vries, 1993.

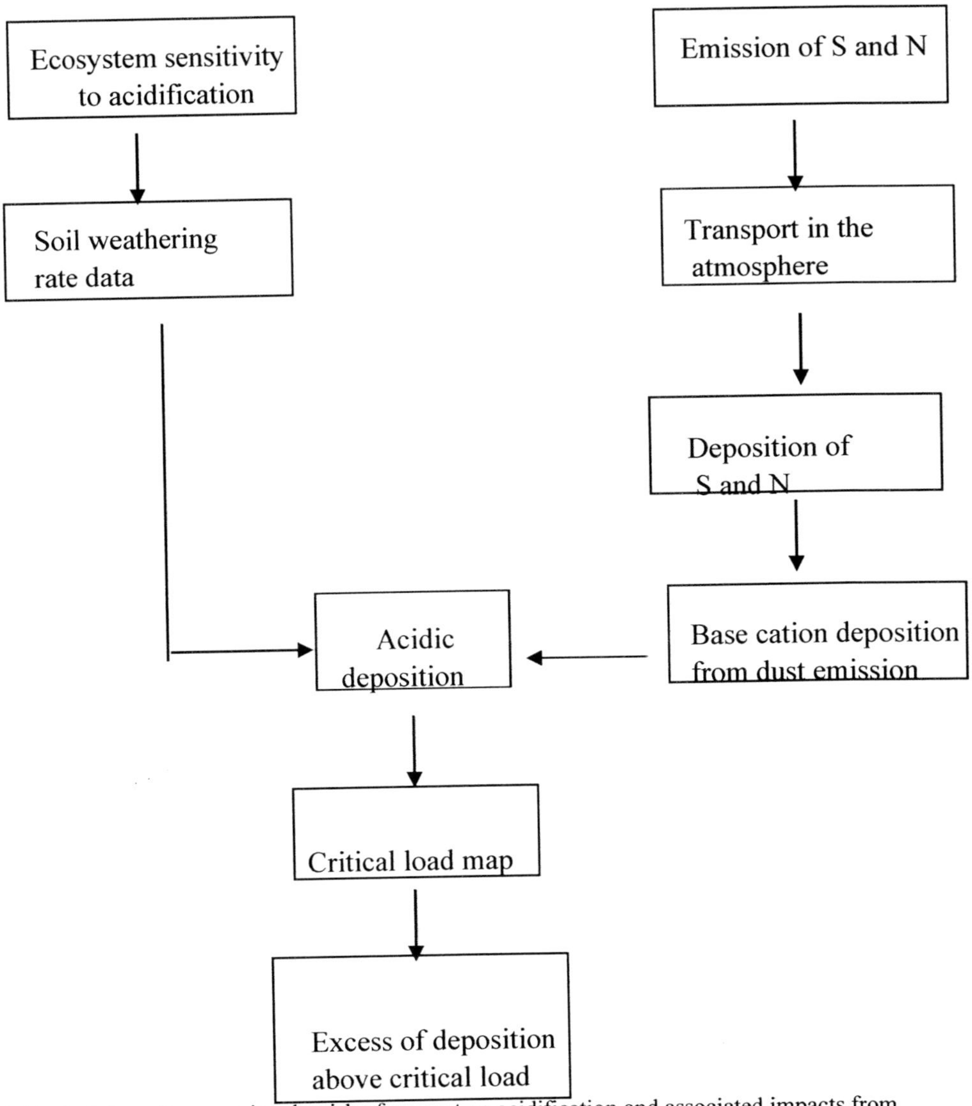

Figure 4. Methodology for assessing the risk of ecosystem acidification and associated impacts from acidifying deposition

Source: Kuylenstierna et al., 2001

Relative Sensitive Method [Chadwick and Kuylenstierna 1991]

Sensitivity may be defined as "the responsiveness of an important variable to an independent forcing variable". The absolute site-specific sensitivity is indicated by the current chemical status of the soil or surface water of a site together with the tolerance limits of species and the regional 'relative sensitivity' compares the ability of one region to withstand the effects of acidic inputs with the ability of another region. Different parts of the ecosystem will respond in different ways to acidic deposition. Regional sensitivity may be worked out from the distribution of ecologically relevant site factors or could be assessed from regionalized chemical data for soils and surface waters relevant to the survival of biota. In this method major environmental factors that determine site (or ecosystem) sensitivity are identified. These factors are weighted and combined together to give an overall index of relative sensitivity. These estimates are used to attribute deposition targets to relative sensitivity classes. The factors which are used to determine the sensitivity of a site to acidic depositions, are:

I. Bedrock lithology- Rock types have been assigned to one of two categories based on their weathering rates. The slow weathering rock types (eg. granite, quartz sandstones, decalcified sandstones etc.) are those with a low ability to neutralize acids because of rate limitation. The other rock types (eg. calcareous sandstones, limestone, coal measures etc.) which have higher ability to neutralize acids constitute the next category.

II. Soil Type: Soil types have been assigned to categories on the basis of soil chemistry and factors which control soil chemistry. Soil types with characteristics which may cause the soil chemistry to approach the aluminium buffer range, under the influence of acidic depositions will increase site sensitivity and are assigned to category I. Those soil types with major acids-buffering systems operating above this value are designated category II. Thus sites with category I soil types have higher sensitivity than those with category II soils.

III. Lands use: The studies done in Europe has shown that coniferous forest vegetation is considered to increase site sensitivity most because of the way in which such trees confer certain hydrological features to a site and because of the characteristic of the typical acid more soil organic layer characteristics formed under coniferous forest stands [Miller, 1985; Mikola, 1985]. Rough grazing and heathland vegetation also produce more humus and so this type of vegetation is also considered to increase site sensitivity, though to a lesser extent than coniferous forest. Deciduous forest vegetation have a lower relative sensitivity than sites with vegetation causing the production of a more humus. It is also assumed that practices such as fertilizers and lime application will artificially maintain pH and base saturation levels on intensively managed arable and rich grazing land and reduce sensitivity accordingly [Bache, 1983].

IV. Rainfall: As amounts of rainfall increase the base cation, aluminium and other acid-ion leaching rates increase because of the enhanced flow of water through the soil. Thus there is a decrease in the potential for acid neutralization by the mineral soil in high rainfall areas. Areas with a mean annual rainfall greater than 1200 mm are categorized as having a higher sensitivity than those receiving less.

The weighting procedure to the above mentioned factors should be such that when factors are combined the resultant relative sensitivity matches field observations and experience of such sites. Weights therefore may have to be modified as new information becomes available. For Europe Table 1 shows how the four site factors have been divided in to categories and also the associated weights used in their combination to arrive at an overall assessment of sensitivity to acidic deposition.

Table 1. Division of site factors into categories and associated weights for use in combination.

Factor	Weight	Category	Weighting
Rock type	2	I. Siliceous, slow weathering rocks	1
		II. Faster weathering rocks	0
Soil type	1	I. Major acid buffering < pH 4.5	1
		II. Major acid buffering > pH 4.5	0
Land use	3	I. Coniferous forest	1
		II. Rough grazing	½
		III. Deciduous forest	1/3
		IV. Arable land	0
Rainfall	1	I. > 1200 mm (annual average)	1
		II. <1200 mm (annual average)	0

Source: Chadwick and Kuylenstierna, 1991

The weighting given to rock type reflects the importance of mineral weathering in the neutralization of acidity whereas the effect on rainfall is not considered to affect sensitivity to the same extent as mineral weathering or land use.

Application of Relative Sensitivity Method to Study Acidification in Asian Developing Countries Including India

A method to derive the relative sensitivity to terrestrial ecosystems was explained and preliminary critical load values were assigned in the study done by Kuylenstierna, 2001. A range of values for critical loads and base cation deposition have been used to investigate in maps depicting the excess of deposition above the critical loads. These show an increasing risk of acidification in 2050 in extended regions of Southern and Eastern Asia in comparison to 1990. Certain Areas, especially in Asia were shown at risk even when high values of critical load and base cation deposition were used. Their method to evaluate relative sensitivity used proxy variables for soil mineral weathering (a rate limited but essentially capacity – unlimited buffering mechanism) representing a long term source of buffering and exchange of base cations on the soil cation exchange complex (capacity limited but rate unlimited representing a short –term buffering mechanism). The ability of cation exchange to buffer soils was taken as cation exchange capacity (CEC) measured at field pH. CEC and saturation data from a reference soil database [ISRIC, 1995] with data from around the world was used to provide average values for each soil type according to FAO, [1995]. This method used a combination of three classes of CEC and five classes of base saturation to define five relative classes of sensitivity (Table 2)

Table 2. Method used to allocate soil types to relative sensitivity classes

		Base Saturation (%)				
		0-20	20-40	40-60	60-80	80-100
CEC (meq 100 g^{-1} at field pH)	< 10	1	1	2	3	5
	10-25	1	2	3	4	5
	>25	2	3	4	5	5

The sensitivity score is: 1-high sensitivity; 5- low sensitivity

Source: Kuylenstierna et al., 2001

Most of the portion of India came in to the category 5 i.e. least sensitive, where as some portion of south west coastal India fall in category 2 and some areas of Eastern coastal areas (Bengal and Orissa) fall in to the category 4. Low sensitivity sites are confined to drier regions and areas where soil is derived from parent material with a high content of high weathering rate minerals. In order to make a quantitative comparison to deposition values critical load need to be assigned to the classes of sensitivity. It is assumed they are correlated as both are based on soil buffering. It is also being assumed that the critical load for a class lie in the same range as sites of the same sensitivity in different parts of the world. The preliminary critical load ranges taken in this study is shown in Table 3 with the understanding that these values will change as more information becomes available, particularly from developing countries.

Table 3. Preliminary critical load ranges assigned to relative sensitivity classes

Sensitivity Class	Critical load range (meq m^{-2} yr^{-1})	
	Low range	High range
1	0-25	25-50
2	25-50	50-100
3	50-100	100-200
4	100-200	200-400
5.	>200 (no critical load)	>400 (no critical load)

Note: The higher value of each range has been used to map critical load exceedance

Source: Kuylenstierna et al., 2001

In the aforementioned study a measure of risk i.e. exceedance of critical loads were derived and mapped to see the degree to which net acidic deposition derived from sulfur emissions (Sulphur deposition minus the base cation deposition rate) exceeds critical loads. In order to examine the extent to which the ranges of base cation deposition and critical load influences the deposition, the exceedances were calculated for the year 1990 and 2050 using a low and high base cation deposition and low and high critical loads as shown in the above Table (3). In 1990 calculations revealed that only some parts of China and South East Asia in

Asian continent show areas of exceedance, India is at no risk at all. In 2050 result where the exceeded areas in Europe and North American decrease in size due to emission decreases achieved by the UN/ECE Protocols [UNECE, 1996; 1997] and clean Air act in comparison to 1990 results, there was an increase in exceedance in many developing countries particularly in China, parts of south and South-East Asia. In India only few areas on South West coast in 2050 shows high exceedance and some part of Himalayas showed low exceedance which could develop if emissions follow the scenario used in this study. This results changes to low exceedance when higher values of critical load and base cation deposition are used as shown in Table 3 in column high range. Thus the choice of critical load and base cation deposition ranges can affect whether areas projected to be exceeded or not e.g. in the Himalayas exceedance vanish when high range of critical load and base cation deposition range are taken and South West coastal areas of India were now projected to fall in low exceedance Area. The exceeded areas therefore indicate ecosystems at risk where acidification and consequent impacts to terrestrial ecosystems may occur.

In the development of the IMPACT module of the RAINS-Asia model [Foell et al., 1995] the role of the sensitivity mapping has been done to assess:

- data inputs to both the method of Relative Sensitivity and the Steady State Mass Balance Model
- include ecological considerations in addition to the biogeochemical basis of the steady state mass
- balance method and
- the geographic distribution of sensitivity as an aid to verify the distribution of critical loads over Asia thus enhancing the reliability of the results.

This study of mapping sensitivity was done to create a distribution of ecosystems that have different relative reactions to a given rate of acidic deposition on terrestrial ecosystems. For the assessment of sensitivity, relative units of buffering rate were used rather than quantitative values of weathering rate, the reason is to accommodate for the variability of the data and uncertainties in weathering rate values. This methodology was based on three main aspects: how to allocate vegetation types to classes that indicate the sensitivity of ecosystems to acidic deposition, how to determine the buffering ability from soil type attributes and climatic factors and finally how to link the vegetation categories to the soil (buffering) characteristics. In this Asian work the combination method has been kept simple to form five final classes of sensitivity to acidic deposition. From the results it was seen that the most sensitive areas are found in South-eastern Asia, parts of the Himalayas range, parts of the Tibetan plateau and in the southern part of the boreal forest that dips into northern China. There are other small areas of high sensitivity such as in parts of Japan and the rain forest strips in southwestern India. The high sensitivity regions are largely found on acidic soils and the ecosystem shows a low ability to buffer acidic deposition in the long term. The insensitive areas cover the mainly dry, desertified region and rich agricultural lands. From the results it was clear that most of India is not sensitive to acidic deposition.

The Steady State Mass Balance Method (SSMB) [Sverdrup et al., 1990]

This method applies to soil, ground water, lakes and surface water runoff from catchments. It use the static mass balance between long term inputs of acidity and long term in soil production of alkalinity. This method is also called Simple Mass Balance (SMB) model: SMB indicates that we are simplifying the description of the biogeochemical process involved, a necessity when considering the large scale application and the lack of adequate data. SSMB indicates that we are dealing with steady state conditions only, leading to considerable simplifications. These models and methods have been elaborated in several tasks force meetings and workshops [Nilsson and Grennfelt, 1988; Downing et al., 1993; Posch et al., 1995]. The basic principle of the method is to identify the long term average sources of acidity and long term average sources of acidity and alkalinity in the system and to determine the maximum acid input that will balance the system at the biologically safe limit. Several assumptions have been made in the steady state calculations. First of all, it is assumed that ion exchange is at steady state, and that there is no net change in base saturation or no net transfer of acid neutralizing capacity (ANC) from soil solution to the ion exchange matrix. It is assumed that for nitrogen there is no net denitrification, adsorption or desorption, the nitrogen cycle is at steady state. Sulphate is also assumed to be at steady state, no sulphide oxidation, sulphate uptake, sulphate permanent fixation or sulphate reduction are considered as significant. A simple hydrology is assumed, where there is straight infiltration through the soil profile. The models assume soil solution equilibrium at all times [Sverdrup and Warfvinge, 1988; de Vries, 1991].

Soil acidification is defined as an increase in total absorbed acidity in the soil relation to the content of absorbed base cations in ion exchange positions, generally interpreted as a decrease in base saturation. Lake and streams water acidification implies a decrease in the solution alkalinity and pH. Initially in an acidification process, a soil may acidify, as it absorbs hydrogen ions in exchange for base cations. This the runoff from the soil which is undergoing acidification may stay semi-natural, due to the in-situ neutralization of percolating solution. Later as the storage of base cations have been exhausted, and the neutralizing capacity with it, the acid percolate will no longer be neutralized. The acid soil will in the acid state be able to acidify a neutral rain by the same ion exchange process. No change in adsorbed acidity or base cations will take place when the total net inputs of acid to the soil are balanced by the long term total net inputs of alkalinity production in the soil system.

In a wider context of time, soil acidification may be defined as the lowering of the total system content of ANC [de Vries and Breeuwsma, 1987]. This is to be interpreted in terms of acid neutralizing capacities stored in ion-exchange positions, organic material and weatherable minerals. In this context the long term soil development , podzolization or depletion of soil storage of easily weatherable minerals, which has occurred since the last ice-age in Northern Europe, is also a type of natural acidification, even if it is very slow and quantitatively small as compared to anthropogenically induced acidification.

The calculation of critical loads for ecosystem discussed in this section is based on the method given by Hettelingh et al., 1995. The method computes equilibrium between acidification increasing processes (sulphur deposition and base cation uptake) and acidification decreasing processes (weathering rate and base cation deposition). This method assumes a time independent steady state of chemical interactions involving equilibrium between the soil solid phase and soil solutions.

CRITICAL LOAD OF S

Critical load of acidity (CL(A)) includes the acidity contributed by both S and N compounds and for its evaluation, both pollutants are considered simultaneously. Critical load of acidity is computed as follows

$$CL(A) = BC_w + H_{crit.} Q + Al_{crit.}Q \qquad (1)$$

where BC_w is base cation weathering rate, H_{crit} is the critical hydrogen leaching, Al_{crit} is critical aluminium concentration and Q is runoff.

The maximum allowable deposition of S, which does not lead to 'harmful effects' in the case of zero N-deposition is given by

$$CL_{max}(S) = BC_{dep} - BC_u + CL(A) \qquad (2)$$

where BC_{dep} is the deposition and BC_u the net growth uptake of base cations. Critical load of acidity is not directly applicable for evaluation of required S emission reduction, therefore the current UN/ECE protocols concentrate on a single compound either S or N. Therefore, the critical load of acidity has been divided between S and N assuming (a) the fraction of S deposition of total acid deposition is used as proxy of the part of the critical load of acidity attributed to S and, (b) N deposition contributes to acidification only when it is not taken up or immobilized by the ecosystem. Thus, a term sulphur fraction (S_f) has been introduced which is defined by

$$S_f = S_{dep}/(S_{dep} + N_{dep} - N_u - N_i) \qquad \text{for } N_{dep}>N_u+N_i \ (S_f = 1 \text{ otherwise}) \qquad (3)$$

where S_{dep} is the S deposition and N_{dep} is the deposition of both oxidized and reduced N, N_u and N_i are net growth uptake and immobilized N. The critical deposition of S, for a given N deposition is given by

$$CD(S) = S_f. CL_{max}(S) \qquad (4)$$

Critical deposition of S has been compared to S deposition in negotiating the sulphur protocol by computing the so-called exceedance of critical deposition. Exceedance of critical deposition is given as the difference between S deposition and critical deposition of S given as

$$Ex(S) = S_{dep} - CD(S) \qquad (5)$$

and $Ex(S) \leq 0$ ensures that the ecosystem is protected.

BesideS, the deposition of N also contributes to the acidification of ecosystems. The excess deposition of S and N can be written as:

$$Ex\ (N_{dep}, S_{dep}) = S_{dep} + (1 - f_{de}).N_{dep} - (1\text{-}\ f_{de})\ (N_u + N_i) - CL_{max}(S) \qquad (6)$$

where f_{de} is denitrification factor, N_{dep} is the deposition of N and $CL_{max}(S)$ is given by eq. (9). If the exceedance of acidity in above equation is negative or zero i.e. $Ex(N_{dep}, S_{dep}) \leq 0$ for the given pair of deposition (N_{dep}, S_{dep}) it can be said that critical loads are not exceeded. However, there are many values of S and N deposition for which exceedance becomes zero i.e. unique critical load values cannot be defined. Furthermore, for

$$N_{dep} \leq N_u + N_i = CL_{min}(N) \qquad (7)$$

The maximum allowable deposition for S is given by $CL_{max}(S)$; and the maximum 'harmless' acidifying deposition of N is obtained by inserting $S_{dep} = 0$ in eq. (13) and solving $Ex(N_{dep}, 0) = 0)$:

$$CL_{max}(N) = CL_{min}(N) + CL_{max}(S)/(1\text{-}f_{de}) \qquad (8)$$

THE EXCEEDANCE OF CRITICAL LOADS

Critical loads are calculated to express the sensitivity of ecosystems to the deposition of sulphur and nitrogen in quantitative terms, which can be directly compared to the deposition of S and N and thus be used in negotiations on emission reductions of these compounds. The comparison between deposition and critical loads has been done by computing the so called exceedance i.e. the difference between deposition and critical loads.

$$Ex\ (S + N) = S_{dep} + N_{dep} - CL\ (S + N) \qquad (9)$$

In Fig 5. let the point E_1 denote the present deposition of N and S, then the amount of S+ N deposition to be reduced to reach non-exceedance (the thick line in Figure) is independent of the path i.e. it is the same for reaching the point Z_1, Z_2, Z_3 since the slope of the line is one. If S_{dep} is greater than CL_{max} (S) the critical load is exceeded, although S_{dep} might be smaller than CL (S + N) and Ex (S + N) might be negative (e.g. the deposition situation represented by point E_2). Reducing N_{dep} substantially, one reaches the point Z_1 and therefore non-exceedance without reducing S; on the other hand one can reach non-exceedance by only reducing S_{dep} (by a smaller amount) until reaching Z_3; finally with a smaller reduction of both S_{dep} and N_{dep} one can reach non-exceedance as well (eg. point Z_2). In practice external factors such as costs of emission reductions will determine which path will be followed to reach zero exceedance. This approach reduces the problem of critical loads and exceedances to mere numbers and facilitates communication with non-technical users (policy makers).

For use in European emission reduction negotiations, the large number of critical loads has to aggregate. Cumulative distribution functions and percentiles have been used to present critical loads of sulphur and acidity in map format. Thus the concept of the percentile function or protective isoline is followed. An example of a protection isoline is shown in Fig 6. All combination of N_{dep} and S_{dep} lying below the p-th percentile function protect 100-p percent of the ecosystems in a given grid cell. Although no unique exceedance can be defined, but one can distinguish five cases:

1. the point (N_{dep} and S_{dep}) lies below the percentile function i.e. no exceedance
2. reductions in N_{dep} or S_{dep} are interchangeable i.e. non-exceedance can be reached by either N reductions or S-reduction alone.
3. Some reduction in S deposition are mandatory
4. Some reductions in N_{dep} are mandatory and
5. Both reductions in N and S are required to achieve non exceedance

A study was done to compute and map critical loads for Asia [Hettelingh et al., 1995] for 31 different vegetation types. The methodology involved the steady state mass balance (SSMB) originally developed for Europe. Results indicated low critical loads (i.e. below 500 eq ha^{-1} yr^{-1}) for south-eastern Asia, parts of the Himalayan range and Tibetan plateau and in parts of the boreal forest in northern China. Other areas with low critical loads (high sensitivity) include the rainforest strips in south-western India. These results do not imply that soils are acidified yet but some areas with low critical loads are subject to a higher risk of acidification than other areas. This map also provides indication of where continued excess of critical loads may, in some areas earlier than in others, ultimately exhaust the buffering capacity of the soils. The dry region in most of India and north western China show relatively high critical loads (exceeding 2000 eq ha^{-1} yr^{-1}). This study had concluding remarks that the calculation of critical loads for much of Asia with its highly varying environmental conditions and ecosystems is a considerable effort. Therefore, extensive validation procedure is required to improve the assessment of risk of damage on local scale.

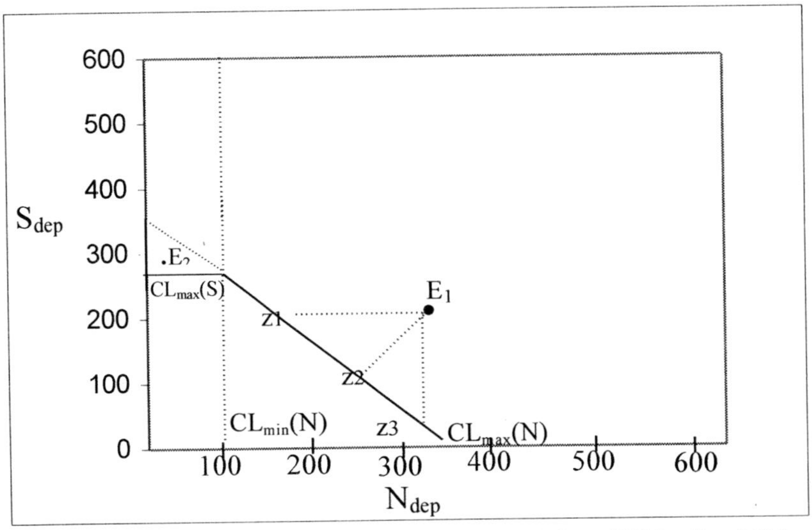

Figure 5. Relationship between N_{dep} and S_{dep} and the critical loads for acidifying S and N for a hypothetical ecosystems. source: Posch et al., 1995

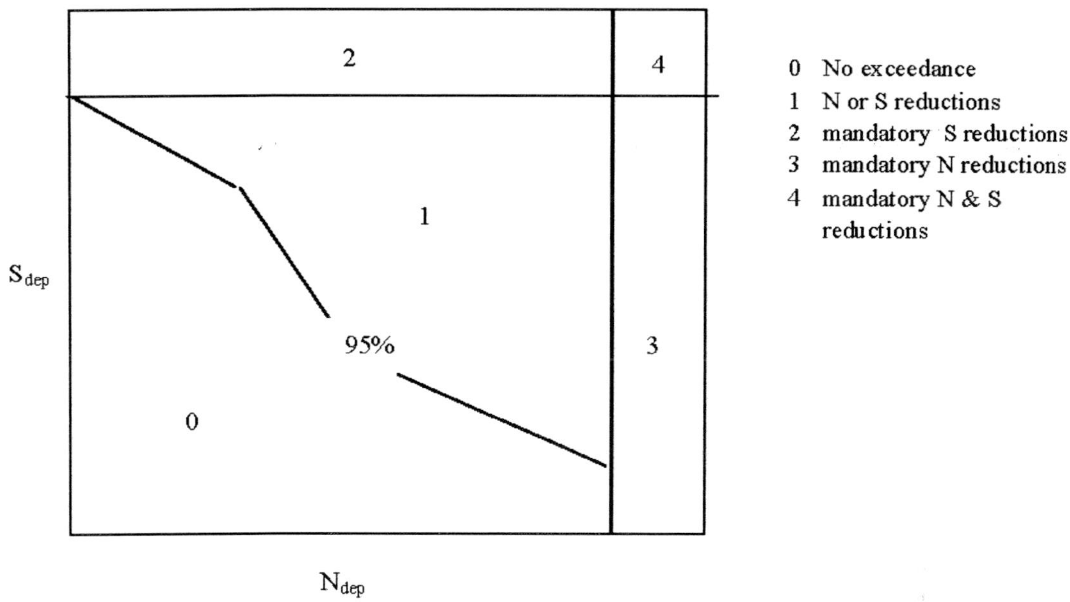

Figure 6. Example of a protection isoline (percentile function)
Source: Posch et al., 1995

Brief Description of Some Case Studies Related to Critical Load in India

Geographical and climatic conditions are advantages with India which provide natural buffering ability due to following reasons:

a) Tropical climatic conditions reduce atmospheric CO_2 dissolution in the rain water.
b) The predominantly alkali rich soil has an additional neutralizing effect
c) The sea and soil derived aerosols lessen acidic deposition
d) Higher temperature and sunlight increase the efficiency of atmospheric chemical reactions which decrease the acidic deposition
e) Indian coal has less sulphur content approximately 0.5 to 2% compared to 3% found in coal of western countries.
f) India has no significant natural sources of sulphur such as volcanoes etc.

The Background Air Pollution Monitoring Network (BAPMoN) data from ten stations in India for the period 1974-90 was analysed by Datar et al., 1996 to determine the background trends. The annual precipitation volume weighted mean (APWM) values of pH were calculated which is shown in Table 4.

Table 4. Mean pH of rainwater in Indian BAPMoN stations (1974-1990)

Station	Observed pH
Allahbad	6.93
Jodhpur	7.42
Kodaikanal	6.28
Minicoy	6.58
Mohanbari	5.98
Nagpur	5.97
Port Blair	6.15
Pune	6.43
Srinagar	7.72
Visakhapatnam	6.01

Source: Datar et al., 1996

The background pH of rain over India lies in the range 5.97 – 7.42. These values are higher than 5.65 for unpolluted rain. The time series of pH for six out of ten stations i.e. Allahabad, Kodaikanal, Mohanbari, Port Blair, Nagpur, Pune and Vishakapatnam show dipping tendencies, which at Vishakapatnam is quite pronounced. The post 1982 epochal mean is significantly lower. A possible explanation for the downward tendency of pH is the increase, over the years in the concentrations of strong acid radicals. A negative correlation of pH with NO_3 which is the general feature of the Indian BAPMoN [Mukhopadhay et al., 1992] indicates that rapidly increasing vehicular traffic could be the primary cause for declining tendencies of the pH.

The alkaline properties of soil particulates have so far sustained the alkaline nature of rain water (pH >5.65) and have been controlling the spread of acid rain in India. Acid rain whereas it has occurred in India is purely a local phenomenon and is restricted within short distances in the upwind and/or downwind of the industrial complexes. Although there is no problem of acid rain at present in India, its occurrence in future cannot be ruled out as air pollution is already severe in metropolitan industrial cities. The scenario may change in the coming decades when the pressure on atmospheric environment will increase due to population growth, changing agricultural practices, increasing number of vehicles, power generator and industrial development.

The RAINS-Asia Model

In development of RAINS-Asia model aforementioned approaches (relative sensitivity and steady state mass model) are used by the scientists. RAINS-Asia (Regional Air Pollution Information and Simulation Model for Asia) model is a user-friendly, integrated, PC based model for regional policy analysis and decision-making regarding energy sector development. It is a tool used to assess and project future trends in emissions, transport and deposition of air pollutants and their potential environmental impacts. RAINS-Asia was developed by a team of Asian, European, and North American Scientist under the leadership of the World Bank and the Asian Development Bank. This model covers 23 countries in East, South and South-East Asia which are divided into 94 regions. The model uses 1990 data as a base and calculates future energy, emissions and environmental parameter through 2020, in ten-year increments. The RAINS-Asia model consists of three modules each addressing a different part of the acidification process. The Regional Energy and Scenario Generator (RESGEN) module estimates energy pathways based on socio-economic and technological assumptions: the Energy and Emission module (ENEM) uses the energy scenarios to calculate sulphur emissions and costs of control strategies. The deposition and critical loads (DEP) module, which consists of the two sub module; Atmospheric Transport and Deposition (ATMOS) submodule and the Environmental Impact and Critical loads (IMPACT) submodule. The atmospheric module (ATMOS) of RAINS-Asia analyses long-range transport and deposition of sulphur in Asia region, their levels and source locations (from ENEM) with meteorological, chemical and physical data is used to calculate the resulting sulphur deposition patterns. The model provides annual average (wet + dry) sulphur (SO_2 and SO_4) deposition values and monthly average SO_2 concentration values for each 1 degree grid cell. The model allows the user to assess the spread of pollution from an individual source or region. The IMPACT submodule assesses the sensitivity of various ecosystems to acid deposition to the deposition data generated by the ATMOS module. The carrying capacity of acid deposition of 31 types of ecosystems is estimated. Values of sulphur deposition are based on a yearly average and are calculated at 1 x 1 degree resolution. This model assesses only the indirect effects of sulphur deposition on soil. The steady state mass balance method is used to calculate critical loads in the RAINS-Asia impact submodule.

Based on the results from RAINS model, it is seen that the energy consumption under the base scenario increase from 15811 peta joules (PJ) to 63621 PJ during the period 1990-2020, thereby showing an increase of 302%. The SO_2 emissions during the same period show an even greater increase of 321% i.e. from 4.48 million tones (Mt) to 18.87 Mt (Table 5).

Table 5. SO_2 emissions and energy consumption by fuel for India

Fuel	SO_2 emissions (thousand tons of SO_2)				Energy consumption (PJ)			
	1990	2000	2010	2020	1990	2000	2010	2020
Coal	2187	3203	5529	9449	4308	6163	10278	17223
Oil	1291	2146	3919	7480	2351	4311	9282	21512
Others	1001	1270	1631	1942	9152	12434	17372	24886
Total	4479	6619	11079	18871	15811	22908	36932	63621

Source: RAINS-Asia 7.02 version

The percentage of SO_2 emissions from coal and oil are 49 and 29% respectively in 1990 and subsequently increase to 50 and 40% in 2020. This is basically due to the share of coal in the total energy consumption being roughly the same (27%) while that of oil increases from about 15% to 34% during the period 1990-2020. Maps were drawn to shown the excess sulphur deposition above critical loads in 1990,2000, 2010 and 2020 under the reference scenario. It was evident from the study [Panwar et al., 1998] that in 1990 the excess sulphur deposition was restricted only to a few locations, however the situation changes significantly in 2020 whereby large regions in northern and eastern India show sulphur deposition level that exceed the critical load in those areas. Many regions experience excess sulphur deposition between 2 to 5 gm m^{-2} yr^{-1}. It was noted here that even under the highest emission scenario considered here, the western and central parts of India do not experience excess sulphur deposition above the critical load mainly due to the alkaline nature of the soil particles.

The RAINS model is capable to simulate SO_2 emission control strategies and explore their costs and their regional environmental benefits. Table 6 shows the SO_2 emissions and the associated control costs under various control Technology Scenario i.e. Best Available Technology (BAT), Advanced Control Technology (ACT), Basic Control Technology (BCT) and Local Application of Advanced Technology (LAAT)

Table 6. Emissions and abatement costs for India under various control scenarios

Scenario	Emissions (thousands tones of SO_2)			Abatement cost (million US$ yr^{-1})		
	2000	2010	2020	2000	2010	2020
Base-BAT	2058	3389	5906	5721	9370	17055
Base-BCT	5074	7970	13054	1887	3424	6213
Base-LACT	5505	8408	13435	1060	2024	3892
Base-ACT	4752	6866	10522	1834	3379	6337
Efficiency BAT	1784	2684	4476	4781	7279	12357

Source: RAINS-Asia 7.02 version

BAT: Best Available Technology; BCT: Basic Control Technology; LACT: Local Applications of Advanced Technology; ACT: Advanced Control Technology

If no counter measurers are taken, an initial analysis with the RAINS model indicates that currently observed trends in energy consumption will impose significant environmental threats to the ecosystems in many parts of India. There are a variety of measures that can be taken to reduce SO_2 emissions and thereby avoid excess deposition. The potential exits for an increase in the cost effectiveness of strategies if measures are focused on specific fuels, technologies, economic sectors, emission sources or ecologically sensitive regions.

RAINS-Asia model was also used [Mohan and Kumar, 1998] to focus attention on south Bihar, parts of West Bengal and Delhi as these areas are comparatively found to be more vulnerable in terms of pollution levels [Kulshrestha et al., 1995; Ravichandran et al., 1994]. The critical load for Delhi for maximum allowable total acidic deposition with respect to the percentage to ecosystem threatened in various scenario is given in Table 7.

Table 7. Exceedance of area from critical level and cost of technology under three scenarios

Options and estimated cost	Percentage exceedance of area from critical level			
	1990	2000	2010	2020
Region: S-E Bihar and West Bengal				
Advanced Control Technology available all over the world	50	55	85	90
Best Control Technology	50	55	85	100
Locally Available Advanced Control Technology	500	55	85	90
Region: Delhi				
Advanced Control Technology available all over the world	0	0	0	0
Best Control Technology	0	0	0	0
Locally Available Advanced Control Technology	0	0	0	0
Cost of application (mio US$ yr^{-1})				
Advanced Control Technology available all over the world	17.61	1834.22	3379.12	6337.29
Best Control Technology	17.61	1887.35	3424.09	6213.49
Locally Available Advanced Control Technology	17.61	1059.98	2023.77	3891.70

Source: Mohan and Kumar, 1998

Due to sulphur deposition exceeding critical load from 0% in 1990 to 15% in 2010 if advanced control technology is used to control the emission. It may rise to 50% if only basic control technology is used for emission control. If the present trend continues unhindered, the ecosystem threatened may be well over 85% by the year 2020 even if the advanced control technology is used to control the emissions in this area. A small area in south Bihar as well as parts of west Bengal was found experiencing stress on their entire ecosystem with percentage exceedance of area from critical level reaching 50% by the year 2020. The main reason for south Bihar and parts of west Bengal facing such treats is due to the location of the largest coal belt and thermal power plants in the region. Further, there are several small-scale enterprises using coal as their energy input not adhering to emission standards. Cost of application of adopting the various control technologies is also included in the Table 7. Therefore it is desired that appropriate measures are take to reverse the process. Among various other measures, proper sitting of future industries and possibly past relocation of existing ones would be required to reduce the impact on the ecosystem by the year 2020.

In the study done in Agra (Bhattacharya, 2002) the SO_2 emissions were calculated for this region using the RAINS-Asia model with Best Control Technology (BCT), Locally Advanced Control Technology (LACT) and Advanced Control Technology (ACT). The results were obtained by keeping the base year as 1990 as kept in the original model and then changing the base year to 2000 i.e. feeding the actual data of SO_2 emission instead of predicted data which was available for this region for year 2000. The difference in percentage between the predicted and actual values by changing the base year from 1990 to 2000 was computed. It was 50.6% for the year 2000 and the predicted difference percent increases to 52.2% by the year 2020 using the Best Control Technology (BCT) and the difference percent between the predicted and actual values SO_2 emission using Locally Advanced Control Technology (LACT) and Advanced Control Technology (ACT) are 35.2 for the year 2000 and the predicted difference percent is 34.3% by the year 2020. The difference percent between the predicted and actual values of critical loads are 22.7% for the year 2000 and the predicted differences percent increases to 37.1% by the year 2010 and in the year 2020 the difference is 22.7% as shown in Table 8.

Table 8. Calculated and predicted critical load for Acid deposition for Agra region taking base year as 1990 and 2000 (from RAINS-Asia model)

Base Year	Critical load (Eq ha^{-1} yr^{-1})		
	2000	2010	2020
1990	1250	1823.5	2516.9
2000	995	1251.9	2000.2

Source: Bhattacharya, 2002

The difference in exceedance of this region from critical level under these scenarios taking base year as 1990 and 2000 were also derived using RAINS-Asia model shown in Table 9.

Table 9. Differences in exceedance

Option	% exceedance of area taking base year 1990			% exceedance of area taking base year 2000		
	2000	2010	2020	2000	2010	2020
Advanced Control Technology	0	15	85	0	0	0
Best Control Technology	0	50	90	0	10	50
Advanced Control Technology (locally available)	0	50	85	0	15	55

Source: Bhattacharya, 2002

Thus this study shows that the future scenario of this region in north India is not as bad as predicted by RAINS-Asia model taking base year 1990. This difference in the predicted and actual results also lead to an important conclusion that models should be validated from time to time.

Critical Loads for the Agra Region in India

A steady state mass balance method (SSMB) was used to calculate the critical load of sulphur and nitrogen for soil. The present load of sulphur (161.1 eq ha^{-1} yr^{-1}), nitrogen (49.9 eq ha^{-1} yr^{-1}) and ammonium (172.4 eq ha^{-1} yr^{-1}) were calculated from wet and dry deposition collected from Agra region. The values of the critical load calculated of S and N for soil with respect to *Cenchrus ciliaris* [Satsangi et al., 1998], *Citrus argentifolium* (lemon) [Bhanti et al., 2003], *Hibiscus rosa-sinensis* (Hibiscus) and black siris (*Albizzia odoratissima*) [Bhattacharya et al., 2004] are shown in Table 10.

Table 10. Calculated critical loads for different species available at Agra region

Common name	Botanical name	Critical load for Soil	
		S (eq ha^{-1} yr^{-1})	N (eq ha^{-1} yr^{-1})
Anjan grass	Cenchrus ciliaris	309.2	381.8
Lemon	Citrus argentifolium	432.4	1123.0
Hibiscus	Hibiscus rosa-sinensis	180.7	253.8
Black siris	Albizzia odoratisima	181.6	254.7

The values in the above table shows that present load are lower than the critical loads in this region. In the above mentioned studies the calculated values of critical load of actual acidity were 432.4 eq ha^{-1} yr^{-1} and 377.1 eq ha^{-1} yr^{-1}. These obtained values are lower than the values of actual acidity assessed by the RAINS-Asia model for this region which is in the range of 500-1000 eq ha^{-1} yr^{-1} (Fig. 9). This analysis will be able to anticipate pollution risk to a particular receptor.

In a recent study [Satsangi et al., 2003] the assessment of the potential for soil acidification in north India using critical load approach and locally derived data for acidic and basic inputs was done. In this study major ions (Cl^-, NO_3^-, SO_4^{2-}, Ca^{2+}, Mg^{2+}, Na^+, K^+ and NH_4^+) were analysed in wet and dry deposition samples collected for 2 years using a polyethylene bottle and funnel collector at Agra in India. The deposition of ionic components (Ca^{2+} and Mg^{2+}) derived from natural sources i.e. soil were higher than those of anthropogenic origin. In rainwater samples, non-sea-salt fraction was found to be 60-90%. Results suggest that most of the acidity, which occurs due to NO_3^-, SO_4^{2-} and Cl^- is neutralized by alkaline constituents, which originate from airborne local soil and dust transported from the Thar desert. Acid neutralizing capacity of soil have also been quantified and found to be 33 x 10^3 neq g^{-1}. Using deposition data, the critical load for acidity of soil with respect to Ashoka and Eucalyptus was evaluated by the SSMB method. The calculated value of critical load of actual acidity was found to be 622.4 eq ha^{-1} yr^{-1}. The relationship between N and S deposition and the exceedance of the critical loads of acidity (i.e. S and N) are illustrated in Fig. 7 $CL_{max}(S)$ and $CL_{max}(N)$ were found to be 2192.4 and 4512.4 eq ha^{-1} yr^{-1}, respectively. The thick line represents the critical deposition, which causes no 'harmful effects' and this function may be termed the critical load function of the ecosystems. It is also evident from the Figure (7) that the current deposition load (represented by dash line (---) in the figure) of S and N, which was estimated from the wet and dry deposition data are much lower than the critical loads of S and N (represented by dark line (——) in the figure). Potential exceedance has also been calculated for soil with respect to Eucalyptus and Ashoka trees. Exceedance of critical deposition was found to be –1709 eq ha^{-1} yr^{-1} which is consistent with the exceedance value mapped by sensitivity approach [Kuylenstierna *et al.*, 2001]. No exceedance of critical loads indicates that the ecosystem is protected and till date there is also no impact of acidification to the ecosystem. Thus, indicating that the acidity, which is formed during the atmospheric reaction, is neutralized by the alkaline substances released from the soil and other sources. But with deterioration of the environmental quality and with the increasing S and N deposition as the trend shows if in future the critical load lies above the thick line then one can plan a path reducing to non-exceedance either by reducing N_{dep} or S_{dep} or by the reduction of both as shown in the Figure 7.

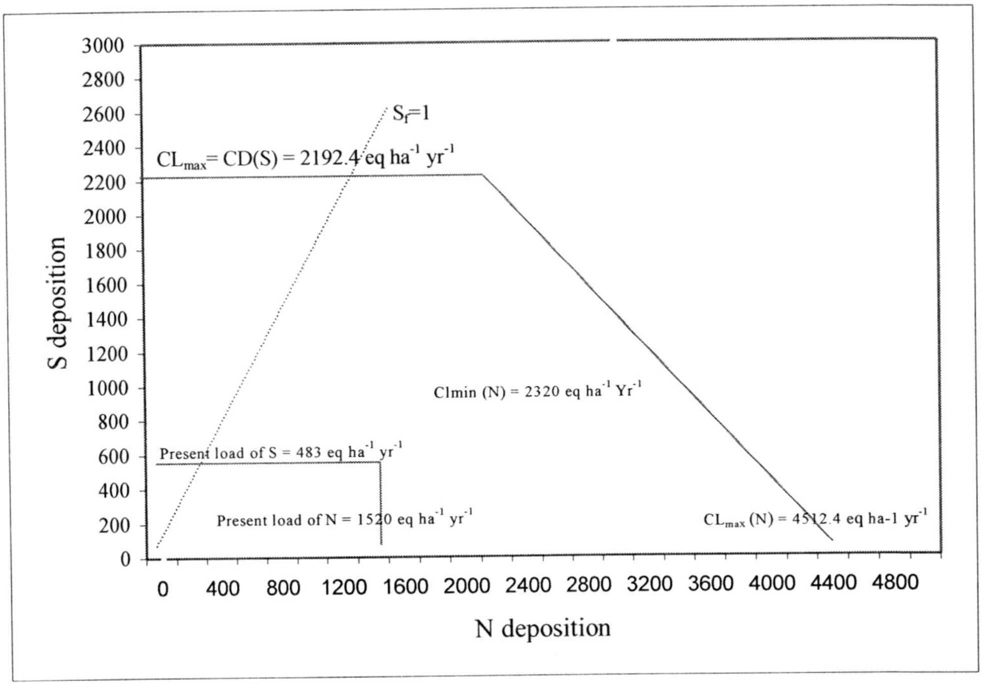

Figure 7. The relationship between N deposition and S deposition and the exceedance of acidity for soil
Source: Satsangi et al., 2003

The present critical load exceedance value calculated for soil indicates that the ecosystem which is enriched with these species is still protected. It is recognized that these are uncertainties in critical load and deposition estimates and there is a need to improve the accuracy of both. However this simple approach used here helps to identify areas where exceedance is most likely for a particular species.

Besides the S and N species, increasing levels of ozone have significant environmental impacts including human health and reduction in crop yields. Ozone is a regional problem alike acid deposition, which will require regional cooperation and emission reduction policies to control. In this respect workshops within UN/ECE Convention on Long range Transport of Air pollution at Egham (U.K.) and Berne (Switzerland) have resulted in a definition of critical levels of ozone for agricultural crops, being an accumulated ozone exposures above a threshold of 40 ppb (AOT 40) of 3000 ppb.h with in the months May, June and July during daylight hours only in Europe. For forest trees, the critical level was defined as an AOT 40 of 10,000 ppb.h in months from April to September integrated 24 hours a day [Smith, 1994]. Recently, Satsangi et al., [2004] used AOT 40 exposure on crop damage in Agra region. As there is no O_3 threshold exposure index established for Asian countries, AOT 40 exposure index used for European countries has been applied in this study. The estimated AOT 40 for summer and winter periods is approximately 3 and 1.5 times lower than the critical levels of O_3 thus showing that the state of affairs in Agra at the present level of emissions of O_3 and its precursors is not serious but if current rate of O_3 exposure persist, the AOT 40 value by 2020 would be 3 to 4 times of its present level and then one might predict the risk of crop reduction from O_3 pollution. However obtained results are certainly preliminary and include uncertainties and must be viewed in the lines of enriching O_3 data and establishment of an O_3 exposure index for Asian countries.

UNCERTAINTIES AND LIMITATION IN CRITICAL LOAD CALCULATION

Critical loads are presented as thresholds and specifying threshold for the natural environment is very problematic for several reasons. The first is that sharp threshold are rather rare, the natural environment is characterized by variation in many dimensions then there is a chance of random variation also. Then secondly there is difficulty of dealing with diversity and variability. Any damage threshold will depend on which species is chosen for protection and how damage is defined.

The uncertainties involved in the average critical load calculation can be divided into there major source of origin which are

a) Critical chemical limit
b) Input data uncertainties due to
   - spatial variability in the field
   - accuracy of parameter estimation
c) Validity of model assumption

The scale of data used is also an important issue. Usually general trends will be picked out and many of the most impacted areas will not be identified. For sensitivity classification uncertainties exist in attributing base saturation (BS) and cation exchange capacity (CEC) data to soil types due to variability within the soil type. By using internationally derived reference data it is possible that drastically altered soil profiles would be avoided (e.g. acidified or fertilized soil). In SSMB method, uptake, immobilization, precipitation and reduction of S are considered insignificant in forest soil. Some of this could be challenge for instance S uptake into plants is perhaps 10% of base cation uptake, sulphate adsorption is also an admittedly significant process etc. An important question is whether the sensitivity assessments and critical loads in absence of important parameters like weathering rates are applicable in tropical and subtropical ecosystems.

As mentioned in the study [Skiffington, 1999] critical load are the best viewed as highly uncertain estimates of relative risk which themselves incorporate political choices rather than precise damage thresholds determined by an objective scientific process. Pollution control science should also think of going beyond critical loads to the prediction and communication of pollution control policies effects on organisms and ecosystems

THE WAY FORWARD

Though subject to all limitations described above, critical loads have undoubtedly been useful in developing emission control policies in Europe. If they are to continue to be useful, policy makers must appreciate their limitation. Problems will arise if they are treated as though they are hard numbers known with absolute accuracy. Barkman [1997] proposed a definition of critical load in terms of risk analysis context: a critical load is "a quantitative estimate of an exposure to one or more pollutant below which the probability of long term unacceptable effects on specified elements of the environment is politically acceptable according to current knowledge". This makes it clear that use of critical loads should be exercise in risk management and that political choices are involved in determining the value of critical loads.

In India, since the sensitivity of terrestrial ecosystems to acidic deposition links different database (soil, vegetation and climate) and focuses on vegetation, monitoring networks should be strengthened, more so in vulnerable areas to estimate the real effect on vegetation and human health. As acid deposition in this country is a regional scale phenomena it should be assessed on the basis of data on wet and dry deposition measured at regionally representative sites. Thus, an integral approach including valid measurements of wet and dry deposition at regionally representative site and estimation of emission source strength is needed. Because of the predominantly rural and agricultural nature of large portions of India, emissions of ammonium associated with livestock and the intensive use of fertilizers to meet the growing demand for food should be addressed properly in deriving critical loads because ammonia in rainwater acts as a base, neutralizing the acids and elevating the pH of precipitation. However after it is deposited on soils biochemical processes cause ammonia to act as a strong acidifying agent. The role of ammonia in this regard is not yet well characterized in this region so its study should be given high priority. Moreover deposition alone is not the only cause of increased risk of damage. Atmospheric concentrations of SO_2, NO_x and O_3 have been shown to cause direct damage to natural ecosystems and crops as well as having health effects. Interaction between pollutants are likely, such as acidity with heavy metals and the possibility of climate change and associated changes in cropping system and vegetation land cover may add to the complexity.

Nevertheless, although the RAINS model has some limitations in its current stage of development, this model can be applied in combination with the critical load approach to support policies which are aimed at reducing acidic emissions such that excess of critical loads can be controlled. This analysis will be able to anticipate pollution risks to a particular receptor. Further involvent of incountry policy makers, managers and scientist is needed to refine the procedures suitable for deriving critical loads for vast country with variable environmental conditions like India.

REFERENCES

Acid News, Environmental fact sheet No. 6, April 1995.

Bache, B.W., The implications of rock weathering for acid neutralization in ecological effects of acid deposition Report PM 1636 (Solna: National Swedish Environment Protection Board), 1983, 175-187.

Barkman, A.; Applying the critical loads concept: constraints induced by data uncertainty; Report in Ecology and Environmental Engineering 1997:1; Univeristy of Lind: Lund, Sweden, 1997.

Bhanti, M.; Lawrence, A. J.; Saini, R.; Satsangi, G.S.; Taneja, A., Proc. 12th National Symposium on Environment, 2003, 542-546.

Bhattacharya, A., Assessment of critical loads of acid deposition. Ph.D. Thesis, Chemistry Department, St. John's College, Agra University, 2002

Bhattacharya, A.; Mudgal, R.; Taneja, A., J. of Hazardous Materials, 2004, 106B, 157-160.

Boumans, L.J.M.; Van Grinsven, J.J.M., Aluminium concentration in the groundwater, Bilthoven, The Netherlands, National Institute of Public Health and Environmental Protection, Repor 724901002, 1991, 43pp.

Chadwick, M.J.; Kuylenstierna, J.C.I., Perspectives in Energy, 1991, 1, 71-93.

Datar, S.V.; Mukhopadhyay, B.; Srivastava, H.N., Atmospheric Environment, 1996, 30(21), 3677-3682.

de Vries, W., Methodologies for the assessment and mapping of critical loads and the impact of abatement strategies on forest soils, Wageningen, The Netherlands, DLO Winand Starring Center for Integrated Lands, Soil and Water Research, Report, 1991, 94, 205.

de Vries, W., Water, Air and Soil Pollution, 1993, 68, 399-434.

de Vries, W.; Breeuwsma, A. Water, Air and Soil Pollution, 1987, 35, 293-310.

Downing, R.J.; Hettelingh, J.P.; de Smet, P.A.M. (Eds.), Calculation and mapping of critical loads in Europe, Copenhagen, Denmark, 1993, 428.

FAO, 1995. FAO Digital soil map of the world. Food and Agriculture Organization, Rome, Italy.

Foell, W.; Amann, M.; Carmichael, G.; Chadwick, M.; Hettelingh, J-P; Hordijk, L; Zhao, D. Rain-Asia: An assessment model for air pollution in Asia. Report on the World Bank Sponsored Project "Acid rain and emission reduction in Asia", 1995.

Henriksen, A.; Kamari, J.; Posch, M.; Wilander, A., Ambio, 1992, 21(5), 356- 361.

Hetteleingh,J-P.; Sverdrup, H.; Zhao, D., Water, Air and Soil Pollution, 1995, 85, 2565-2570.

Heywood, V.H., UNEP, Global Biodiversity Assessment (Ed. Cambridge Univ. Press, Cambridge), 1995.

IPCC (Intergovernmental Panel on Climate Change), Climate Change, 1995, Cambridge Univ. Press, Cambridge, 1996, 9-49.

ISRIC, 1995. World Inventory of Soil Emission Potentials. International Soil Reference & Information Center, Wageningen The Netherlands, 1995.

Kulshrestha, U.C.; Sarkar, A.K.; Srivastava, S.S.; Parashar, D.C., Proceedings of the International Conference on Acidic Deposition, Kluwer Academic Publishers, Dordrecht, 1995.

Kuylenstierna, J.C.I.; Rodhe, H.; Cinderby, S.; Kevin, H., Ambio, 2001, 30(1), 20-28.

Lubchenco, J., Science, 1998, 279, 491-497.

Mikola, P., The effects of trees species on the biological properties of forest soil report 3017 (Solna: National Swedish Environmental Protection Board) 1985.

Miller, H.G. Soil Use and Management, 1985, 1, 28-29.

Mohan, M.; Kumar, S., Current Science, 1998, 75(6), 579-593.

Mukhopadhyay, B.; Datar, S.V.; Srivastava, H.N. Mausam, 1992, 43, 249-258.

Nilsson, J. (Ed.) Critical loads for nitrogen and sulphur Report 1986:11 (Copenhagen: Nordic Council of Ministers), 1986.

Nilsson, J; Grennfelt, P. (Eds.) Critical loads for nitrogen and sulphur Report 1988:15 (Copenhagen: Nordic Council of Ministers), 1988.

Panwar, T.S., Acid rain- case study for India (including RAINS-Asia model) Acid rain and emission reduction in Asia, Phase II, Project Report, TERI, 1998.

Posch, M.; de Smet, P.A.M.; Hettelingh, J.P.; Downing, R.J., Calculation and mapping of critical thresholds in Europe: Status Report 1995, RIVM Report No. 259101004.

Ravichandran, C.; Padmanabhamurty, B., Atmospheric Environment, 1994, A28, 2291-2297.

Satsangi, G.S.; Lakhani, A.; Kulshrestha, P.R.; Taneja, A., J. of Atmospheric Chemistry, 2004, 47(3), 271-286.

Satsangi, G.S.; Lawrence, A.J.; Lakhani, A.; Taneja, A., Chemosphere, 2003, 53, 1011-1021.

Satsangi, G.S.; Taneja, A.; Khare, P.; Singh, S.P.; Lakhani, A.; Kumari, K.M.; Srivastava, S.S., The Science of the Total Environment, 1998, 119, 222.

Skiffington, R.A., Environmental Policy analysis, 1999, 33 (11), 245A-252A.

Smith, R.I.; Andreson, C.W.; Fowler, D., J. Res. Natl. Inst. Stand. Technol., 1994, 99, 353.

Sverdrup, H.; de Vries, W.; Henriksen, A., Mapping critical loads, Nordic Council of Ministers, Geneva, Miljorapport 1990:15, Nord 1990:98, 1990.

Sverdrup, H.; Warvinge, P., Assessment of critical loads of acids deposition on forest soil, in J. Nilsson (ed.). Critical loads for sulphur and nitrogen, Nordic Council Ministers and the United Nations Economic Commission for Europe (ECE), 1988, 81-130.

UN-ECE, 1996, 1997, Convention on long range Transboundary Air Pollution and its Protocol, ECE/EB. AIR/50. United Nations Economic Commission for Europe, Geneva, Switzerland, 1996-1997.

Warfvinge, P.; Sverdrup, H., Report in ecology and environmental engineering 1995:5 ISSN 1104-2877

In: Trends in Agriculture and Soil Pollution Research
Editor: James P. Livingston, pp. 191-204
ISBN: 1-59454-325-9

Chapter 8

URBAN EXPANSION: A THREAT TO CROP LAND. A GIS BASED STUDY FROM A PREDOMINANTLY AGRARIAN ECONOMY (INDIA)

Shahab Fazal
Department of Geography; Aligarh Muslim University; Aligarh 202002, INDIA;
email: shahabfazal@rediffmail.com

ABSTRACT

India is passing through a transitional phase where the urban centers are the focal points of change. India is traditionally an agrarian country with three-quarter of its population living in rural areas. But with the phenomenal growth of population the agricultural sector is unable to sustain it in the rural areas with no other alternate occupation is available. There is a mass migration of population to urban centers in search of secondary and tertiary occupations. As the population increases in urban areas, the demand of land for various urban activities also increases. This results in urban expansion and also has its bearing on urban land use.

The present study makes an attempt to measure the urban expansion, the changes in land use, to study the pace and direction of land transformation and the loss incurred to the agricultural land in a growing urban center of India. Aligarh city is chosen as the study area which is one of the class I city situated in the western part of the most populous state of Uttar Pradesh (India). Aligarh City is basically known for the world famous seat of education and its industrial base. But the city also has well developed predominantly agricultural hinterland.

The study is based on remotely sensed data, which were processed in GIS environment using ILWIS software. The result shows that significant expansion and land transformation has taken place in Aligarh City mainly due increase in population and change in occupational structure, which raised the demand as well as value of land in and around the city. The city is expanding at a rapid rate swallowing the fertile agricultural land converting first in to vacant land to be used for construction afterwards.

INTRODUCTION

In recent years, human activities have been recognized as a major force shaping the biosphere. Human actions rather than natural forces are the source of most contemporary changes in the state and flows of the biosphere. Understanding these actions and the social forces that drive them is crucial to understanding, modelling and predicting local, regional as well as global environmental change and also for managing and responding to such change. Land transformation is one of the most important fields of human induced environmental transformation, with an extensive history dating back to antiquity (Wolman and Fournier,1987). Since Neolithic times, the modification of the earth by human action mainly involved impacts on the soil and biotic resources. Land transformation did not abate but, rather, accelerated and diversified with the onset of the Industrial Revolution, the globalization of the world economy and the expansion of population and technological capacity. Forests were cleared, grasslands plowed or grazed, wetlands drained and crop lands and settlements expanded, yet never so rapidly as in the last few years. Almost all of the world's lands are now used and managed, albeit in widely varying degrees of intensity (Richards,1990). Land transformations, although localized, contribute to wider reaching, including globally systemic, processes.

Global demand for the products of the land is likely to continue growing in the foreseeable future. Maintaining the capacity of the land to sustain that demand will remain of fundamental importance. The level of concern elicited by current trends reflects the possibility that much land transformation in some sense constitutes land degradation (Blaike and Brookfield,1987). Changes in the land have often been seen as improvements by some or all land users. Alteration is nearly inseparable from human occupation and use, and the goal is to encourage improvement and to counter forces that encourage degradation. To assess what effect a particular land transformation and its consequences will have involves difficult issues of forecasting (future resource demands and opportunities as affected by technical and socio – economic change) and evaluation (distribution and the rights of future generations). However, we now possess abetter scientific knowledge of the physical extent, character and consequences of land transformation, which serves as the foundation for any such assessment (Turner and Meyer,1994).

Settlement refers to the occupation of land for human living space. As land cover, settlement represents the most profound human alteration of the natural environment through the imposition of structures, buildings, paved surfaces and compacted bare soils on the ground surface. Until quite recently, nearly everyone lived in rural areas. In 1800, only 3 per cent of the world's population lived in urban centres of 5,000 or more and many of these behaved like large villages in their socio-economic activities (Fazal,2000). By 1900, 14 per cent of the world's population was living in urban centres and this proportion increased rapidly after 1950 to reach close to 50 per cent today. While the percentage of people in urban areas has been growing, the size and number of urban settlements is also increasing rapidly.

Urban settlements cover a relatively small part of the earth's surface. One estimate suggested that urban areas cover only 1 per cent of the total land surface (Miller,1988). Globally, it has been forecast that 24 million hectares of crop land will be transformed to urban use by the year 2000. This is only about 2 per cent of the world's total crop land but it provides the equivalent present-day food supply for some 84 million people

(Simmonds,1989). The loss of agricultural land to urbanization is most severe in low and middle-income nations. One estimate suggested that by 2000, more than 476,000 hectares of land a year will be built up in low and middle-income nations (WRI,1988). Some specialists have expressed considerable concern about the loss of agricultural land to urban development while others complain that planning controls restrict the supply of land for new urban developments. The situation varies greatly from one country to another and, in general, the expansion of settlements over agricultural land is better managed in high income nations. The loss of agricultural land to human settlements is far more serious in India; about 1.5 million hectares of land (mostly agricultural) went to urban growth between 1955-1985 and a further 800,000 hectares were expected to be transformed between 1985 and 2000 (Chabra,1985). With rapid urbanization and the extension of urban areas, combined with continuing population growth, both agricultural and social scientists have long expressed a concern as to whether India will be able to feed its population (Fazal,2000). Notwithstanding its large urban population, most of India's population lives in villages with fewer than 5,000 inhabitants. Agriculture still employs around two-thirds of the labour force even though the agricultural sector only contributes 27 per cent of the nation's gross domestic product. Throughout the 1980s and 1990s, value added in industry and services grew twice as fast as value added in agriculture (World Bank,1999). Until the 1960s, agricultural growth in India stood at about 3.2 per cent, marked mainly by an expansion in the area under cultivation. In the 1970s, a growth rate of about 2.2 per cent was due mainly to increases in yields. From 1980-81 to 1993-94, a further acceleration in the growth rate to around 3.4 per cent was achieved mainly through further yield increases. But in last five years, this growth has not been smooth because increases in yield have peaked (Fazal,2001). Overall, the situation is worrying given the great population pressure on land, with an average agricultural landholding of only 1.57 hectares (and reducing rapidly), most which is still rainfed (Ranade and Dev,1998). In 1998, the food grain demand in the country was 193 million tonnes, which was barely achieved due to favourable monsoons. By the year 2015, the demand will be around 350 million tonnes. To meet such food requirements, production needs to increase at a rate of 3 per cent per annum; but in the last five years, the growth rate in food grain production was only 1.6 per cent (Fazal,1998). Achieving these targets is even more challenging since agricultural scientists have already suggested a decline in crop yields, and there is not much scope for further expansion in the agricultural area.

In India, urban population is increasing through migration of population to urban centers from rural areas. The main reason for this migration is lack of employment opportunities in the rural areas. The main occupation in rural areas of India is agriculture, which is unable to bear the burden of increasing population in rural areas. The agricultural land remains unchanged, the dependent population is increasing leading to high land-man ratio. Demographic growth stimulates structural change through multiplier effects and behavioural changes attract immigrants but the basis of urbanization is economic change and in particular the growth of large-scale production and expansion of infrastructure. Urban areas have become centers of all the activities, they are exploding in size, land use is changing quickly and constantly intensifying urban lands and large-scale land transformation is taking place due to substantial increases in urban land value. This research topic was chosen with these circumstances in mind and Aligarh, a Class I city in western Uttar Pradesh (India), was chosen as the study area.

The present study aims to study the land use pattern of Aligarh City, its expansion, to find out the pace and direction of land transformation and the resultant loss of agricultural land in the city. Aligarh city was chosen as the study area, which is one of the class I city (population of 100000 persons or more) in the western part of Uttar Pradesh. It is known for the world famous seat of education (Aligarh Muslim University) and its well-developed industrial base. But the city also has well developed predominantly agricultural hinterland. But lately the city has started losing its agrarian characteristics and more and more of its population now shedding agrarian sector and switching to secondary and tertiary sector. The development in secondary and tertiary occupation sector is luring the population not only from the city's hinterland but also from far off places for providing better job opportunities

Data and Methodology

The present study is based on both primary and secondary data. Urban land use map was prepared using remotely sensed data (satellite imageries). These data along with extensive field checks and surveys provide accurate and up to date data. The advantages of these data are many, because urban areas are very dynamic and complex in nature and conventional data source were unable to cope with these data, as the information were too much and they too get changed in a very short period of time. On the other hand their collection took a very long period. So there was an urgent need for an alternate data source, which can solve this problem. Satellite imagery has shown great utility in urban studies especially IRS – 1 D, which has the spatial resolution of 5.8 meters.

The study is spread over a time span of twenty-six years (1974 – 2001). The urban sprawl mapping was done with the help of Aligarh city *Guide map* prepared by Survey of India (1974) on 1:20000 scale and IRS 1D *geocoded panchromatic satellite analogue imagery* on 1: 12500 scale acquired in 2001. The minimum delineation unit for both maps were 0.5 cm x 0.5. Further the data were processed in GIS environment as GIS along with remotely sensed data provide the means for the rapid production of accurate and up to date data for planning. The inherent flexibility of GIS techniques can make important contribution for improving the speed and qualitative performance of urban and regional planning. GIS techniques also allow quite complex quantitative and spatial relationships to be made and displayed in graphical form, and improve the clarity of the planning and decision making process. The present study has been carried out using ILWIS software (The Integrated Land and Water Information System).

Discussion

Aligarh is an old city, which grew from an uplifted part *(Kol),* witnessed a long and cherished history including *Mughal's, Maratha's and Britisher's rule*. Aligarh city situated in the western part of Uttar Pradesh (Figure1), between *Ganga* and *Yamuna* rivers at 27°53′ north latitude and 78°35′ east longitude. It is district headquarter and is famous as educational and industrial city. This city has gained importance because of its proximity to the national capital (New Delhi) and industrial cities of Uttar Pradesh (Kanpur and Ghaziabad). Systemic and planned development of Aligarh city would serve dual purpose – it would reduce the pressure on these cities, as it would provide employment opportunities in the city and it can even supply raw materials to these cities. But due to some social and economic problems the desired level of progress has not been attained.

Urban Land Use of Aligarh City

The present study is spread over 15398 hectares of land, which is much beyond the actual urban limit of Aligarh city. This extended area covers expansion of the city in all directions. From the perusal of table 1 and figure 2, it is evident that Aligarh city has witnessed large-scale changes in land use in and around the city. The statistics shows that urban area has increased from 2276.8 hectare in 1974 to 5445.8 hectare in 2001 (an increase of 139 per cent). Similarly built up area has increased from 1832.7 hectare to 4534.9 hectare (147 per cent). All these increases are at the expense of non – built up area, which shrunk from 13566.1 hectare to 10863.6 hectare (a decrease of 20 per cent) during the study period. The city has witnessed major land use changes among classes such as residential area, an increase of 2501.2 hectare (244 per cent increase), commercial area, an increase of 145.7 hectare (258 per cent increase), vacant land, an increase of 465.6 hectare (114 per cent increase) and industrial area an increase of 78.3 hectare (96 per cent increase). These changes were basically swallowed the agricultural land of the surrounding areas. Agricultural land has lost 3078.3 hectare (a decrease of 23 per cent) to the urban sprawl of Aligarh city. Among the other land use classes not much change has been observed. Area under villages has marginally reduced as some of these villages were swallowed by urban expansion. Similarly 361-hectare land was occupied by the Aligarh Muslim University and 17.6 hectare by Aligarh fort. Both of these are located in the northern part of the study area and remained unchanged.

Table 1. Aligarh City, Area under different land Uses (1974 and 2001)

Land Use / Year	Area (1974)	Area (2001)	Change in Area	%age Change
Residential	1025	3562.2	+2501.2	244.01
Villages	135.4	107	-28.4	-20.8
Commercial	56.3	202	+145.7	258.79
Government Institutions	106.4	110.8	+4.4	4.3
University	361.1	361.1	--	--
Social & Recreational	70.2	74.2	+4.2	5.69
Industrial	78.3	153.6	+78.3	96.16
Aligarh Fort	17.6	17.6	--	--
Vacant Land	406.6	872.2	+465.6	114.5
Tree Plantation	172.9	67.4	-105.5	-61.02
Agricultural Land	12924	9845.7	-3078.3	-23.82
Water bodies/logged	45	60.7	+15.7	34.88
Total	**15398.8**	**15398.8**	--	--
Total Built up Area	1832.7	4534.9	+2702.2	147.4
Total Non Built up Area	13566.1	10863.6	-2702.5	-19.93
Total Urban Area	2276.8	5445.8	+3169	139.18

Note: Area in hectare.

Source: Based on Aligarh city, Guide map (Survey of India) and IRS 1D Satellite Imagery.

Table2. Aligarh City, Land Transformation (1974 and 2001)

Land Use	1 Res.	2 Vil.	3 Com	4 Govt.	5 Univ.	6 Soc.	7. Ind.	8. Fort	9. Vac.	10 Tree	11 Agr.	12 Water.
1. Residential Area	**1025**	28.4							425.6	105.5	2047.4	
2. Villages		**135.4**										
3. Commercial Area	105.7		**56.3**								40	
4. Government Institutions	4.4			**106.4**								
5. University					**361.1**							
6. Social and Recreational						**70.2**					4	
7. Industrial Area							**78.3**		83.7		70	
8. Aligarh Fort								**17.6**				
9. Vacant Land									**406.6**		1030.9	
10. Tree Plantation										**172.9**		
11. Agricultural Land											**12924**	
12. Water bodies/Logged							15.7					**45**

Note: 1. Area in hectare.
2. Figures in bold letters (diagonally) are area under that particular land use in 1974, while figures in the same column is shift in area to other classes and figures in same row are increase in area, captured from other land use classes.
Source: Based on Aligarh city, Guide map (Survey of India) and IRS 1D Satellite Imagery.

The salient characteristics, which emerges from table 1 and figure 2 are as followed. Aligarh city has recorded a significant increase in built up area and this is mainly because of population growth and development in secondary and tertiary occupation structure. But this built up expansion was haphazard and without any planning. Even most of the recent residential colonies are developed without the provisions of basic amenities, resulting in severe congestion in road traffic, lack of open spaces and over flowing of drains. The urbanization process in Aligarh city is characterized by rapid growth in population followed by increase in built up area but not much change in utilities and services of the city.

During the study period Aligarh city has under gone substantial land transformation, but only in some specific areas, which is the out come of social, economic, religious and cultural factors. The newer part of the city has grown in the north eastern part of railway line, which runs almost from the middle part of the city. Initially, the newer part consists of University and the civil lines but their pull factor and relatively lesser congestion in this part have drastically transformed it. This part now dominated by residential class and commercial uses, this has resulted in substantial increases of land value. The study reveals that with time, the older part of the city has become more and more congested while on the other hand the city has expanded outward in a haphazard manner and swallowed the agricultural land in the fringe area. The urban land use of Aligarh city is dominated by residential class, of which, most of it area is unplanned. The second largest area is occupied by vacant land, which is the outcome of rapid transformation of occupational structure and emergence of secondary and tertiary activities. The main reason for the increase of vacant land in the city was that initially when industrialisation took place, there was shortage of open land for construction. This demand has led to increase in the land value and owners of land started to get more money instantly, which they were not getting in agricultural sector. In anticipation more and more people started to offer their land for construction resulting in phenomenal transformation of land from agricultural uses to vacant land. Later this demand due to stagnation and saturation subsided, resulting in increase of vacant land, which is neither used for urban development nor is it useful for agricultural sector. The city has 153.6 hectares of land under industries, which is quite high by traditional Indian standard. Commercial area in the city occupies 202 hectares of land. There are two distinct commercial zones in the city. One is located in the older part of the city and second is of recent origin to cater the demand of newer part of the city. The newer commercial area is in the north eastern part of the city and it is characterized by showrooms and offices of big organizations. However both of these are along the major roads. Interestingly area under city's recreational and utilities, services covers almost negligible area, which indicates that, not much attention is paid to these facilities despite significant increase in population.

The land use pattern of Aligarh City is different from the general land use pattern of western cities. The existence of a mono-economic system and an advanced transportation technology of the western cities had led to a maximum specialisation of land use which has most frequently taken the form of concentric circles of relatively homogeneous economic usage. From the perusal of Figure 2 and 3, no such clear zonation characterises the land use pattern of Aligarh city. It is hard to generalise about the patterns of residence. The rapid growth of population of the city associated with the socio-economic changes, which are creating an emergent middle class, have caused a proliferation of squatter settlements in the interstices and fringes of the city. The dominance of various social and religious groups in distribution of various residential pockets is also quite a prominent feature. At present Aligarh city shows little conformity with the western cities where direct relationship is observed between social class and distance from the centre of the city with working class living closest to the city core and upper class in the city fringe. The pattern of residence in Aligarh city includes a mixture of poor and rich elements of the city in both the core and outer areas within their social and religious residential area. It represents some transitory phase of development between the patterns of pre-industrialism and post-industrialism. There were two distinct commercial zones in Aligarh city, one in older part and another in newer part. These zones have peculiar radial dimension along the main roads. Some other commercial areas are scattered, but located along the roads and their concentration declines with distance. These commercial areas are crowded and a picture of congestion and traffic jams. The industrial land use of the city resembles the western cities in some way. The larger industrial units are located in the fringe area but the main difference from western cities is that it is not confined to specialised estates but are scattered. The other land use classes especially like vacant land is governed by local factors. The same trend is not necessarily found in other cities of India. Another significant feature of Aligarh city is lack of public space like parks, play ground etc. and even public utilities and services have negligible occupance in the city. It indicates lack of planning and co-ordination among various government authorities for developing these facilities, which are under severe stress from rapidly increasing population. The over all spatial organization in Aligarh city represents indigenous characteristics, which stem from indigenous living conditions and the indigenous ways of adaptation with them. This pattern confirms the Gerald Breese's opinion for Indian urbanization where he termed it as subsistence urbanization (Breese, G., 1969).

Land Transformation in Aligarh City

Land transformation is the process of identifying differences in the state of any phenomenon by observing it at different time periods. Land is in a continuous state of transformation as a result of various natural and man-made processes. The study of land transformation, therefore, requires a comprehensive understanding and monitoring of all the factors that cause the land transformation. During the study period, Aligarh city has not only expanded from its original size but there was significant interchange of land among various land use classes. Table 2, and figure 3, which are prepared applying G.I.S. technique, shows the changes in various urban land use class of the city. These changes are because of the development of city resulting in increased demand of land for residential, commercial, industrial etc. purposes. This demand of land along with site attraction, functional convenience, functional magnetism and the land value of that particular area ultimately influence the pace and direction of urban land use change.

The residential area has increased by 2501 hectare, capturing land mainly from agriculture (2047.4 hectares), vacant land (425 hectares), tree plantation (105 hectares) and villages (28.4 hectares). Residential area also lost area for commercial development (105 hectares). Commercial area has increased from 56 hectares to 202 hectares gaining land from residential area (105 hectares) and vacant land (40 hectares) in the city. Industrial area has increased from 78 hectares to 153 hectares capturing mainly the agricultural lands (70 hectares) because this development is mostly away from the city centre. Plantation area has reduced by 105 hectare and the entire area was used for residential. The city has recorded a massive increase from 406 hectares in 1974 to 872 hectares in 2001 under the land use class vacant land. It captured land from agriculture class (1030 hectares) while lost area mainly for residential (425 hectares) and commercial (40 hectare) development. During the study period all the city expansion is mainly on agricultural lands, which decreased from 12924 hectares to 9845 hectares resulting in heavy losses of fertile agricultural lands.

The study also tried to study the landowner's decision-making process. Decision about the use of land involves a multiplicity of factors: the objectives of land user; the process by which he reaches on decision and the socio – economic factors that consciously or unconsciously influences the decision. The sample survey conducted on 48 landowners, where land transformation (agriculture to other uses) has taken place during the study period. The sampled landowners were divided in to three categories on the basis of land holdings.

Large: landowners disposed more than 2000 square meters of agricultural land, 12 sampled landowners belonged to this category.

Medium: landowners disposed 500 – 2000 square meters of agricultural land, 28 sampled landowners belonged to this category.

Small: landowners disposed less than 500 square meters of agricultural land, 18 sampled landowners belonged to this category.

To know the factors which lead to their decision of disposing agricultural land. They were asked to give three important factors each, which are presented in table 3. The result shows shift from agriculture to other activity was the most prominent followed by getting instant money, protecting future and high status of non-agriculture occupation. Large landowners were influenced most by change of occupation and low status of agriculture in society. Medium class of landowners were influenced by change in employment and to protect the future they sell out their agricultural lands. Interestingly small landowners were attracted by instant money by selling off their land, which was uneconomic and their sustenance was difficult. The sample confirms that land use is the product of human decision, which is greatly influenced by environment and economics.

Land transformation in Aligarh city is influenced greatly by distraction from agricultural activity. People are switching from agricultural activity due to lesser profit margins and also the activity has lower status in the society, uncertainty of profit margins, operational problems etc. This raises a very crucial issue in a predominantly agrarian economy that is endowed by fertile agricultural setting. The people are shedding their agricultural land and getting instant money but becoming landless, giving rise to unemployment problem. The urban sector is unable to cope with large population looking for jobs. This ultimately results in emergence of different socio – economic problem.

Table 3. Reasons for land transformation among sampled land owners in Aligarh city (2001)

Type of Landowner	1. C.E.	2 I.M.	3. P.F.	4. S.S.	5. M.I.	6. U.P.	7. O.P.	8. U.L.	9. Fr.	10. Oth.	Total
Large (land disposed more than 2000 square meters)	**11** *(20.4)*	7 *(12.9)*	6 *(11.1)*	**10** *(18.6)*	6 *(11.1)*	6 *(11.1)*	6 *(11.1)*	--	--	2 (3.7)	**54**
Medium (land disposed 500 – 2000 square meters)	**21** *(25)*	**12** *(14.4)*	**16** *(19)*	9 *(10.7)*	**10** *(11.9)*	7 *(8.3)*	7 *(8.3)*	--	--	2 (2.4)	**84**
Small (land disposed less than 500 square meters)	**3** *(8.3)*	**10** *(27.8)*	**1** *(2.8)*	**3** *(20.4)*	**4** *(8.3)*	**1** *(2.8)*	--	9 *(25)*	**4** *(11.1)*	1(2.8)	**36**
Total	**35** *(20.1)*	**29** *(16.7)*	**23** *(13.2)*	**22** *(12.6)*	**20** *(11.5)*	**14** *(8.0)*	**13** *(7.5)*	**9** *(5.2)*	**4** *(2.3)*	**5** (2.9)	**174**

Note: Based on field survey.

Figures in bracket are percentages to the total of land owner category.

1. **C.E. – Change in employment:** Urbanization has generated employment opportunities with increased and assured remuneration and land owners have obtained jobs out side agricultural sector so they are now disposing off their lands to get extra money.
2. **I.M. – Instant Money:** Agricultural activity is being depressed activity, so land owners are disposing off their lands to get instant money.
3. **P.F. – Protecting Future:** Economic returns from the agricultural sector are lower, so land owners dispose off their land when real estate prices are high and thus assured of future.
4. **S.S. – Social Status:** Social Status is higher in jobs in urban area than in the agricultural activity.
5. **M.I. – Maximizing Income:** Since returns in agriculture sector returns are lower so land owners to maximize their income dispose land in phases to get maximum income from their lands.
6. **U.P. – Uncertainty of Profit Margins:** Uncertainty of profit margins due to variation in natural – social – economic conditions also led to land transformation.
7. **O.P. – Operational Problems:** Due to urban shadow effect land owners faces several operational problems in agricultural activity in the urban fringe.
8. **U.L. – Uneconomic Landholding:** Small land owners are unable to sustain their family from the small land holding so they dispose their land work as labourers in urban area.
9. **Fr. – Forced:** Property dealers and land developers pressurise small land owners to dispose their land as these people needs big areas for making colonies.
10. **Oth. – Other Factors:** Some other personal reasons can also result in land transformation from agriculture to built up urban area.

Loss of Agricultural Land

The above discussions very clearly points towards the development of Aligarh city, which has instigated migration to the city, adding to already high population of the city. This resulted in increased demand for land, resulting phenomenal expansion of the city swallowing fertile agricultural land. The present study attempted to measure the loss of agricultural land during the study period using GIS techniques with crucial inputs also from the field surveys. Figure 4 and table 4 shows that during the study period agricultural land was reduced by 3078 hectare (a decline of 23 per cent). The agricultural land in the study area is declining at an alarming rate of 114 hectare annually and the estimates indicates that by year 2025 another 2850 hectares of fertile agricultural land would be swallowed by the city expansion. This loss of agricultural land seems more alarming when the out put from land is considered. A loss of 3078 hectares means that the city is loosing approximately 10,000 quintals of food grain production annually. By year 2025, this loss would be of 17,7000 quintals. This much of production would be sufficient to meet the requirements of the increasing population of the city (city population has recorded 3 per cent growth) which otherwise needs to fulfilled externally.

Table 4. Rate of urban expansion and loss of agricultural land (1974 to 2001)

Year	Urban Area (in hectare)	Change		Time Span	Mean Change	
		Area (in hectare)	%age increase		Hectare / year	%age change / year
1974	2276.8	3169	139.2%	27	117.3	5.15%
2001	5445.8					
Year	**Agricultural Land (in hectare)**	**Change**		**Time Span**	**Mean Change**	
		Area (in hectare)	%age loss		Hectare / year	%age change / year
1974	12924	3078.3	23.82%	27	114	0.9%
2001	9845.7					

Note: Area in hectare.

Source: Based on Aligarh city, Guide map (Survey of India) and IRS 1D Satellite Imagery.

CONCLUSIONS

Some of the main findings of the present study are as follows:

- The study area has recorded a significant increase in residential area to accommodate the rapidly growing population of the city but not much change has occurred among utilities, services and recreational facilities etc., which indicates haphazard city expansion.
- Agricultural land is being converted into non-agricultural land (built up and non-built up urban expansion) because of rapid expansion in Aligarh City.
- Ribbon like urban development along main roads is a noticeable phenomenon in the city. The urban expansion is taking place in a peculiar way. In the first phase agricultural land is converted into vacant land along the main roads. In second phase this land is used for constructing buildings. During the third phase the intervening peripheral land between the roads are used for urban development. This process of urban development can be termed as frontal expansion, which is characterized by piecemeal accretion of the built up areas.
- Vacant land in and around Aligarh City has increased substantially. The main reason for its increase is the shortage of open land for construction. This demand has led to increase in the land value and owners of land started to get more money instantly, which they were not getting in agricultural sector. So in anticipation more and more people started to offer their land for construction resulting in phenomenal transformation of land from agricultural uses to vacant land.
- Urbanization pattern of Aligarh city is characterized by poverty, acute shortage of housing and civic amenities, dismal filth and diversified occupance of space, which resembles like a large village. Here in lies the major departure of the urbanization pattern in Indian cities from the western cities. Urbanization may be an indicator of development in western countries but in India, especially in growing small towns it is difficult to accept. Here, the population is increasing rapidly, the demand for land also increases. But civic authorities have little control over urban development and expansion, which results in violation of basic rules and regulations in municipal area.
- There is an urgent need to look into the agricultural system in India. The rate, at which agricultural land is being destroyed, needs serious thinking on part of planners and policy makers. The farmers are distracted from agricultural activity due to poor productivity, uncertainty of production, low returns, lower social status and operational problems due to their shift to other occupation in the city. These reasons for land transformation need to be taken seriously. India has a predominantly agrarian economy, where people are distancing themselves from agricultural activity due to inferior social status is unfortunate. Government needs to encourage the agricultural sector and protect the interest of farmers.
- Unlawful conversion of agricultural land needs to be restricted because the region is gifted with best of natural setting for agricultural activity. The

important fact is that ***'land is finite'*** and the converted land can not be recovered back in future.

- A more significant factor intensifying the problem of loss of cropland is associated with the cost of such land conversion. These costs are twofold, and are related to the actual cost of replacing the lost farmland, as well as to government programs that prompt urban expansion and encourage loss of cropland. When the prime farmland is lost, the value of the crop production is immediately foregone. The cost to replace that land is fairly substantial. The government makes effort to bring some additional land for agricultural uses by reclaiming wastelands but the question is "how much and at what cost"?
- Land use is not static, but rather a dynamic interacting system. Changes are occurring in each of the categories of land use, which have an impact on the residents of the nation both at local as well as regional level. Urban sprawl with its attendant destruction of prime cropland seems far severe than popularly supposed. Although urban expansion cannot be restricted, but with proper management and planning we can restrict and direct this expansion in desired and sustainable way, protecting the fertile agricultural lands. It needs stricter implementation of agricultural land conversion laws. The farmers should be provided some incentives to remain in farming activity, as this is the only way to control migration of rural population to urban centres. Even they should be encouraged to go back to farming activity and this is the only way to restrict conversion of cropland into vacant land.

The basic objective of land use planning is to integrate the essential use of land. Every effort should be made to find the optimal use of land in the national interest. Naturally every claimant puts forth his demand as most important. One of the important tasks of land use planning is to strike a balance between various interests.

REFERENCES

[1] Alam, M. and Pokshishevsky, V.V. (1976) ' *Urbanization in Developing Countries'* Asia Publishing House, Hyderabad.

[2] Alam, M. (1965) 'Hyderabad-Secundrabad: *A Study in Urban Geography'* Asia Publishing House, Hyderabad.

[3] Blaikie, P. and Brookfield, H.C. (1987) ' *Land Degradation and Society'*. Methuen, London.

[4] Borah, J. (1985) *'Spatial Structure of Urban Influence in the Neighbouring Areas of Gauhati City'* Concept Publisher, New Delhi.

[5] Bose, A. (1988) *'India's Urbanisation, 1901-2001'* Tata McGraw Hill Publication, New Delhi.

[6] Breese, Gerald (1969) *'Urbanization in Newly Developing Countries'* Prentice-Hall of India Pvt. Ltd. New Delhi.

[7] Chabra, R (1985), *'India: Environmental Degradation, Urban Slums, Political Tension'*. Draper Fund Report.

[8] Deshpande, C.D. et al., (1980). *'Impact of a Metropolitan City on the Surrounding Region'* New Delhi.

[9] Fazal, S. (2001) 'Land Reorganization Study Along Major Roads'. *Land Use Policy*, Elsevier Science Publication, Vol. 18, No.2. United Kingdom.

[10] Fazal, S. (2001) 'Application of Remote Sensing and GIS Techniques in Urban Sprawl and Land Use Mapping. A Case Study of A Growing Urban Centre in India'. *Asian Profile*, Vol. 29. No.1. Canada.

[11] Fazal, S. (2001) 'Dynamics of Urban Land Transformation in Developing Economy'. *Indian Journal of Regional Sciences, Kolkata*, Vol. XXXIII, No.1.

[12] Fazal, S. (2000) 'Urban Expansion and Loss of Agricultural Land. A GIS Based Study of Saharanpur City, India'. *Environment and Urbanization*, Vol.12. No.2, IIED, London.

[13] Fazal, S. (1998) ' *Agricultural production and price mechanism'*. Chugh Publications, Allahabad.

[14] Johnson. J.M. (1984) *'Sub-Urban Growth'* John Willey & Sons., London.

[15] Miller, G.T. (1988) *'Living in the Environment'* Wadsworth, Belmont, California.

[16] Misra, R.P. (1998) *'Million Cities of India'* Vikas Publishing House Pvt. Ltd., New Delhi.

[17] Prakasha Rao, V.L.S. (1983) *'Urbanisation in India. Spatial Dimensions'* Concept Publication, New Delhi.

[18] Ranade, A. and Dev, M.S. (1998) *'Agriculture and Rural Development'* in Parikh, K.S. (editor), India Development Report, Oxford University Press, New Delhi.

[19] Richards, J.F. (1990) ' *Land Transformation'* in Turner, B.L. et al, - The Earth as Transformed by Human Action. Cambridge University Press, USA.

[20] Simmonds, I.G. (1989) *'Changing the Face of the Earth: Culture, Environment, History'*. Blackwell, Oxford, UK.

[21] Surekha, Pandit (1988) *'Urbanization the Indian Way'* ed. Yadav, C.S. 'Perspectives in Urban Geography' Volume 3, Concept Publishing Company, New Delhi.

[22] Turner, B.L. and William, B. Meyer, (1994) *'Changes in Land Use and Land Cover: A Global Perspective'*. Cambridge University Press, USA.

[23] Verma, S.S. (1995) *'Urbanization and Regional Development in India'* Chugh Publications, Allahabad.

[24] Wolman, G. and Fournier, F.G.A. (1987) ' *Land Transformation in Agriculture'*. John Wiley and Sons, UK.

[25] World Bank (1999) *'Entering the 21st Century: World Development Report, 1999-2000'*, Oxford University Press.

[26] WRI (World Resource Institute) (1988) *'World Resources 1988 – 89'*, Basic Books, New York.

[27] Yadav, CS. (1988): *'Comparative Urbanization: City Growth and Change'* Concept Publishing House, New Delhi.

In: Trends in Agriculture and Soil Pollution Research
Editor: James P. Livingston, pp. 205-229
ISBN: 1-59454-325-9

Chapter 9

ECO-FRIENDLY MOLLUSCICIDES, PISCICIDES AND INSECTICIDES FROM COMMON PLANTS

Ajay Singh*[*]*, S. K. Singh, R. P. Yadav, V.K. Srivastava, D. Singh and S. Tiwari
Natural Product Laboratory; Department of Zoology,
D.D.U. Gorakhpur University, Gorakhpur- 273 009 U. P. INDIA

ABSTRACT

Mosquitoes and snails are still the world's number one vectors of human and animal diseases and are conspicuous nuisance pest as well. Mosquitoes are potential vectors of malaria, filaria, yellow fever, brain fever, dengue fever etc. and many aquatic snails act as vectors for the larvae of trematodes and thereby cause two diseases i.e. fascioliasis and schistosomiasis. Presence of predatory fishes in fish culture ponds is also a serious problem due to their faster growth rate and better utilization of cultured carp habitats and food. Due to their carnivores nature they engulf the fingerlings of cultured carps and adversely effect the aquaculture production. The WHO and FAO have tested several thousands of synthetic compounds for the control of these vectors and predatory fishes. Though effective, these pesticides have so far not proved themselves to be entirely satisfactory. With a growing awareness of environmental pollution, efforts are being made to find out molluscicidal, piscicidal and insecticidal products from plant origin. Being products of biosynthesis, these are potentially biodegradable in nature. Several groups of compounds present in various plants have found to toxic to target organisms. Thus, saponins, tannins, alkaloids alkenyl phenols, glycosides, glycoalkaloids, flavonoids, sesquiterpenes lactones, terpenoids and phorbal esters have been found to be poisonous to target organisms at acceptable doses ranging from < 1 – 100 ppm. These compounds come from 56-60 families of angiosperms. Although, at present a very little literature available on the control of vector snails, mosquitoes larvae and predatory fishes through plant origin pesticides. An attempt has been made by this review to present

[*] Dr. Ajay Singh; E-mail: ajay_s@sancharnet.in

current information on products of plant origin, which might be useful for the control of snails, mosquitoes and predatory fishes.

INTRODUCTION

The intensive application of pesticides in modern agriculture and public health operation systems has resulted in serious environmental problems (Minelli and Rebeiro 1996; Dua et al., 1998; Waliszewski et al., 1999; Singh et al., 2004 b). The pesticides that have received most attention include Pentachlorophenol (PCPs), Polychlorinated biphenyls (PCBs), Atrasine (S-triazines), Organochlorines (OCs), organophosphates (Ops) and Carbamates, since they are widely used and are highly persistent compounds. The freshwater bodies adjoining to agriculture fields are continuously being contaminated by the toxic wastes of chemical pesticides (Bourgeois et al., 1993; Nayak et al., 1995) and pose a potential direct threat to freshwater organism, particularly to sensitive animals, such as fishes and prawns (Saravanan et al. 2003; Selvarani and Rajamanickam, 2003 and Park et al., 2004).

Basic research for over more than fifty years in biology and biochemistry has made it possible to envisage not only how new pesticides may be synthesized but also a completely new approach for the control of vectors using secondary plant products which may be toxic to a specific vectors yet harmless to non-target organisms. In recent years, considerable attention has been directed to the research and application of molluscicides, insecticides, larvicides, insect growth regulators of juvenile hormones analogus in protection of human being and their domestic animals. These substances have been used successfully on large scale in vectors management programme, also (Srivastava et al., 2003). More than 2,000 plant species belonging to different families and genera have been reported to contain toxic principles, which are effective against insects. Among the well represented plant pesticides is 'Pyrethrums' obtained from *Chrysanthmum cinariaefolium*, which is mainly used as a domestic insecticide because it is non-toxic to man and warm-blooded animals and is highly sensitive to light.

In recent time, the use of plant products has gained unprecedented impetus, all over the world. The people of North – Eastern region of India, in particular the rural and tribal people living in some remote areas, primarily depends upon the folk and traditional medicine. Indigenous knowledge of these plants used for the different purpose in different areas. A large number of plant families have furnished many classes of product, which may vary with the degree of pesticidal activity. Several countries have promoted the use of plant product with a wide range of ideal properties, are as

- High pesticidal properties.
- Easily availability.
- Easily biodegradable with less hazard of environmental contamination.
- Low mammalian toxicity.
- Solubility in water.
- Low cost.
- Grow abundantly in endemic areas and safe for operator.

The above-mentioned properties of plant product have opened a new vista. In reference of India, our country possess a rich biodiversity of medicinal plants are used for many purposes. The plant products (extracts also) have been used from the immemorial. The Vrikshayurveda is the branch of Ayurveda that deals with plant health and recommends drugs possessing specific qualities of treatment of insect attack. During the co-evolution of plants and insects, plant has bio-synthesized a number of secondary metabolites to serve as defense chemicals against pest attack. Although only 10,000 secondary metabolites have been chemically identified so far, the total number may exceed 400,000 (Swain, 1977).

A list of various plants/ products have been tested during past on a decade and have been shown to possess molluscicidal, piscicidal, insecticidal, larvicidal, growth inhibitor chemosterilant and repellent effect against mollusc, fishes and mosquitoes. The present article deals with botanical products, which have demonstrate their efficacy in the management of disease vectors as alternative of synthetic pesticides

Molluscicides of Plant Origin

Mollusc is the second largest group of invertebrate animals, in variety, after insect, occupying all possible habitats except aerial. Although the exact number of existing molluscan species is still a matter of speculation, Abbott (1954) has estimated a total of about 1,10,000 living species, 80,000 amongst which are gastropods, 10,000 bivalves, and 5,000 belonging to the other three classes of mollusca. Godan, (1983) on the other hand believe the number of living species is about 1,20,000.

Terrestrial snails and slugs cause considerable damage to both cultivated and useful non-cultivated plants. The animals can make their appearance in any damp area, but damage can occur also during relatively dry weather. Along with slugs, terrestrial snails also cause considerable damage to vegetable gardens, agricultural crops and fruits orchards. Singh and Agarwal, (1981) reported that *Pila globosa* an amphibious snails causes damage to paddy crops in northern part of India. In freshwater, the larvae of parasite trematodes also pass part of their life. Many aquatic snails act as vectors for the larvae of trematodes and there by, cause a number of diseases. Schistosomiasis is caused by *Schistosoma*, it is a devastating disease of mankind second only to malaria in its deleterious effect (Lambert, 1966; Jobin, 1973; Bali et al., 1986).

Fascioliasis caused by *Fasciola hepatica*, the large liver fluke, common in sheep, cattle, goat and other herbivorous animals throughout the World. Froyed (1975) reported that about 21% cattle and 7% sheep were infected with liver fluke in Great Britain. In India, the freshwater snails *Lymnaea acuminata* and *Indoplanorbis exustus* are the intermediate host of *Fasciola hepatica* and *Fasciola gigantica* (Hyman, 1970), which causes immense harm to domestic animals of this country.

The best method of controlling both the diseases such as schistosomiasis and fascioliasis is chemotherapy. Using orally-administered drugs for individuals with moderate or severe level of infection. The disadvantage of this approach is that it does not eliminate the infection entirely, the cost of recurrent treatment may become prohibitive and drug resistance may become a problem. A sure way to tackle the problem of schistosomiasis and fascioliasis is destroy the carrier snails and remove an essential link in the life cycle of the flukes. This can be accomplished in a number of ways including the use of many synthetic or plant molluscicides (Agarwal and Singh, 1988; Ndamba et al., 1995; Singh et al., 1996b).

Several plants of family leguminosae have compounds highly toxic to harmful snails (table 1). Of the 153 crude extracts of Panamanian plants of different families, *Hymenaea coubaris* (Leguminosae) is most effective against *Biomphalaria glabrata* (Marston et al., 1996). *Ambrosia maritima* (Asteraceae) have significant molluscicidal activity against snail species *Biomphalaria, Bulinus* and *Lymnaea,* examined in fields (El-sawy et al., 1981). Sesquiterpene lactones are responsible for the molluscicidal activity of Brazilian spp. Vernonia (Callegari lopes, 1991). Shoeb et al., (1993) screened many plants of family aga-vaceae for their molluscicidal activity against freshwater snails. *Agave attenuata* powder shows effective molluscicidal activity against *Biomphalaria alexandrina, Bulinus truncatus* and *Lymnaea cailliaudi* (Shoeb et al., 1993). Members of the family Guttiferae are shown to contain compounds highly effective against *Biomphalaria glabrata* (Cepleannu et al., 1994). *Anacardium occidentale* and *Spondias mombin* belongs to family Anacardiaceae shows strong molluscicidal activity against freshwater snails *Biomphalaria glabrata.* (Laurens et al., 1997). Aqueous and alcoholic extracts of *Asparagus racemosus* (Chifundera et al., 1993) and *Uriginia epigea* leaves (Amusan et al., 1997) of the liliaceae family, exhibit high mortality rate (100%) against *Lymnaea natalensis* (LC_{50} 1.0 mg/L) and *Bulinus africans* (LC_{50}, 50-100 ppm), respectively.

Pestoban, a herbal pesticide manufactured by Indian Herbs, Research and Supply Co. Pvt. Ltd, India, is a liquid concentration of *Cedrus deodara* Linn oil and *Azadirachta indica* oil extracted from their seed, and the powder of seeds of *Embelia ribes,* dissolved in a non-ionic emulsifier, srichaskshar has high molluscicidal activity against harmful snails *Lymnaea acuminata* and *Indoplanorbis exustus* (Singh and Singh, 1994; Singh et al., 1995). In another study Singh et al., (1998) studied the synergism of an oil of *Azadirachta indica,* a powdered of *Allium sativum* bulbs and an oleoresin of *Zingiber officinale* rhizome by piperonyl butaoxide and MGK-264 against the snails *Lymnaea acuminata* and *Indoplanorbis exustus.* The response of snails to the synergised mixtures was both time and dose dependent.

Singh and Agarwal (1984a) observed that the crude latex of *Euphorbia royleana* caused 100% snail mortality within 24h at concentrations as low as $2.7x10^{-5}$ (v/v), the snail mortality was dose-dependant and the toxic moiety of the latex was stable even at 100°C. Snails exposed to latex of *Euphorbia royleana* exhibited typical symptoms of nerve poisoning and death took place within 24h. It was shown that the latex was an acetylcholinesterase inhibitor and its anti-AChE activity in the snail *Lymnaea acuminata* was very high in comparison to any synthetic organic pesticides. In another study, Singh and Agarwal (1984b) also observed that the latex of *Euphorbia royleana* reduced the level of 5-hydrodxy-tryptamine (5-HT) and dopamine in the nervous tissues of *Lymnaea acuminata.* Singh and Agarwal, (1992 a) reported that, the latices of several euphorbious plants significantly reduced the alkaline and acid phosphatase activity in nerve tissue of *Lymnaea acuminata.* Cheng, (1971) and Amin, (1972) have recorded the molluscicidal properties of *Thea olesosa, Croton tiglium, Sehima argenta* and *Jatropha* spp. Adewunmi and Morquis, (1980) studied the molluscicidal properties of methanolic extracts of the fruit of *Jatropha gossypifolia* and *Jatropha podaqrica.*

Table 1. A list of plants having molluscicidal activity.

Plant	Plant part Tested	Class of active moiety	Extracts	Species	References
ACANTHACEAE					
Brillantaisia vogelinae	S	Unknown	Water	*Bulinus globosus*	Adewunmi & Sofowora, 1980
Crossandra flava	F	Unknown	Water	-	-
Lankesteria elegans	L	Unknown	Water	-	-
Nelsonia canescens	L	Unknown	Methanol	-	Okunji & Iwu, 1988
AGAVACEAE					
Dracaena arborea	L	Sterols, Saponins	Water	*Lymnaea natalensis*	Okunji & Iwu, 1988
D. manni	S, F		Methanol	*B. globosus*	Adewunmi & Sofowora1980
Sanseviera trifasciata	L	Unknown	Water	*B. globosus*	
Yucca aloifolia	I	Spirostonal	Ethanol	*Biomphalaria glabrata*	Kishor & Sati, 1990
ANACARDIACEAE					
Anacardium occidentale	C	Alkenyl phenols	Water	*B. glabrata*	Sullivan et al, 1982
ANNONACEAE					
Annona senegalensis	S	Unknown	Water	*B. globosus*	Adewunmi & Sofowora, 1980
Cleistopholis patens	R	Unknown	Water	*B. globosus*	Adewunmi & Sofowora, 1980
APOCYANACEAE					
Alstonia booneide	L, S	Unknown	Methanol	*B. globosus*	Johns et al, 1982
Hunteria umbellata	L	Unknown	Ethanol	*B. globosus*	Fronczek et al, 1984
Alstonia scholaris	La, S	Glycosides	Water	*L. acuminata*	Singh & Singh, 2003a
Nerium indicum	La	Unknown	Water	*L. acuminata*	Tomassini & Matos, 1979
Rauvolfia caffia	S, R	Unknown	Water	*B. globosus*	Liu & Nakanishi, 1982
R. vomitoria	L, F	Alkaloids, sterols, Triterpenes	Water	*B. globosus, L. natalensis*	Adewunmi & Sofowora, 1980
Voacanga africana	S	Unknown	Methanol	*B. globosus*	Adewunmi & Sofowora, 1980
Thevetia peruviana	La, S	Thevetin	Water	*L. acuminata, I. exustus*	Singh & Singh, 2003a, b, c, Singh et al. 2000
ARACEAE					
Culcasia scandens	L	Unknown	Water	*B. globosus*	Adewunmi & Sofowora, 1980

Table 1. A list of plants having molluscicidal activity. (Continued)

Plant	Plant part Tested	Class of active moiety	Extracts	Species	References
ARALIACEAE					
Hedera helix	-	Triterpenes, Saponins Saponin	Methanol	*B. globosus*	Adewunmi & Sofowora, 1980
Polyscias dichroostachya	L	-	Methanol	*B. globosus*	Adewunmi & Sofowora, 1980
ASTERACEAE					
Ambrosia confertiflora	W	Sesquiterpene lactones	Water	*B. glabrata*	Vagas et al, 1986
A. maritima	-	Sesquiterpenes	Hexane	*B. glabrata, Bulinus quernei*	Belot et al, 1991
Baccharis trimera	-	Diterpenes	Water	*B. globosus*	Fibo et al, 1980
Eupatorium odoratum	F	Unknown	Ethanol	*B. globosus*	Adewunmi & Sofowora, 1980
Heliopsis longipes	R	Amide (affin)	Hexane	*Physa occidentale*	Johns et al, 1982
Podachaenium emiens	-	sesquiterpenes	Hexane	*B. globosus*	Fronczek et al, 1984
Wedelia scaberrima	L	Diterpenes	Hexane	*B. globosus*	Tomassini & Mattos, 1979
BALANITACEAE					
Balanites aegyptiaca	R	Spirostanol, Saponins	Methanol	*B. glabrata*	Liu & Nakanshi, 1982
BIGNONIACEAE					
Kigelia africans	F	Unknown	Water	*B. globosus*	Adewunmi & Sofowora, 1980
BOMBACACEAE					
Bombax costatum	S, R	Unknown	Water	*B. globosus*	Adewunmi & Sofowora, 1980
CAESALPINACEAE					
Cassia singuena	S	Unknown	Water	*B. globosus*	Adewunmi & Sofowora 1980
Delonix regia	S	Unknown	Water	*B. globosus*	Adewunmi & Sofowora, 1980
Detarium senegalense	R	Unknown	Water	*B. globosus*	Adewunmi & Sofowora, 1980
Dialium guineense	F	Unknown	Water	*B. globosus*	Adewunmi & Sofowora, 1980
Tamarindus indica	S	Unknown	Water	*B. globosus*	Adewunmi & Sofowora, 1980
CANNACEAE					
Canna indica	L, R	Unknown	Water	*B. globosus*	Adewunmi & Sofowora, 1980
CHENOPODIACEAE					
Artriplex nummularia	F	Saponins	Water	*B. globosus*	Christensen, 1985
CLUSPIACEAE					
Garcinia kolo	S	Serols/triterpene, tannins	Water	*B. glabrata*	Okunji & Iwu, 1988

Table 1. A list of plants having molluscicidal activity. (Continued)

Plant	Plant part Tested	Class of active moiety	Extracts	Species	References
COMBRETACEAE					
Combretum delichoprtalum	L	Unknown	Water	*B. globosus*	Adewunmi & Sofowora, 1980
C. ghasalense	S, R	Unknown	Water	*B. globosus*	Adewunmi & Sofowora, 1980
Terminalia avicennioides	S	Unknown	Water	*B. globosus*	Adewunmi & Sofowora, 1980
T. macroptera	S	Uknown	Water	*B. globosus*	Adewunmi & Sofowora, 1980
T. mollis	S	Unknown	Water	*B. globosus*	Adewunmi & Sofowora, 1980
EBENACEAE					
Diospyros usambarensis	-	Naphthoquinones	Methanol	*B. glabrata*	Marston et al, 1984
ERICACEAE					
Arctostaphylos uva-ursi	L	Tannins	Water, Methanol	*B. glabrata*	Schaufelberger & Hostettmann, 1983
EUPHORBIACEAE					
Acalypha ornata	L, R	Unknown	Methanol	*B. globosus*	Adewunmi & Sofowora 1980
Alchornea cordifolia	F	Alkaloids, Sterols	Water	-	Okunji & Iwu, 1988
Bridelia atroviridis	S	Unknown	Methanol	*B. globosus*	Adewunmi & Sofowora 1980
B. ferruginea	L	Sterols, Saponins	Water	*B. glabrata, B. pfeifferi*	Okunji & Iwu, 1988
Cryptogonone argentea	R	Unknown	Water	*B. globosus*	Adewunmi & Sofowora 1980
Euphorbia antisyphlitica	La	Unknown	Water	*L. acuminata*	Singh & Agarwal, 1987
E. lactea criststa	La	Unknown	Water	*L. acuminata*	Singh & Agarwal, 1988
E. neutra	La	Unknown	Water	*B. globosus*	Adewunmi & Sofowora 1980
E. pulcherima	R	Unknown	Water	*B. globosus*	Adewunmi & Sofowora 1980
E. royleana	La	Unknown	Water	*L. acuminata*	Singh & Agarwal, 1984a.
Jatropha gossypifolia	La	Jatrophane	Water	*B. globosus*	Adewunmi & Sofowora 1980
Manihot glaziovii	S	Unknown	Water	*B. globosus*	Adewunmi & Sofowora 1980
Schima argenta	-	-	Water	*B. globosus*	Cheng, 1971
Thea olesosa	-	-	Water	*B. globosus*	Amin. 1972.
Uapaca guinensis	S	Sterols, Saponins	Methanol	*B. glabrata, B. pfeifferi*	Okunji &Iwu, 1988

Table 1. A list of plants having molluscicidal activity. (Continued)

Plant	Plant part Tested	Class of active moiety	Extracts	Species	References
Croton tiglium	La, S	Saponins, Tanins	Water	*L. acuminata, I. exustus*	Yadav & Singh 2002, Yadav et al. 2004b
Codiaeum variegatum	La, S	Saponins, Tannis	Water	*L. cuminata, I. exustus*	Yadav & Singh, 2001, 2003, Yadav et al. 2004a
Euphorbia pulcherima	La, S	Ellagic acids	Water	*L. cuminata, I. exustus*	Singh et al, 2004a
Euphorbia hirta	La, S	Ellagic acids	Water	*L. cuminata, I. exustus*	Singh et al, 2004a
FABACEAE (PAPILIONACEAE)					
Amphimas pterocarpiodes	S	Tannins, Sterols	Water	*B. glabrata, B. pfeifferi*	Okunji & Iwu, 1988
Dioclea reflexa	R	Unknown	Methanol	*B. globosus*	Adewunmi & Sofowora 1980
Erythrina senegalensis	F	Unknown	Water	*B. globosus*	Okunji & Iwu, 1988
Indigofera kerstingii	L, R	Unknown	Water	*B. globosus*	Adewunmi & Sofowora 1980
I. secundiflora	S, R	Unknown	Water	*B. globosus*	Adewunmi & Sofowora 1980
I. spicata	S	Unknown	Water	*B. globosus*	Adewunmi & Sofowora 1980
Stylosanthes biscosa	S	Unknown	Water	*B. globosus*	Adewunmi & Sofowora 1980
Tephrosia bogelii	L	Flavonoids	Petrolium ether	*B. glabrata*	Marsaton et al 1996.
FAGACEAE					
Quercus spp.	S	Tannins	Water, methonal	*B. glabrata*	Schaufelberger & Hostettmann, 1983
HAMANELIDACEAE					
Hammelis virginiana	L	Tannins	Methanol	*B. glabrata*	Schaufelberger & Hostettmann, 1983
HIPPOCASTANACEAE					
Aesculus indica	S	Triterpines, glycosides	Petrol	*B. glabrata*	Okunji & Iwu, 1988
HIPPOCRATEACEAE					
Hippocratea welwitschii	R	Unknown	Ethanol	*B. globosus*	Sati & Rana, 1987.
ICACINACEAE					
Icacina trichanta	L	Unknown	Methanol	*B. globosus*	Okunji & Iwu, 1988
Pyrenacantha staudtii	L	Unknown	Ethanol	*B. globosus*	Okunji & Iwu, 1988
KRAMERIACEAE					
Krameria triandra	R	Tanins	Methanol	*B. glabrata*	Schaufelberger & Hostettmann, 1983

Table 1. A list of plants having molluscicidal activity. (Continued)

Plant	Plant part Tested	Class of active moiety	Extracts	Species	References
LAMIACEAE (LABIATAE)					
Hyptis pectinata	L	Unknown	Water	*B. globosus*	Adewunmi & Sofowora 1980
H. suaveolens	L	Unknown	Water	*B. globosus*	Okunji & Iwu, 1988
Ocimum canum	L	Unknown	Water	*B. globosus*	Adewunmi & Sofowora 1980
LAURACEAE					
Cassytha filiformis	Ap	Unknown	Ethanol	*B. globosus*	Okunji & Iwu, 1988
LECYTHIDACEAE					
Napoleona vogelii	L	Unknown	Water, methanol	*B. globosus*	Okunji & Iwu, 1988
LILIACEAE					
Allium sativum	B	Allicin	Water	*L. acuminata*	Singh & Singh, 1993
Asparagus curillus	-	Spirostanol saponins	Water	*B. glabrata*	Sati et al, 1984
Eriospermum abyssinicum	R	Unknown	Water	*B. glabrata*	Adewunmi & Sofowora 1980
LONGANIACEAE					
Anthocleista djalonensis	L	Unknown	Methanol	*B. trunctatus*	Okunji & Iwu, 1988
MELIACEAE					
Azadirachta indica	S	Tanin	Methanol	*B. glabrata*	Ayoub & Yankov, 1986
Ekebergia senegalensis	L	Unknown	Water	*B. glabrata*	Adewunmi & Sofowora 1980
MENISPERMACEAE					
Dioscoreophyllum cumminssi	L	Alkaloids, saponins, tannin	Water	*B. glabrata*	Okunji & Iwu, 1988
D. tenerum	L	Alkaloids, saponins, tannin	Methanol	*B. glabrata*	Okunji & Iwu, 1988
Rhigiocarya racemifora	L	Alkaloids, saponins, tannin	Ethanol	*B. glabrata*	Okunji & Iwu, 1988

Table 1. A list of plants having molluscicidal activity. (Continued)

Plant	Plant part Tested	Class of active moiety	Extracts	Species	References
MIMOSACEAE					
Acacia catechu	W	Tannins	Water	*B. glabrata*	Schaufelberger & Hostettmann, 1983
A. dudgeoni	L, S	Unknown	Water	*B. globosus*	Adewunmi & Sofowora 1980
A. nilotica	L, S	Tannins	Water	*B. globosus*	Ayoub, 1985.
Albizza adianthiofolia	L	Unknown	Methanol	*B. globosus*	Okunji & Iwu, 1988
A. amara	S	Tannin	Methanol	*B. pfeifferi*	Ayoub & Yankov, 1986
A. coriaria	S	Tannin	Methanol	*B. trunctatus*	Ayoub & Yankov, 1986
A. lebbeck	S	Tannin	Methanol	*B. trunctatus*	Ayoub & Yankov, 1986
A. zygia	L	Tannin	Methanol	*B. trunctatus*	Ayoub & Yankov, 1986
Culpurnia aurea	L	Alkaloids	Methanol	*B. trunctatus*	Marston & Hostettmann,1985
Dalbergia nitidula	S	Tannins	Water	*B. glabrata*	Schaufelberger & Hostettmann, 1983.
Dichrostachys glomerata	L	Unknown	Water	*B. globosus*	Adewunmi & Sofowora 1980
Entada africana	S	Tanin	Water	-	Ayoub & Yankov. 1986.
Tetrapleura tetraptera	F	Unknown	Water	*B. globosus*	Adewunmi & Sofowora 1980
MORACEAE					
Ficus exasperata	L	Unknown	Methanol	*B. globosus*	Okunji & Iwu, 1988
F. glumosa	S	Unknown	Water	*B. globosus*	Adewunmi & Sofowora 1980

S= stem bark, L= leaf, O= oil cake, F= fruit, R=root, W=whole plant, La= latex, I=inflorescence, C= cashew nut shell, Se= seed, A=aerial part

Pharmacological action of *Croton tiglium* is due to the presence of alkaloids (Rizk, 1987). The alkaloids are naturally occurring organic bases, which contain at least one nitrogen atom either in the heterocyclic ring or linked to an aliphatic skeleton. They are found in vascular plants and rarely occur in gymnosperms, cryptogams and monocotyledons. Okunji and Iwu, (1988) screened several plants of different families for molluscicidal properties and suggested that the toxic properties of these plants may be due to the presence of alkaloids. Toxicity in codiaeum spp. is due to the presence of tanin in the latex (Wealth of India, 1985). Tanin are complex phenolic compounds, divided in to two groups (i) The hydrolysable tanins, which are esters of garlic acid and also glycosides of these esters and (ii) The condensed tanins, which are polymers derived from various flavonoids. The molluscicidal activity found to be related to the free phenolic groups of the tanins. Ayoub and Yankov (1986), screened several tanin-bearing plants of different families for their molluscicidal activity. On assumption was made by him, that the molluscicidal activity of the tanin-bearing plants is proportional to the amount of tanin present in the various morphological parts.

Several species of family Euphorbiaceae, which contain diterpenes, show molluscicidal properties. Those with known molluscicidal properties are *Euphorbia royleana, E. antisyphilitica, E. lacteal acristata, E. pulcherima, E. neutra* and *Jatropha gossypifolia, Croton tiglium,* and *Codiaeum variegatum,* (Singh and Agarwal 1984 a, 1984b, 1987, 1988, 1990, 1991, 1992 a 1992 b, Singh, 1991, Singh and Singh, 2003a, Singh and Singh, 2003b, Singh, 2000, Singh and Singh, 2004, Yadav, 2000, Yadav and Singh, 2001, 2002,). Recent work has demonstrated that the toxicological actions of the latex can be attributed to a new class of diterpenes, which are esters of phorbol (12-deoxyphorbol, 12-deoxy-16-hydroxy-phorbol, ingenol, 5-deoxy-ingenol, resiniferotoxin and tinyatoxin) (Kinghorn and Evans, 1975). It has been reported that phorbol esters interact with and activate the recently discovered protein kinase-C (Takai et al., 1977).

A list of various plants / products, which have been shown to possess molluscicidal activity against harmful snails is given in table 1.

PISCICIDES OF PLANT ORIGIN

Fishing with the aid of plant toxins was formerly very common. Today this easy method of fishing is still practiced in remote areas. The method is simple – the poisonous ingredients are pounded and thrown in to a pool or dammed up sections of a small river. After a short time the fish begin to rise to the surface and can then readily be taken by hand. The fish can be eaten without health problems (Singh, 2001). Several bio-cidal plants have been in use for fish catching practices by the tribal communities in large numbers including Tharu, Bhotia, Kol, Gond, Kharwar and Korwas that inhabit remote villages and forest areas of the State Uttar Pradesh and Bihar (Prakash and Singh, 2000).

According to Neuwinger (1994), 258 fish poisonous plants are present in Africa, based on 25 years of field research by the author in tropical Africa and evaluation of herbarium notes. 10-20 percent fishing poisonous are probably still unknown. They are spread among 167 plant genera and 60 families. The evaluation shows a clear dominance of the leguminosae (Caesappiniaceae, Mimosaceae, Papopmaceae) in the hierarchy of fish poisoning plants. It also remarkable that a great proportion are in euphorbiales.

The latices of several genera of the Euphorbiaceae and in particular of different species of *Euphorbia* have been used extensively by fisherman in different countries as fish poison of high biological activity (Watt and Breyer–Brandijk, 1976; Novock et al., 1980). The rhizome of *Euphorbia biglandulosa* are pounded in order to release the latex in stagnant water of rivers. The rapidly dissolving poison first paralyses and then kills the fish. The sap though causes irritantation in the human skin and has no toxic effect on the people who handle the latex or eat the fish. *Argemone mexicana* belong to family–Papaveraceae and commonly known as pila kantaila. The leaf, seed and fruits of this plant have strong piscicidal activity (Dutta Munshi et al., 1991). Bhatt and Dhyani, (1990), have studied the lethality of pesticidal compounds of plant origin on some freshwater animals like water beetle, water scorpion, back swimmer, water bug, freshwater fish (*Bsrillus bendelisi*, Hom) and different stage of tadpole larvae of *Buffo melanostictus* (Boul.).

Piscicidal activity of *Madhuca indica* (Sapotaceae) has been compiled by Bhakuni et al., (1969); Talwar and Jhingaran, (1991); Satyavati and Gupta (1987). Choubey et al., (1989) evaluate the relative toxicity of oil cake of *Madhuca indica* in different species of animal viz. *Paratelphusa jacquomontii* (crab), *Pheretima posthuma* (earthworm), *Lebistus reticulates* (fish), *Pila globosa* (snail) and tadpole of *Rana tigrina*. Tadpoles were found to be most sensitive to aqueous extracts of oil cake of this plant to the presence of sterol glycoside, saponins and flavonol glycosides. A terpene, 4-deoxyphorbal the active component of plant *Euphorbia tirucalli* is highly poisonous to fish (Kamat and Muthe, 1995).

The toxic effect of ripe fruit pulp of hingan, *Balanites roxyburghii*, containing saponin on fishes *Glossogobius giurius*, *Chanda nama*, *Sarotherodon mossambica* and *Channa marulius* have been evaluated in the laboratory. *Channa marulius* was found more resistant to the toxic effect of hingman fruit than other three fishes. The hingman fruit pulp used for eradication of weed and unwanted fishes from the culture ponds. Same result was also found in case of karanj, *Pongamia pennata* seed on different fishes i.e. *G. giuris*, *Chanda nama*, *Oreochromis mossambicus* (Mohapatra and Nayak, 1998; Mohapatra and Sovan sahu, 2000). The piscicidal compounds came from 20 families of angiosperms summarised in table 2.

Table 2. A list of plants having piscicidal activity.

Plant	Plant parts tested	Class of active moiety	Extracts	Fish spp.	References
Apocynace					
Nerium indicum	L	Oleandrin	Water	*Channa punctatus*	Tiwari and Singh 2004
Nerium indicum	L	Oleandrin	Acetone	*C. punctatus*	Tiwari and Singh 2003
Nerium indicum	L	Unknown	Water	*C. punctatus*	Singh and Singh 2000
Anacardiaceae					
Lannea coromandelica	S	Unknown	Water	*Ophiocephalus punctatus*	Kulakkattolickal, 1989
Semecarpus anacardium	F	Unknown	Water	*O. punctatus*	Kulakkattolickal, 1989
Anonaceae					
Miliusa veluntina	S	Unknown	Water	*O. punctatus*	Kulakkattolickal, 1989
Miliusa veluntina	L	Unknown	Water	*O. punctatus*	Kulakkattolickal, 1989
Cannabinaceae					
Cannabis sativa	L	Unknown	Water	*O punctatus*	Kulakkattolickal, 1989
Combretaceae					
Terminalia chebula	F	Unknown	Water	*O. punctatus*	Kulakkattolickal, 1989
Terminalia alata	R	Unknown	Water	*O. punctatus*	Kulakkattolickal, 1989
Euphorbiaceae					
Codieum varigatum	S	Taraxerol	Ether	*Channa punctatus*	Yadav et al. 2004a
Euphorbia royleana	S	Unknown	Water	*C. punctatus*	Singh and Singh 2000
Euphorbia tirucalli	S	4-deoxy phorbol	Water	*C. punctatus*	Tiwari et al. 2001
Euphorbia tirucalli	La	4-deoxy phorbol	Water	*C. punctatus*	Tiwari et al. 2001
Glochidion velutinum	S	Unknown	Water	*Ophiocephalus punctatus*	Kulakkattolickal, 1989
Mallotus philippensis	S	Unknown	Water	*O. punctatus*	Kulakkattolickal, 1989
Manihot esculenta	R	Unknown	Water	*O. punctatus*	Kulakkattolickal, 1989
Labiatae					
Colebrookea oppositifolia	L	Unknown	Water	*O. punctatus*	Kulakkattolickal, 1989
Lecythydiaceae					
Careya arborea	R	Unknown	Water	*O .punctatus*	Kulakkattolickal, 1989
Leguminosae					
Acacia catechu	S	Unknown	Watewr	*O. punctatus*	Kulakkattolickal, 1989
Acacia pennata	S	Unknown	Water	*O. punctatus*	Kulakkattolickal, 1989
Dalbergia sp.	S	Unknown	Water	*O. punctatus*	Kulakkattolickal, 1989

Plant	Plant parts tested	Class of active moiety	Extracts	Fish spp.	References
Lythraceae					
Lagerstroemia parviflora	S	Unknown	Water	*O. punctatus*	Kulakkattolickal, 1989
Meliaceae					
Azadirachta indica	S	Nimbisidine	Water	*O. punctatus*	Kulakkattolickal, 1989
Myrtaceae					
Syzygium cumini	S	Unknown	Water	*O. punctatus*	Kulakkattolickal, 1989
Papilionaceae					
Derris elliptica	S	Unknown	Water	*O. punctatus*	Kulakkattolickal, 1989
Indigofera sp.	S	Unknown	Water	*O. punctatus*	Kulakkattolickal, 1989
Polygonaceae					
Pterospermum hydropiper	S	Unknown	Water	*O. punctatus*	Kulakkattolickal, 1989
Polygonum neplense	W	Unknown	Water	*O. punctatus*	Kulakkattolickal, 1989
Polyonaceae					
Polygonum hydropiper	L, B, R	Plygodial	Water	*Heteropneustes fossilis*	Barnes and Lodder 1962
Rosaceae					
Artybotria sp.	R	Unknown	Water	*O. punctatus*	Kulakkattolickal, 1989
Rubiaceae					
Adina cordifolia	S	Unknown	Water	*O. punctatus*	Kulakkattolickal, 1989
Catunaregam spinosa	F	Unknown	Water	*O. punctatus*	Kulakkattolickal, 1989
Rutaceae					
Aegle marmelos	L	Unknown	Water	*O. punctatus*	Kulakkattolickal, 1989
Sapotaceae					
Madhuca indica	O	Rotenone	Water	*Lebistus reticulatus*	Choubey et al. 1989
Madhuca indica	O	Rotenone	Water	*Channa punctatus*	Perschbacher and Sarkar,1989
Solanaceae					
Nicotiana tabacum	L	Unknown	Water	*Ophiocephalus punctatus*	Kulakkattolickal, 1989
Solanum xanthocarpum	F	Unknown	Water	*O. punctatus*	Kulakkattolickal, 1989

S= stem bark, L= leaf, O= oil cake, F= fruit, R=root, W=whole plant, La= latex

INSECTICIDES OF PLANT ORIGIN

In all existing insect pests, mosquitoes are greater public enemy (Chandler and Read, 1981) and responsible for more human illness than any other group. About 3000 species of mosquitoes have been identified all over the world (Hati, 1991). Out of them 400 species of mosquitoes belonging to 19 genera and 42 sub-genera are recorded in India. Of these, only four genera, viz; *Anopheles, Culex, Aedes, Mansonia* are the vectors of various mosquito born diseases (Tewari, 1995). They alone carry Malaria, Yellow fever, Dengue, and Bancroftian filariasis. They also carry several types of encephalitis and occasionally involved in the mechanical transfer of Tularemia and Anthrax (Stefferud, 1952).

Malaria, the great disabler, prevails through out the tropics and temperate regions. Species of *Anopheles* mosquitoes are the carriers of human malaria. Dengue was considered to be urban disease in India but in the recent past several epidemics have also occurred in rural areas (Mourya and Joshi, 2002). *Aedes aegypti* is the main vector of dengue and yellow fever. Elephantiasis a disfiguring malady of peoples in the tropic and sub- tropic caused by *Wuchereria bancrofti* is transmitted by *Culex quinquefasciatus* mosquito (Menon and Rajagopalan, 1980). Japanese encephalitis (JE) being primarily a zoonosis, vertebrate hosts such as pigs and birds play an important role in the maintenance and amplification of the virus (Pant, 1979). In 1973 large outbreak of Japanese encephalitis occurred in the state of West Bengal and this was the first recognized epidemic of JE in India (Chakrovarthy et al., 1975). Japanese encephalitis virus was first isolated from mosquitoes of culicine species including *Culex tritaeniorhynchus* in Japan (Mitamura et al., 1938).

In recent years, there has been revival of interest in plant- derived insecticides, which are selectively toxic, do not bio-accumulate, and exhibit relatively short persistence in the environment (Shanker and Solanki, 2000). Attempts have been made to search for new classes of insecticides derived from plants owing to low toxicity and less persistence in nature (Marston and Hostettmann, 1985; Singh et al., 1996b; Kloos and Mc Cllough, 1987). Some of plant products like alkaloids, vegetable oils, plant extracts, triterpenoids, rotenone and azadiractine have lead to discovery of new molecule from botanicals as alternative pest control agents (Shankar and Solanki, 2000). Plant products can be obtained either from the whole plant or from a specific part by extraction with different types of solvents such as aqueous, methanol, chloroform, hexane etc., depending on the polarity of the phytochemicals, Studied carried out so for have shown that some chemical act as general toxicant (insecticidal / larvicidal) against both adult as well as larval stages of mosquitoes, while others interfere with growth and development (growth inhibitors) or with reproduction (chemosterilant) or produce olfactory stimuli thus acting as repellent or attractant.

Deshpanday and Tipnis, (1971) reported that the essential oils of *Ocimum basilicum* have been found to be insecticidal and larvicidal towards mosquito. Recently Bhatnagar et al., (1993) identified methyl chvicol the major constituents of *Ocimum basilicum* oil as more effective than eugenol or component of *O. sanctum* oil which was found to active against three species of mosquitoes *Anopheles stephensi, Aedes aegypti and Culex quinquefasciatus.* Kumar and Dutta (1987), reported toxicity of many essential oil towards 4th instars larva of *Anopheles stephensi,* the oils are *Cedrus deodara, Cymbopogon nardus, C. flexuosus, C. martini, Lavandula officinale (L. angustifolia), Mentha arvensis, Ricinus communis, Eucalyptus globules, Melia azadirachta, Caryptophyllus (Syzygium aromaticum).* A large number of essential oils have evaluated for their repellent activity against mosquitoes, the most effective oils being those of *Daerydum franlint* (Huonpine wood), oil of leaves of *Bakhousia myritifolia, Melaleuca bractia and Zieria smithi* (Sharma et al., 1994; Singh et al., 2001). In fields trails *Cinnamomum camphora* oil protected from various *Anopheles* for about 12 hours (Ansari and Razdan, 1995; Bhatnagar et al., 1993).

The effect of essential oils upon insects depends on several parameters including chemical composition and species susceptibility of insects as the physiological responses of each insect species towards the same crude plant extract or essential oil was not same (Bhatnagar et al., 1993). Available reports reveal that lipophilic nature of compounds, alkyl side chain, free phenolic, hydroxy or methylene dioxy group are significantly for insecticidal activity, while aliphatic straight chain Ketone and aryl Ketone compound exhibit strong repelling activity, further importance of specific position of side chain in few molecules enhance the activity (Singh et al., 2001).

A large number of plant extracts have been reported to have mosquitocidal larvicidal, growth inhibitor chemosterilant and repellent activity against mosquito vectors given in Table 3.

Table 3. A List of plants producing repellant, larvicidal and insecticidal activities against mosquitoes.

Name of the Plants	Properties	References
Allium sativum	Larvicidal	Amonkar and Benerjee, 1971
Angelica	Insecticidal	Singh et. al. 2001
B. myrstifolid	Repellant	Sharma et al., 1994
Cedrus deodara	Insecticidal	Singh et. al. 2001
C.camphora	Repellant	Ansari and Razdan 1995; Bhatnagar et al. 1993
C. lyrata	Insecticidal, Repellent	Sharma, 1991
C.citratus	Repellent, larvicidal	Singh et. al. 2001; Selander et al. 1974
C.nardus	Repellent	Singh et. al. 2001
C. domestica	Larvicidal, Repellent	Singh et al. 2001
C. martini varsofia	Repellent	Singh et. al. 2001
C. martini var motia	Repellent	Singh et. al. 2001:Selander et al.1974
Daucas carata	Insecticidal	Purohit et al. 1983
Duerydium franklint	Repellent	Sharma et al.1994
Denettia tripetala	Insecticidal	Maradufu, 1978
Eucalyptus species	Repellent	Sharma et al. 1994
Eucalyptus globules	Larvicidal	Singh et. al. 2001
E. ugenia	Larvicidal	Singh et. al. 2001
Larix decidua	Repellent	Trigg,1996
L. angusifoliat	Repellent	Singh et. al. 2001
Linaloe	Repellent	Sharma et al. 1994
Melaleuca bractea	Repellent	Yusufoglu and Hasdemir, 1996
Melia azadiracta	Repellent, Larvicidal	Singh et. al. 2001
Mentha arvensis	Larvicidal	Singh et. al. 2001
Nepta catane	Repellent	Iwvabe et al. 1981
Ocimum sanctum	Repellent, larvicidal	Hayashi, 1995; Hebalkar et al. 1992
Origanum majorana	Larvicidal	Singh et. al. 2001:Selander et al.1974
Ricinus communis	Larvicidal, Repellent	Singh et. al. 2001
Rose	Repellent	Singh et. al. 2001
Tridax procumbers	Repellent, Insecticidal	Nayak et al.1995
Vitex negundo	Repellent	Klock et al.1987
Zingiber officinale	Repellent	Singh et. al. 2001
Euphorbia roleyana	Insecticidal	Srivastava et al. 2003
Nerium indicum	Insecticidal	Srivastava et al. 2003

ACTIVE COMPOUNDS

Plants are the richest source of renewable bioactive organic chemicals. The total number of plant chemicals may exceed 40,000 of these 10,000 are secondary metabolites whose major role in the plants is reportedly defensive (Swain, 1977; Cooper and Johnson, 1984). Numerous defensive chemicals belonging to various categories (terpenoids, alkaloids, glycosides, phenols, tannins, etc.) that cause behavioural and physiological effects on pests have already been identified. Some important compounds are as given below:

Saponins- Saponins are naturally occurring plant glycosides, which form a soapy lather with water they consist of a sugar moiety and an aglycone unit. Monodesmosidic saponins (Sugar moiety only at position C-3) possess toxic activity whereas biodesmosidic saponins (sugar moiety both at C-3 and C-28 are inactive).

Alkaloids- Alkaloids are naturally occurring organic bases, which contain at least one nitrogen atom either in the hetrocyclic ring or linked to an aliphatic skeleton. They are usually colourless, crystalline, non-volatile solids, slightly soluble in water but soluble in ethanol, ether and chloroform.

Flavonoids- The term flavonoids embraces all compounds whose structure is based on flavone. Flavonoids are C_{15} compounds (exclusive of O-alkyl groups and secondary substituents), which are composed of two phenolic nuclei connected by three Carbon unit.

Diterpeneoids- The diterpenoids from a group of compounds having general molecular formula $C_{20}H_{32}$. They are not steam volatile and are usually obtained from plants. A new class of diterpenes which are esters of phorbol (12-deoxyphorbol, 12-deoxy-16 hydroxy-phorbol, ingenol, 5-deoxy-ingenol resiniferotoxin and tinyatoxin) possess highly toxic activity against pests.

Monoterpenoids: Monoterpenoids are made up of two isoprene units ($C_{10}H_{16}$) and are the chief constituents of essential oils.

Sesquiterpenes lactones: Sesquiterpenoid lactones are those compounds, which possess a sesquiterpene skeleton having an additional lactone ring.

Iridoids: Iridoids are those monocyclic monoterpenoids, which possess a lactone ring instead of having the ρ-menthane skeleton.

Furanocoumarins: Furanocournarins are a group of compounds in which the 1, 2 benzopyrene skeleton is fused with a furan ring.

Tanins: Tanins are complex phenolic compounds which can be divided into two groups, (i) the hydrolysable tannins, which are esters of gallic acid and also glycosides of these esters and (ii) the condensed tannins, which are polymers derived from various flavonoids.

CONCLUSION

There are a very large number of plants, which contain compounds lethal to target as well as non-target organism at doses, which are much below those for synthetic pyrethroids (Marston et al., 1996; Singh et al., 1996a; Amusan et al., 1997; Singh, 1991; Singh and Singh, 2003a,b; Singh and Singh, 2004; Singh et al., 2004 b). Use of such products has the additional advantage that these contain biodegradable compounds, which are less likely to cause environmental contamination. After all such compounds are not only confined to the plants in which they are found but also possibly gets distributed in the environment. We strongly feel that if these herbaceous products are used as molluscicides, piscicides and insecticides they would not only control the vector snails, predatory fishes and mosquitoes but would also have the advantage of easy availability, low cost, easy biodegradability and greater acceptance amongst the users. Further more we feel that with further progress in biotechnology, such products could be raised form sources other than those plants in which they are currently found. Production of plant pesticides could, in long run also becomes an important industry using biotechnological methods.

Most of the studies cited in this article however suffer form one common drawback. While there is much information on the toxicity and lethal doses of these plant pesticides, very little literature available on their mode of action and effect on other organisms. We are hopeful that pesticides of plant origin will find practical use as many of the plants having molluscicidal, piscicidal or mosquitocicidal activity have been used by mankind for Integrated Pest Management Programme (IPM) without polluting their environment.

REFERENCES

Abbott, R. T. (1954). The marine molluscs of Grand Cayman Island. Monogr. Acad. *Natur. Sci,* Philadelphia. Pp 11.

Adewunmi, C. O. and Morquis, V.O. (1980). Molluscicidal evaluation of some *Jatropha* sps., grown of the plant tested. *J. Crud. Drug. Res.* 18: 141-145.

Adewunmi, C. O. and Sofowora, E. A. (1980). Preliminary screening of some plant extracts for molluscicidal activity. *Planta Med.* 39: 57-65.

Agarwal, R. A. and Singh, D. K. (1988). A Harmful gastropods and their control. *Acta Hydrochim. Hydrobiol.* 16: 113-138.

Amin, M. A. (1972). Preliminary report on the molluscicidal properties of Habat EI-Mallok, *Jatropha* sp. Trans. Roy. Soc. Trop. Med. Hyg. 66: 805-806.

Amonkar, S. V. and Benerjee, A. (1971). Isolation and charactrization of larvicidal principal of garlic. *Sci.* 174: 1343-1344.

Amusan, O. O. G. Msonthi, J.D. and Makhuba, L. P. (1997). Molluscicidal activity of *Urginia epigea. Fitot.* 68: 185-186.

Ansari, M. A. and Razdan, R. K. (1995). Repellent action of *C. martini* staf var sofia against mosquito. *Ind. J. Malariol.* 31: 95-102.

Ayoub, S. M. H. (1985). Flavanol molluscicides from the Sudan Acacias. *Int. J. of Crud Drg. Res.* 23: 87-90.

Ayoub, S. H. M. and Yankov, L. K. (1986). The molluscicidal factor of tannin-bearing plants. *Int. J. of Crud. Drug. Res.* 24: 16-18.

Bali, H. S. Singh, S. and Sharma, S. (1986). The distribution and ecology of vectors snails of Punjab. *Ind. J. Ecol.* 13: 31-37.

Barnes, C. S. and Lodder, J. W. (1962). The structure of Polygodial: a new sequiterpene dialdehyde from *polygomium hydropipes*. L. Auset. J. Chem. 15: 322-327.

Belot, J. Bornarel, P. Diouf, M. and Polderman, A. M. (1991). *Ambrosia maritima* L. Molluscicidal effects on the land snails *Lymnaea natalensis*, *Bulinus forskalii*, *Bulinus globosus* and *Biomphalaria pfeifferi* from Senegal. *Plant Sci.* 74: 167-170.

Bhakuni, D. S. Dhar, M. L. Dhar, M. M. Dhawan, B. N. and Mehrotra, B. N. (1969). Screening of Indian plants of biological activity. *Ind. J. Exp. Biol.* 7: 250.

Bhatt, J. P. and Dhyani, P. P. (1990). Lethality of piscicidal compounds of plant origin to some freshwater animals, *J. Fresh. Biol.* 2 (1): 5-12.

Bhatnagar, M. Kapur, K. K. Jellees, S. and Sharma (1993). Laboratory evaluation of insecticidal properties of *Ocimum basilicus* L. and *O. sanctum* L. plants essential oils and their major constituents against vector mosquito species. *J. Entomol. Res.* 17: 21-26.

Bourgeois, D. P., Gaudet, J., Deveau, P. and Maller, N. (1993). Micro extraction of organophosphorous pesticides from environmental water and analysis by gas chromatography. *Bull. Envrion. Contam. & Toxicol.* 50: 433-440.

Callegari-lopes, J.L. (1991). Sesquiterpenes lactones from vernonia. *Mem. Inst. Oswaldo Cruz,* 86: 227-230.

Cepleannu, F. Hamburger, M.O. Sordat, B. Msonthi, J.D. Gupta, M.P. Saadou, M and Hostettmann K. (1994). Screening of tropical medicinal plants for molluscicidal, larvicidal, fungicidal and cytotoxic activity and brine shrimp toxicity. *Int. J. Pharmacy,* 32: 294-307.

Chakrovarthy, M.S. Mukherjee, M.K. Das, B.C. and Hati, A.K (1975). The first epidemic of Japanese encephalitis in India virological studies. *Ind. J. of Med. Res.* 63: 77-82.

Chandler, A.C. and Read, C. P. (1981). *Introduction to parasitology.* John-Wiley and Sons Inc., New York. pp 882.

Cheng, T.H. (1971). Schistosomiasis in mainland China: A review of research and control programmes since 1949. *Amer. J. Trop. Med. Hyg.* 20: 26-53.

Chifundera, K. Baluka, B. and Mashimango, B. (1993). Phytochemical screening and molluscicidal potency of some Zairean medicinal plants. *Pharmacol. Res.* 28: 333-340.

Choubey, S. M. Mahajan, R. T. Jawale, S. M. and Karnik, A. G. (1989). Toxic effect of oil cake of *Madhuca indica* (Gmel) in different species of animals. Environmetal series, Vo.II in "*Recents trend in toxicology*" ed. by R. Prakesh, pp.189-193.

Christensen, S. B. (1985). *Atriplex nummularia*, a source for the two molluscicides saponins: Hederagenin-3-0-α-D-glucuronopyranoside and calenduloside *J. of Nat. Prod.* 48: Pp161.

Cooper, M. R. and Johnson, A. W. (1984). In reference book 161. *Ministry of Agriculture, Fisheries and Food*, London.

Deshpanday, R.S. and Tipnis, H.P. (1971). Insecticidal activity of *Ocimum basilicum* L. Pesti. 11: 11-12.

Dua, V. K. Pant, C. S. Sharma, V. P. and Pathak, G. K. (1998). Determination of HCH and DDT in finger-prick whole blood dried on filter papers and its field application for monitoring concentrations in blood. *Bull. Environ. Contam. & Toxicol.* 56: 50 - 57.

Dutta Munshi, J. Sinha, M. K. and Lal, B. K. (1991). Phytochemical screening and quantitative analysis of certain piscicidal plants of Santhal Parganas. *J. Fresh. Biol.* 3 (2): 177-181.

El-Sawy, M. F. Bassiouny, H. K. and El-Magdoub, A. I. (1981). Biological combat of schistosomiasis. *Ambrosia maritima* (damsissa) for snail control. *J. Egypt Soc. Parasitol.* 11: 99-117.

Fibo, D. S. D. Sarti, S. J. Vichnewski, W. Bulhoes, M. S. and Filha, de F. L. H. (1980). *Ret. Fac. Form. Odontol. Ribeiro Poreto* (Uni. Sao Paulo). 17: 43.

Fronczek, F. R. Vargas, D. Fischer, N. H. and Hostettmann, K. (1984). The molecular structure of 7-Alpha-Hydroxy-3-Desoxy-Zaluzanin-c, a molluscicidal sesquiterpene lactone. *J. of Nat. Prod.* 47: 1036.

Froyed, G. (1975). Liver fluke in Great Britain: A survey of affected livers. Vet. Rec. 97: 492-495.

Godan, D. (1983). *Pest slugs and snails, Biology and control* (ed. Dora Godan) translated by Sheila gruber. Springer- Verlog, Berlin Heidelberg, New York.

Hati, A. K. (1991). Malaria West Bengal State Book Board, Calcutta. pp 316.

Hayashi, S. (1995). *Repellency and insecticidal activity of Japanese mint oil.* Aromat. 13: 80-81.

Hebalkar, D. S. Hebbalkar, C. D. Sharma, R. N. Joshi, V. S. and Bhat, V. S. (1992). Mosquito repellent activity of oils from *Vitex negundo* L. leaves. *Ind. Med. Res.* 95: 202-203.

Hyman, L. H. (1970). The invertebrate. Vol. 6. Mollusca I. Mc Graw Hill, New York.

Iwvabe, M. O. E. Osisiogu, I. W. U. and Agbakuru, E. O. P. (1981). Dennetia oil. A potential new insecticide. Tests with adults and nymphs of *Periplanata Americana* and Zonocerus variegates. J. Med. Entomol. 74: 249-252.

Jobin, W. R. (1973). Environmental control of bilharzias snail in small reservoir. J. of the Irri. and Drain. Div. 99: 365-373.

Johns, T. Graham, K. and Towers, G. H. N. (1982). Molluscicidal activity of affinin and other isobutyllamides from the Asteraceae. Phytochem. 21:2737-2738.

Kamat, D. V. and Muthe P. T. (1995). Poisonous effect of *Euphorbia tircalli* on fish. J. of Ani. Morphol. and Physiol. 42 (1-2): 65-68.

Kinghorn, A. D. and Evans, F. J. (1975). A biological screen of selected species of the genus euphorbia for skin irritant effects. Planta. Med. 28: 325 - 335.

Kishor, N. and Sati, O. P. (1990). A new molluscicidal spirostanol glycoside of *Yucca aloifolia* J. of Nat. Prod. 33: 1557-1559.

Klock, J. A. Darlingaton, M. V. and Balandrin, M. F. (1987). 1,8 cineole (eucalyptol), a mosquito feeding and ovipositional relellent from volatile oil of *Hemizonia fitchii* (Astreaceae). J. Chem. Ecol. 3: 2131-2141.

Kloos, H. and Mc Cullough, F. S. (1987). Plants with recognized molluscicidal activity. In Plant molluscicides (eds. KE Mott) NewYork, UNDP/World Bank/WHO, 45-108.

Kulakkattolickal, A. T. (1989). Piscicidal plants of Nepal: ripe fruit of Catunaregam spinosa (Thunb.) (Rubiaceae) and leaves of *Polygonum hydropiper* L. (Polygonaceae) as fish poisons. Aqua. 78: 293-301.

Kumar, K. and Dutta, G. P. (1987). Indigenous plants oils as larvicidal agents against *Anopheles stephensi* Curr. Sci. 16: 959-960.

Lambert, C. R. (1966). Bilharziasis CIBA symposium, 14: 3-14.

Laurens, A. Fourneau, C. Hocqemiller, R. Cave, A. Bories, C. and Loiseau, P. M. (1997). Anivectorial activities of cashewnut shell extracts from *Anacardium occidentate* L. Phyto. Res. 11: 145-146.

Liu, H. W. and Nakanshi, K. (1982). The structures of Balanitins, potent molluscicides isolated from *Balanites aegyptiaca*. Tetrah. 38: 513-519.

Maradufu, A. (1978). Isolation of (5E)-Ocimenone, a mosquito larvicide from *Tagetes minuta* Uoydia 41: 181-183.

Marston, A. Msonthi, J. D. and Hostettmann, K. (1984). On the reported molluscicidal activity from *Tephrosia vogelli* leaves. Phytochem. 23: 1824-1825.

Marston, A. and Hostettmann, K. (1985). Plant molluscicide. Phytochem. 24: 639-652.

Marston, A. Dudan, G. Gupta, M.P. Sois, P.N. Correa, M.D. and Hostettman, K. (1996). Screening of Panamanian plants for molluscicidal activity. Int. J. of Pharmacol. 34: 15 - 18.

Menon, P. K. B. and Rajagopalan, P. K. (1980). Relative importance of different types of breeding habits in contributing to the population of *Culex pipens fatigans* in Pondicherry. Ind. J. Med. Res. 71: 725-733.

Minelli, E. V. and Rebeiro, M. L. (1996). DDT and HCH residues in the blood serum of malaria control sprayer. Bull. Environ. Contam. & Toxicol. 57: 691 - 696.

Mitamura, T. Kitaoka, M. Mori, K. and Okubo, K. (1938). Isolation of the virus of Japanese encephalitis form mosquitoes caught in nature. Tokyo Shinshi, 62: 820-824.

Mohapatra, B. C. and Nayak, G. B. (1998). Assessment of toxicity of ripe fruit pulp of Hingan, *Balanites roxburghii*, on different fishes. J. Aqua. 6: 19 - 21.

Mohapatra, B. C. and Sovan sahu (2000). Toxicity of Karanj, *Pongamia pinnata* seed on different fishes. The 5th Indian Fisheries Forum. 17-20 January 2000. CIFA Bhubneshwar.

Mourya, D. T. and Joshi, V. (2002). Horizontal and vertical transmission of dengue virus in *Aedes aegypti* mosquitoes and its persistence in successive generations. 6th Int. Sym. On virus and vector Borne diseases. Reg. Med. Res. Cen. Bhubanswar.

Nayak, A. K. Raha, R. and Das, A. K. (1995). Organochlorine pesticides in middle stream of the Ganga river, India. *Bull. Environ. Contam. & Toxicol.* 54: 68-76.

Ndamba, J. Molgaard, P. Lemmich, E. Chandiwana, S. K. and Furu, P. (1995). Response of the molluscicidal berry plant (*Phytolacca dodecandra*) to different climatic and edaphic conditions. *Trop. Agric.* (Trinided), 72: 135-140.

Neuwinger, H. D. (1994). Fish poisoning plant in Africa. *Botanica Acta.* 107 (4): 263- 270.

Novock, E. Crea, A. E. G. and Falsone, G. (1980). Inhibition of mitochondrial oxidative phosphoryaltion by 4-deoxyphorbol tiresters, a poisonous constituent of latex of *Euphorbia biglandulosa. Toxicon*, 18: 165 – 174.

Okunji, C. O. and Iwu, M. M. (1988). Control of schistosomiasis using Nigerian medicinal plants as molluscicides. *Int. J. Crude Res.* 28: 246-252.

Pant, C. P. (1979). Control of vectors of Japanese encephalitis. WHO/VBC/79: 733.

Park, D. Minor, M. D. and Propper, C. R. (2004). Toxic response of endosulfan to breeding and non-breeding female mosquito fish. *J. Environ. Biol.* 25: 119-124.

Perschbacher, P. W. and Sarkar, J. (1989). Toxicity of selected Piscicides to the snake head *Channa punctatus. Asi. Fish, Sci.* 2 (2): 249-254.

Prakash, A. and Singh, K. K. (2000). Observation on some high valued ethno medicinal plants among the tribal of Uttar Pradesh. J. Med. & Arom. Plants (in press).

Purohit, P. Mustafa, M. and Osmami, J. (1983). Insecticidal properties of plant extract of *Cuminum cyminum Lnn. Sci. Cult.* 49: 101-103.

Rizk, 1987. The chemical constituents and economic plants of the euphorbiaceae. *Bot. J. of the Linean Soci.* 94: 293-326.

Saravanan, T. S. Mohamed, M. A. Chandrasekar, R. and Sundramoorthy, M. (2003). Freshwater fishes as indicators of Kaveri River pollution. *J. Environ. Biol.* 24: 381-389.

Sati, O. P. and Paul, G. and Hostettmann, K. (1984). Potent molluscicides from *Asparagus.* Pharmazie, 39: Pp 581.

Sati, O. P. and Rana, U. (1987). A new molluscicidal triterpenic glycoside from *Aesculus indica*. *Int. J. of Crud Drug Res*. 25:158-160.

Satyavati, G. V. and Gupta, A. K. (1987). Medicinal plants of India. I. C. M. R. New Delhi.

Schaufelberger, D. and Hostettmann, K. (1983). On the molluscicidal activity of tannin containing plants. *Planta Med*. 48: 105-107.

Selander, J. Kalo, P. Kangus, E. and Perttunem, V. (1974). Olfactory behavior of *Hylobius abietes* L (Col Curlionidae). *Response to several terpenoid fractions isolated from Scoats pine pheoem. Ann*. Entomol. Fenn. 110: 109-115.

Selvarani, D. and Rajamanickam, C. (2003). Toxicity of PCB 1232 on mitochondria of fish *Arius caelatus* (Valenciennes). Ind. J. Exp. Biol. 41: 336-340.

Shanker, C. and Solanki, K. (2000). Botanical Insecticides: A historical perspective. *Asi. Agric. Hist*. 4 (3): 221-232.

Sharma R. N. Desphanday, S. G. Tungikar, V. B. and Toseph, M. (1994). Toxicity of natural essential oils to mosquito *A. aegypti* and *culex fatigans*. Geobios 21: 162-165.

Sharma, R. N. (1991). Proceeding of International seminar on newer trends in essential oils and flavour RRL Jammu (J & K) India. 21-23.

Shoeb, H. A. EL-Sayed, M. M. EL-Wakel, A. A. and Abd-EL-Ghany, L. (1993). The molluscicidal properties of *Thymelaea hirsuta* and *Agave attenuata*. Egypt. J. Bilaharz. 14: 77-89.

Singh, A. and Agarwal, R. A. (1988). Possibility of using latex of euphorbiales for snail control. The Sci. of the Total Environ. 77: 231-236.

Singh, A. and Agarwal, R. A. (1990). Molluscicidal properties of synthetic pyrethroids. J. of Med. & Appl. Malacol. 2: 141-144.

Singh, A. and Agarwal, R. A. (1991). Kinetics of acetylcholinesterase inhibition by latex of euphorbiales in the snail *L. acuminata*. J. Med. & Appl. Malacol. 3: 101- 105.

Singh, A. (1991). Molluscicides of plant origin against harmful snails. Ph.D. thesis Gorakhpur University, Gorakhpur U.P India.

Singh, A. and Agarwal, R. A. (1992a). Toxicity of the latex of euphorbiales. Effect on acid and alkaline phosphatase of the snail *Lymnaea acuminata*. *Biol. Agric. & Hortic*. 8: 211-219.

Singh, A. and Agarwal, R. A. (1992b). Molluscicidal activity of euphorbiales against the snails *Indoplanorbis exustus*. *Acta. Hydrochim. hydrobiol*. 20: 262-264.

Singh, A. and Singh, D. K. (1994). Pestoban, A potent herbal molluscicides. Biol. Agric. & Hortic. Vol: Pp 175-178.

Singh, A. Singh D. K. Mishra, T. N. and Agarwal, R. A. (1996a). Molluscicides of plant origin. *Biol. Agric. & Hortic*. 13: 205–252.

Singh, A. K. Tripathi, A. K. Bindra, R. L. Verma, N. and Kumar, S. (2001). Essential oils and isolation for controlling household insects: Housefly, Cockroach and Mosquito. *J. Med. & Aroma. Plant Sci*. 22/4A & 23/1A, :159-166.

Singh, D. and Singh, A. (2000). The acute toxicity of plant origin pesticides into the freshwater fish *Channa punctatus*. Acta hydrochim. hydrobiol. 28 (2): 92 – 94.

Singh, D. (2001). Studies on toxicological and biochemical effects of phytopesticides on non-target freshwater fish *Channa punctatus*. Ph.D. thesis DDU Gorakhpur University, Gorakhpur U.P India.

Singh, D. and Singh, A. (2003). Effect of stem-bark extract of some common plants on non-target freshwater fish *channa marulius* (Ham). Ind. J. Fish. 50 (4): 525-532.

Singh, D. and Singh A. (2004). Biochemical stress responses in tissues of fish *Channa punctatus* due to lattices of *Nerium indicum* and *Thevetia peruviana*. J. Appl. Toxicol. (in press).

Singh, D. K. and Agarwal, R. A. (1984a). Correlation of the anti-cholinesterase and molluscicidal activity of the latex of *Euphorbia royleana* on the snail *Lymnaea acuminata*. J. Nat. Prod. 47: 702-705.

Singh, D. K. and Agarwal, R. A. (1984b). Alteration of biogenic level in the snail *Lymnaea acuminata* by latex of *Euphorbia royleana*. Toxicol. Lett. 21: 309-314.

Singh, D. K. and Agarwal, R. A. (1987). Latex of *Euphorbia antisyphlitica*, a new potent molluscicides having anticholinesterase activity against the snail *Lymnaea acuminata*. The Sci. of Total Environ. 61: 211-215.

Singh, D. K. and Singh, A. (1993). *Allium sativum* (Garlic), A potent new molluscicide.. *Biol. Agric. Hortic.* 9: 121-124.

Singh, K. Singh, A. and Singh, D. K. (1995). Molluscicidal activity of different combinations of the plant products used in the molluscicides pestoban. *Biol. Agric. Hortic.* 12: 253-261.

Singh, K. Singh, A. and Singh, D. K. (1996b). Molluscicidal activity of neem (*Azadirachta indica*). J. of Ethano. 52: 35-40.

Singh, K. Singh, A. and Singh, D. K. (1998). The use of piperonyl butaoxide and MGK-264 to improve the efficacy of some plant-derived molluscicides *Pesti. Sci.* 54: 145-149.

Singh, O. and Agarwal, R. A. (1981). Toxicity of certain pesticides to two economic species of snails in Northern India. *J. of Econ. Entomol.* 74: 568-571.

Singh, S. K. (2000). Studies on molluscicidal properties of some local plants of eastern Uttar Pradesh against harmful snails. Ph.D. thesis DDU Gorakhpur University, Gorakhpur U.P India.

Singh, S. K. Yadav, R. P. and Singh, A. (2000). Molluscicidal activity of *Thevetia peruviana*, a common medical plant of India. J Med. and Arom. Plant Sci. 22/4A & 23/1A: 113-116.

Singh S. K. and Singh A. (2003a). Molluscicidal and Anti-cholinesterase activity of *Alstonia scholaris* plant against freshwater snail *Lymnaea acuminata*. Pak. J Biol. Sci. 6 (16): 1442-1446.

Singh S. K. and Singh A. (2003b). Effect of the plants *Thevetia peruviana* and *Alstonia scholaris* (Family:Apocynaceae) on acetylcholinesterase activity of *Lymnaea acuminata* snails. Egypt J. Schist. & Infe. & End. Dise. 25: 31-40.

Singh S. K. and Singh A. (2003c). Toxic effect of *Thevetia peruviana* and *Alstonia scholaris* latices on the freshwater snail *Lymnaea acuminata*. Iberus, 22 (2): 19-27.

Singh, S. K. Yadav, R. P. Singh, D. and Singh, A. (2004a). Toxic effect of two common euphorbiales lattices on the freshwater snail *Lymnaea acuminata*. Environ. Toxic. & Pharmacol. 15: 87-93

Singh, S. K. Tripathi P. K. Yadav, R. P. Singh, D. and Singh, A. (2004b). Toxicity of Malathion and Carbaryl Pesticides: Effects on Some Biochemical Profiles of the Freshwater Fish *Colisa fasciatus*. Bull. Environ. Contam. & Toxicol. 72 (3): 592-599.

Srivastava, V. K. Singh, S. K. Rai, M. and Singh, A. (2003). Toxicity of *Nerium indicum* and *Euphorbia royleana* lattices against *Culex quinquefasciatus* mosquito larvae. *Nigerian J. of Med. and Nat. Prod.* Vol 7: 61-64.

Stefferud, A. (1952). Insects (The Year book of Agriculture). Oxford and I B H Publishing House, Washington, D.C.

Sullivan, J. T. Richards, C. S. Lloyd, H. A. and Krishna, G. (1982). Anacardic acid: molluscicide in cashew nut shell liquid. *Planta med.* 44: 175-177.

Swain, T. (1977). Secondary compounds as protective agents: *A ann. Revi. of Plant Physiol.* 28: 479 - 501.

Takai, Y. Kishimoto, A. Inoue, Y. and Nishizuku, Y. (1977). Studies on a cyclic-nucleotide independent protein kinase and its proenzyme in mammalian tissues i.e. purification and characterisation of an active enzyme from bovine cerebellum. *J. Biol. Chem.* 252: 7610 – 7616.

Talwar, P. K. and Jhingaran, A. G. (1991). Inland fishes of India and Adjacent countries. Vol. 1. Oxford and I.B.H. Publishing Co. Pvt. Ltd. New Delhi.

Tewari, S. C. (1995). Mosquitoes of medical importance in Andaman and Nicobar island. Proc. Of Int. Symp. On vector born disease. Pp. 246.

Tiwari, S. Sahani, S. K. and Singh, A. (2001). Control of predatory and weed fishes through environmentally safe plant origin pesticides. In: *Proc. of Int. Cong. Chem. Environ.* (Ed. S.L. Gargh), Pp. 170-174.

Tiwari, S. and Singh, A. (2003). Piscicidal activity of active compound extracted from *Euphorbia royleana* latex through different organic solvent. In: Proc. of 1st Nat. Int. Meet on Med. and Arom. Plants. Eds by A.K. Mathur, S. Dwivedi, D.D. Patra, G.D. Bagchi, A. Sharma and S.P.S. Khanuja), Pp. 330-336.

Tiwari, S. and Singh, A. (2004). Toxic and sub-lethal effects of Oleandrin on biochemical parameters of fresh water air breathing murrel, *Channa punctatus* (Bloch.). *Ind. J Exp. Biol.* 42: 413-418.

Tomassini, T. C. B. and Matos, M. E. O (1979). On the natural occurrence of 15-tiglioyloxy-kaur-16-En-19-oic acid. *Phytochem.* 18: 663-664.

Trigg, J. K. (1996). Evaluation of eucalyptus based repellent against Anopheles sps. In: Tanjania. *J. Anex. Mosquito Control Asso.* 243-246.

Vagas, D. Fronczck, F. R. and Fischer, N. H. (1986). The chemistry of confertiflorin and the molecular structure of confertiflorin and allodesacetylconfertiflorin, two molluscicidal sesquiterpenes lactones. *J. of Nat. Prod.* 49:133-138.

Waliszewski, S. M. Aguirre, A. A. Benitez, A. Infanzon, R. M. Infazon, R. and Rivera, J. (1999). Organo-pesticides residues in *Human blood serum of inhabitants of Veracuz*, Mexico. Bull. Environ. Contam. & Toxicol. 62: 397-402.

Watt, J. M. and Breyer-Brandijk, M. G. (1976). Medicinal and poisonous plants of Southern and Eastern Africa, E&S. Livingston Ltd. Edinburg London.

Wealth of India, (1985). A dictionary of Indian raw materials and Industrial products. Pubs. and Information directorate, *CSIR*, New Delhi. : Pp 201-204.

Yadav, R. P. (2000). Studies on molluscicidal properties of some common plants of family Euphorbiaceae and their environmental impact on non-target organisms. Ph.D. thesis D.D.U Gorakhpur University, Gorakhpur U.P India.

Yadav, R. P. and Singh, A. (2001). Environmentally safe molluscicides from two common euphorbiales. *Iberus*. 19 (1): 65-73.

Yadav R. P. and Singh, A. (2002). Toxic effects of latex of *Croton tiglium* on *Lymnaea acuminata* and *Channa punctatus* Iberus. 20 (2): 31-44.

Yadav, R.P. and Singh, A. (2003). Effect sub-lethal concentrations of *Codiaeum variegatum* latex on freshwater target snail *Lymnaea acuminata* and non-target fish *Channa punctatus*. Nig. J Nat. Pod. & Medi. (in press).

Yadav, R. P. Tiwari, S. and Singh, A. (2004a). Toxic effects of taraxerol extracted from *Codiaeum variegatum* stem-bark on target vector snail *Lymnaea acuminata* and non-target fish. Iberus. (In press).

Yadav, R. P. Singh, D. Singh S. K. and Singh A. (2004b). Toxic effect of stem bark of *Croton tiglium* on metabolism of freshwater snail *Lymnaea acuminata*. Amer. Malaco. Bull. (In press).

Yusufoglu, A. and Hasdemir, B. (1996). Essential oils from Turkey *Pelargonium graveolens* L. Heard as natural mosquito repellent. *Chem. Acta Turnica*, 24: 105-109.

INDEX

absorption, 4, 84, 85, 87, 88, 155, 161
acetic acid, 14
acetonitrile, 99
acidity, 169, 172, 175, 176, 177, 185, 186, 188
activation energy, 99
activation entropy, 110
activation parameters, 105
adenosine, 20, 32
adsorption, 16, 24, 39, 47, 51, 52, 53, 57, 59, 106, 107, 108, 113, 114, 115, 121, 122, 123, 124, 125, 126, 127, 128, 129, 130, 131, 132, 133, 134, 135, 136, 137, 138, 139, 140, 141, 142, 146, 161, 162, 163, 175, 187
aerosols, 165, 179
agriculture, vii, 193, 198, 199, 200, 206
air pollutants, 166, 180
Al, 123, 125, 128, 146, 149, 163, 166, 175
aldehydes, 33
algae, 3, 125
aliphatic compounds, 54
alkane, 68, 71
alkylation, 4
alkylation reactions, 4
amino acids, 11, 147
aniline, 97, 98, 101
animals, 79, 80, 81, 89, 92, 206, 207, 216, 223
antibiotics, 18, 20
aqueous solutions, 46, 56, 57, 71, 76, 97
aquifers, 46, 71, 74, 76
aromatic compounds, 54, 68
aromatic hydrocarbons, 73, 76, 77
aromatics, 68
atmospheric deposition, 89, 166, 167, 169
atmospheric pressure, 41
ATP, 14, 15, 19, 20, 32, 33, 34
bacteria, 3, 4, 6, 7, 8, 14, 15, 17, 20, 21, 22, 24, 25, 26, 28, 29, 30, 33, 36, 37, 38, 39, 40, 41
base catalysis, 111
benzene, 4, 39, 54, 57, 63, 65, 68, 74, 75, 76, 78
bioaccumulation, 80, 122
bioavailability, 1, 24, 39, 87, 142, 146
biochemistry, 206
biodegradability, 222
biodegradation, 3, 6, 44, 48, 67, 69, 71, 73, 77
biological activity, 69, 76, 80, 216, 223
biological systems, 41, 166
biology, 206
biomass, 19, 33
biosensors, 2, 27, 29, 38, 39, 40
biosynthesis, 205
biotechnology, 38, 222
biotic, 47, 72, 192
blood, 223, 225, 228
breakdown, 3, 95, 97
Britain, 207, 224
burning, 46
calibration, 9
Canada, 2, 4, 204
carbon, 14, 33, 45, 86, 95, 96, 97, 99, 107, 109, 112, 114, 115

carboxylic acids, 111
carboxylic groups, 110, 116
carriers, 219
catalyst, 96, 110, 111, 115, 116
catalysts, 95, 105, 110, 111
catalytic properties, 97, 116
catchments, 175
cation, 3, 4, 22, 99, 146, 152, 157, 161, 170, 171, 172, 173, 175, 176, 187
C-C, 145, 155
cell, 6, 14, 16, 17, 28, 29, 30, 31, 33, 34, 38, 40, 47, 178, 181
CF, 84
CH_2, 98, 101
CH_3COOH, 95, 99, 103, 105, 110
characterisation, 75, 228
chemical interaction, 176
chemical properties, 44, 98, 115
chemical reactions, 179
chemiluminescence, 38, 40
chemistry, 40, 121, 143, 166, 171, 228
chlorine, 81, 82
chlorobenzene, 55, 162
chloroform, 219, 221
chloroplasts, 96
chromatograms, 54, 64, 67, 69, 71
chromatography, 4, 38, 41, 55, 223
classification, 187
climate change, 166, 188
climatic factors, 174
CMC, 148, 153
CO_2, 97, 179
coal, 54, 75, 171, 179, 181, 183
coatings, 121, 122, 123, 124, 125, 126, 127, 128, 129, 130, 131, 132, 133, 134, 135, 136, 137, 138, 139, 140, 141, 142
combustion, 76
composite, 77
composition, 19, 24, 40, 44, 48, 54, 56, 63, 64, 68, 69, 71, 72, 77, 86, 122, 142, 220
compounds, 1, 2, 3, 4, 6, 7, 10, 13, 16, 20, 21, 22, 37, 38, 39, 41, 46, 53, 61, 63, 68, 69, 71, 73, 76, 79, 80, 91, 96, 100, 101, 110, 166, 167, 168, 169, 176, 177, 205, 206, 208, 215, 216, 220, 221, 223, 228
conductivity, 58, 59, 75
congeners, 80, 81, 82
conversion, 33, 97, 110, 202, 203
copper, 36, 162
Copyright, iv
covalent bond, 127
crop production, 203
CT, 92, 150
cultivation, 31, 193
culture conditions, 27, 33
damage, 6, 165, 167, 168, 178, 186, 187, 188, 207
database, 172, 188
decomposition, 110
defense, 207
degradation, 45, 46, 62, 63, 68, 69, 71, 72, 95, 96, 97, 99, 100, 101, 102, 103, 104, 107, 108, 110, 111, 116, 117, 147, 192
degradation process, 96
degradation rate, 116
denitrification, 175, 177
Denmark, 189
density, 15, 18, 28, 29, 30, 31, 34, 40, 48, 65, 87, 127, 128
deposition rate, 87, 173
desorption, 47, 76, 108, 137, 138, 145, 146, 147, 148, 149, 152, 153, 154, 155, 157, 159, 160, 161, 162, 163, 175
destruction, 203
dibenzo-*p*-dioxins (PCDDs), 79, 80, 81, 82
dielectric constant, 116
diffusion, 45, 49, 75
dioxin, 79, 81, 82, 85, 86, 87, 91, 92
displacement, 54
dissociation, 138
dissolved oxygen, 15
distilled water, 56, 77
diversity, 187
DNA, 6, 11, 19
DO, 20
drugs, 207
dry matter, 87
drying, 44, 75
ecology, 167, 190, 222
ecosystems, 2, 165, 166, 167, 168, 169, 172, 174, 177, 178, 181, 182, 185, 187, 188

efficiency, 4, 8, 10, 104, 145, 147, 148, 159, 179
electric charge, 138
electrode, 121, 127, 128, 131, 142, 148, 161
electrolytes, 46, 56, 62, 72, 76
emission, 4, 9, 10, 19, 34, 40, 81, 150, 168, 170, 174, 176, 177, 182, 183, 186, 188, 189, 190
emulsions, 47
energy, 33, 43, 44, 71, 166, 180, 181, 182, 183
engineering, 167, 168, 190
enthalpy of activation, 100
entropy, 100, 110
environment, vii, 2, 37, 39, 40, 43, 44, 46, 47, 48, 50, 56, 57, 63, 72, 73, 74, 75, 78, 80, 82, 93, 98, 115, 119, 127, 137, 146, 166, 167, 168, 180, 187, 188, 191, 192, 194, 199, 219, 222
environmental contamination, 206, 221
environmental impact, 38, 180, 186, 229
Environmental Protection Agency, 82, 93
environmental threats, 182
enzymes, 7
ethanol, 20, 100, 101, 102, 107, 221
European Community, 2
evaporation, 59, 123
extrapolation, 108
farmland, 203
fast Fourier transform infrared (FTIR), 145, 151, 155
fibers, 8, 30
filtration, 100
fish, 77, 89, 166, 167, 205, 215, 216, 224, 225, 226, 227, 229
fixation, 175
flame, 4, 37
fluid, 14, 44, 48, 49, 73
fluorescence, 36
food industry, 41
forecasting, 192
France, 1, 2, 14, 21, 22, 24, 25, 40
free energy, 97
frequency, 8, 30, 85
freshwater, 3, 46, 47, 71, 75, 76, 125, 126, 206, 207, 208, 216, 223, 226, 227, 229
FTIR, 145, 151, 155
fuel, 43, 46, 146, 181
fuels, 43, 182
fusion, 6, 10, 11
future, vii, 2, 34, 166, 168, 180, 183, 184, 186, 192, 199, 200, 203
gasoline, 50, 59, 64, 74
gene, 6, 7, 11, 12, 39
Germany, 2
glucose, 13, 14, 16, 17, 18, 19, 20, 26, 28, 31, 32, 33, 34, 35, 36
grains, 54
gravitation, 139
grazing, 91, 171, 172
Grignard reagents, 4
groundwater, 43, 46, 189
habitats, 7, 205, 207
halogen, 21
health effects, 188
heat, 43, 127, 128
herbicide, 95, 96, 97, 98, 107, 113, 116
herbicides, 95, 96, 97, 110, 112, 116
heterogeneity, 24, 106, 108
hexane, 4, 219
hydrocarbons, 3, 43, 44, 46, 47, 50, 51, 54, 55, 56, 57, 61, 62, 68, 69, 70, 71, 72, 73, 74, 75, 76, 77, 78
hydrogen, 111, 175, 176
hydrolysis, 95, 96, 97, 101, 104, 105, 106, 109, 110, 115, 116, 117, 139, 156
hydrophobicity, 112, 116
hydroxide, 97, 102, 110, 112
hydroxyl, 139
India, v, 40, 77, 162, 165, 166, 172, 173, 174, 178, 179, 180, 181, 182, 184, 185, 188, 190, 191, 193, 194, 195, 196, 198, 201, 202, 203, 204, 206, 207, 208, 215, 219, 223, 225, 226, 227, 228, 229
inhibition, 1, 39, 226
inhibitor, 207, 208, 220
inoculum, 22
integrated circuits, 40
interface, 47, 48, 49, 71
ionization, 100, 101
ions, 56, 110, 127, 139, 141, 147, 148, 149, 153, 157, 161, 175, 185

IR, 20
isoprene, 221
isotherms, 113, 124, 126, 128, 132, 133, 136, 139, 146
kinetic constants, 97, 109, 110
kinetic parameters, 97
kinetics, 51, 52, 148
KOH, 99
lakes, 122, 167, 175
land use, 168, 172, 191, 192, 193, 194, 195, 196, 197, 198, 199, 203
ligands, 121, 122, 137, 142
light, 7, 9, 12, 13, 15, 19, 21, 22, 28, 30, 34, 40, 41, 61, 68, 69, 206
linear dependence, 110
links, 188
liquid chromatography, 4, 41
liquid phase, 48, 64, 134
liver, 207
livestock, 188
lubricants, 43
luminescence, 6, 7, 8, 10, 11, 18, 20, 24, 25
macromolecules, 56, 110, 113
macropores, 50
magnetism, 198
mass, 4, 24, 41, 47, 48, 49, 50, 61, 64, 67, 68, 74, 75, 77, 87, 100, 101, 106, 108, 117, 174, 175, 178, 180, 184, 191
medicine, 39, 206
melting, 25
membrane, 40, 99, 100
metabolism, 19, 229
metal extraction, 148
metals, 24, 121, 122, 123, 124, 125, 126, 127, 128, 131, 132, 133, 134, 137, 138, 139, 142, 145, 147, 148, 161, 162, 163, 188
methanol, 100, 213, 219
microorganism, 123
microorganisms, 3, 45, 122, 128
microscope, 123
mitochondria, 226
mixing, 47, 71, 75, 77, 82
MM, 14
modeling, 143
moisture, 44, 45, 50, 51, 52, 57, 58, 59, 63, 72
moisture content, 50, 51, 52, 57, 58, 59, 63, 72
molar volume, 56
molecular biology, 7
molecular structure, 96, 100, 224, 228
molybdenum, 8
monitoring of processes, 1, 3
monomers, 153
MS, 4, 37, 38, 40, 41
NaCl, 14, 54, 56, 145, 150, 157, 158
naphthalene, 54, 61, 74, 76, 162
Netherlands, 76, 189
nitrogen, 99, 111, 166, 169, 175, 177, 184, 190, 215, 221
NMR, 101
nonane, 68
nonionic surfactants, 147
nuclei, 221
nutrients, 45, 62, 69
oil, 44, 54, 76, 77, 181, 208, 214, 216, 218, 219, 220, 223, 224
oils, 77, 219, 220, 221, 223, 224, 226, 229
optical density, 9, 17, 20, 26, 28, 31, 32, 33, 34
optical fiber, 8, 30, 40
optimization, 45, 71
organic chemicals, 75, 221
organic compounds, 45, 46, 47, 53, 56, 61, 71, 76, 78, 100, 147, 161
organic matter, 50, 61, 71, 72, 96, 97, 98, 100, 110, 115, 116, 139
orientation, 11
oxidation, 7, 147, 175
Oxides, 122, 123, 125, 127, 128, 136, 137, 138, 139, 143, 146, 147, 162
oxygen, 7, 14, 17, 19, 26, 28, 29, 96, 123, 128
ozone, 166, 186
particles, 58, 79, 82, 83, 85, 91, 182
PC, 180
PE, 40
peat, 64, 66, 67
permeability, 64, 150

pesticide, 74, 121, 124, 135, 136, 137, 142, 208
pesticides, 73, 119, 205, 206, 207, 208, 222, 223, 225, 226, 227, 228
pH, 14, 15, 16, 17, 18, 19, 21, 22, 23, 26, 28, 29, 34, 45, 95, 99, 100, 101, 102, 103, 110, 111, 112, 121, 122, 123, 124, 126, 135, 137, 138, 139, 142, 145, 146, 147, 149, 150, 151, 155, 156, 157, 161, 162, 163, 165, 166, 171, 172, 173, 175, 179, 180, 188
phenols, 205, 209, 221
photodegradation, 2
photodetectors, 38, 41
photosynthesis, 96
physical properties, 50, 53, 57, 69
pine, 226
plasma, 4, 41
PM, 10, 28, 29, 30, 189
polar groups, 116
polarity, 147, 219
pollutants, 2, 30, 38, 45, 47, 71, 73, 79, 80, 81, 122, 167, 168, 176, 188
pollution, vii, 2, 27, 30, 37, 43, 45, 73, 76, 166, 167, 180, 181, 182, 185, 186, 187, 188, 189, 205, 225
polyethylene, 185
polymer, 136
polymers, 74, 146, 215, 221
polypropylene, 123, 124
polysaccharides, 56
porosity, 48, 50
precipitation, 139, 156, 161, 165, 179, 187, 188
prediction, 187
productivity, 202
protein, 11, 33, 40, 215, 228
proteins, 19
purification, 148, 228
quality, 4, 72, 185
radicals, 180
rain forest, 174
rate constant, 95, 99, 102, 103, 104, 105, 108, 109, 110, 112, 117
rate constants, 99
raw materials, 195, 228
reactant, 28, 117
reaction rate, 95, 99, 100, 111, 115, 117
reagents, 40
reduction, 2, 14, 59, 63, 107, 141, 175, 176, 177, 178, 186, 187, 189, 190
regression, 99, 121, 124, 125, 126, 140
reinforcement, 141
reproduction, 219
resistance, 6, 39, 207
resonance, 96
responsibility, iv
RNA, 19
room temperature, 24
runoff, 175, 176
salinity, 46, 47, 77
sediments, 1, 2, 3, 6, 23, 37, 38, 39, 40, 45, 46, 76, 77, 122
seeds, 208
sensitivity, 2, 4, 8, 9, 21, 24, 170, 171, 172, 173, 174, 177, 178, 180, 185, 187, 188
sensors, 29, 37, 39, 41
services, iv
sewage, 2, 46, 61, 63, 69, 71, 72
sheep, 207
silica, 4, 162
sodium dodecyl sulfate (SDS), 148
soils, 39, 44, 45, 48, 50, 53, 58, 61, 64, 65, 69, 71, 73, 74, 75, 76, 77, 97, 99, 100, 106, 107, 112, 113, 114, 115, 116, 146, 147, 148, 151, 158, 161, 162, 163, 167, 171, 172, 174, 178, 188, 189, 192
solid matrix, 24, 36
solid phase, 25, 44, 45, 48, 49, 50, 51, 59, 72, 122, 133, 134, 176
solubility, 31, 46, 47, 53, 56, 61, 62, 63, 68, 71, 72, 73, 75, 77, 78, 113, 136, 146
solution, 9, 14, 21, 25, 26, 43, 44, 46, 47, 53, 54, 55, 56, 57, 63, 68, 97, 100, 101, 106, 107, 108, 109, 110, 116, 117, 121, 122, 123, 124, 126, 129, 130, 131, 132, 133, 136, 137, 138, 139, 141, 142, 145, 147, 149, 150, 151, 152, 153, 155, 157, 159, 160, 163, 175
solvent, 75, 228
solvents, 4, 46, 61, 219

sorption, 51, 52, 53, 58, 73, 76, 95, 96, 97, 106, 107, 112, 115, 116, 145, 146, 147, 148, 149, 150, 151, 152, 153, 155, 156, 157, 159, 161, 162, 163
sorption isotherms, 95, 106, 115, 151
sorption process, 107
Spain, 46, 76
specific adsorption, 122, 141, 142
specific surface, 122
steel, 8, 28
streams, 47, 71, 167, 175
strength, 56, 99, 101, 102, 106, 121, 122, 123, 124, 140, 141, 142, 150, 157, 163, 188
stretching, 145, 155
styrene, 59
substrates, 69, 71, 98
sugar, 221
sulphur, 169, 175, 176, 177, 179, 181, 183, 184, 190
surface area, 85, 124, 151
surface modification, 162
surface properties, 48
surfactants, 146, 147, 148, 161, 162
Switzerland, 186, 190
temperature, 19, 20, 28, 29, 34, 40, 45, 54, 77, 99, 102, 105, 136, 179
tetrachlorodibenzo-*p*-dioxin (TCDD), 79, 80
thermodynamics, 139
toluene, 4, 54, 59, 74
toxic substances, 96
toxicity, 2, 20, 22, 24, 30, 36, 37, 38, 122, 143, 206, 216, 219, 222, 223, 225, 226
toxicology, 223
transformation, 86, 137, 191, 192, 193, 194, 197, 198, 199, 200, 202
transport, 44, 47, 48, 53, 58, 59, 61, 62, 69, 71, 72, 74, 77, 122, 142, 147, 180
treatment, 4, 38, 59, 62, 64, 69, 71, 72, 207
trees, 167, 171, 185, 186, 190
United Kingdom, 204
USA, 41, 46, 77, 118, 204
UV, 3, 100
UV radiation, 3
validation, 27, 178
vegetation, 87, 88, 166, 167, 171, 174, 178, 188
viscosity, 75
volatilization, 44, 47, 49, 59, 63, 64, 65, 66, 67, 69, 74, 76, 77
wastewater, 2, 20
water, vii, 1, 2, 3, 6, 8, 20, 21, 22, 23, 24, 25, 26, 27, 36, 37, 38, 39, 44, 45, 46, 47, 48, 49, 50, 51, 52, 54, 55, 56, 57, 58, 59, 61, 62, 63, 65, 68, 70, 71, 72, 73, 74, 75, 76, 77, 83, 84, 85, 86, 88, 91, 95, 97, 98, 99, 100, 101, 107, 108, 109, 110, 112, 113, 115, 116, 117, 121, 122, 123, 124, 136, 138, 142, 144, 149, 150, 154, 155, 157, 159, 160, 161, 166, 167, 170, 171, 175, 179, 180, 206, 216, 221, 223, 228
water policy, 2, 3, 38
water quality, 46
wetlands, 192
wettability, 48, 50
wetting, 44
wood, 2, 219
x-ray diffraction, 96